Introduction to the Quantum Theory of Semiconductors

Introduction to the Quantum Theory of Semiconductors

MARVIN M. COHEN
Research Physicist
Harry Diamond Laboratories
Washington, D. C.

Adjunct Professor
American University
Washington, D. C.

GORDON AND BREACH SCIENCE PUBLISHERS
New York London Paris

Gordon and Breach, Science Publishers, Ltd.
12 Bloomsbury Way
London W.C.1

Editorial office for the United States of America

Gordon and Breach, Science Publishers, Inc.
150 Fifth Avenue
New York, N.Y. 10011

Editorial office for France

Gordon & Breach
7-9 rue Emile Dubois
Paris 14e

Library of Congress catalog card number 79-123485. SBN 0 677 02980 2.

To my wife and our parents

Preface

THIS BOOK GREW out of the notes which I used to teach a short course in semiconductor theory at the American University, Washington, D.C. The classes were made up of first year graduate students in physics and engineering, many of whom were already working in semiconductor-related fields they had entered so-to-speak by the "back door," i.e. none had a formal background in semiconductors. It was our intention to develop a course in which these students and those planning to enter the field of semiconductors would get a mathematically oriented introduction to the methods, tools, and concepts which are fundamental to the theory of solids in general, with emphasis on applications to semiconductors.

The manuscript is aimed at students who have had an undergraduate modern physics course and an introductory course in thermodynamics. Chapters 1 and 2 are intended to serve as a review aud not as a complete introduction. I have utilized many of the arguments and figures from a variety of sources and in many instances have modified the figure labeling to correspond to presently accepted values and notation.

I would like to thank Herbert D. Curchack for much encouragement, Drs. Fernand Bedard, Charles Davis, Clyde Morrison, and Lynwood P. Randolph, and students Daniel J. Lanigan and James D. Penar for editing portions of the manuscript. In particular, I thank my wife, Hilde, for typing and editing the entire manuscript.

Washington, D.C. M. M. COHEN

Contents

CHAPTER 1

Review of Quantum Theory

1.1 THE STATE OF A QUANTUM MECHANICAL SYSTEM

UNDERSTANDING of a subject or technology can be achieved only after an expenditure of effort to learn the language by which it may most conveniently be discussed. Because solid state physics is to a large measure based on the quantum theory, the theory of solids may be discussed by utilizing any of the quantum mechanical *representations*. For the student to work effectively in the field of solids, he must have a working knowledge not only of the wave-mechanical Schrödinger representation, but of all of the methods by which solids may be described. For this reason we shall start with a review of the abstract quantum theory as proposed by Dirac[1] from which all representations may be derived.

We shall assume that there exists a hypothetical many-dimensional space each point of which fully describes the state of our quantum mechanical system at time t. The state is assumed to be fully described by the coordinates of the particular point in the space or equivalently by a vector generated from the origin to the point. The many-dimensional vector that describes our state is called a *ket* vector and is written as $|A\rangle$. Thus $|A\rangle$ represents the state A of a quantum mechanical system. For example, as shown in Figure 1.1, the three dimensional velocity vector

$$|\mathbf{V}_{(t)}\rangle = a_{(t)}\,|\hat{V}_x\rangle + b_{(t)}\,|\hat{V}_y\rangle + c_{(t)}\,|\hat{V}_z\rangle \tag{1.1.1}$$

where $|\hat{V}_x\rangle$, $|\hat{V}_y\rangle$, and $|\hat{V}_z\rangle$ are the orthogonal unit velocity vectors (or basic vectors) which define our allowed velocity space, completely defines the state of the velocity of a system at any time t.

The Dual Space

In our study of semiconductor physics we shall make frequent use of the fact that for every vector space a unique dual space may be defined. We shall call the vectors in our dual space *bra* vectors and write them as $\langle B|$. We shall assume that for every ket vector $c|A\rangle$ in our original space there corresponds a bra vector $\langle A|c^*$, where c^* is the complex conjugate of the constant c. In general, however, the ket vector is a complex quantity from which one may not be able to separate the real from the imaginary parts. Thus we call the bra $\langle A|$ the complex imaginary of $|A\rangle$ or $\overline{|A\rangle}$.

We shall define the scalar product between two N dimensional vectors in the same space as

$$\begin{aligned} |A\rangle &= \sum_{i=1}^{N} a_i\,|A_i\rangle \\ |B\rangle &= \sum_{j=1}^{N} b_j\,|A_j\rangle \end{aligned} \tag{1.1.2}$$

where a_i and b_j are constants and

$$\langle A|B\rangle = \sum_{k=1}^{N} a_k^* b_k \tag{1.1.3}$$

Thus defined, the *scalar product* between two ket vectors generates a number. Of course, the scalar product of the real velocity vector $|V_{(t)}\rangle$ with any of the basic ket vectors $|\hat{V}_{(x)}\rangle$, $|\hat{V}_{(y)}\rangle$ or $|\hat{V}_{(z)}\rangle$ yields the component of $|V_{(t)}\rangle$ in the direction of the basic ket (see Figure 1.1), e.g.

$$\begin{aligned}\langle \hat{V}_{(x)}|V_{(t)}\rangle &= a_{(t)}\\ \langle \hat{V}_{(y)}|V_{(t)}\rangle &= b_{(t)}\\ \langle \hat{V}_{(z)}|V_{(t)}\rangle &= c_{(t)}\end{aligned} \tag{1.1.4}$$

In particular we shall make use of a dual space made up of basic vectors which are defined only by their scalar products with the basic ket vectors of the original space.

For example, choose a set of numbers, the ith one of which we designate by ϕ_i, and each of which is a linear function of a single ket vector $|A_i\rangle$.

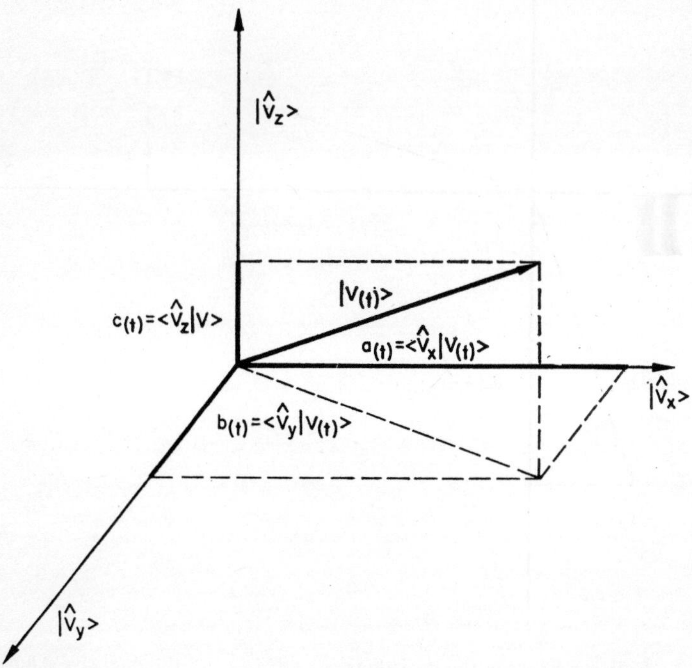

Figure 1.1 The velocity of a system represented as a vector in velocity space.

Each ϕ_i may be looked on as the result of the scalar product of the ith basic ket vector with some bra vector $\langle B_i|$, i.e.,

$$\phi_i = \langle B_i|A_i\rangle \tag{1.5}$$

The set of bra vectors thus generated is the dual space. Our dual space is a linear function of the original ket vectors. As an example, Figure 1.2 shows the relation between a real two dimensional space defined by the basic kets $|\hat{x}\rangle$ and $|\hat{y}\rangle$ and its dual space defined by the basic bras $\langle\hat{B}_1|$ and $\langle\hat{B}_2|$.

The length of a ket vector is as usual defined as the square root of its scalar product with its conjugate bra

$$\langle A|A\rangle^{\frac{1}{2}} = \text{length } |A\rangle \tag{1.1.6}$$

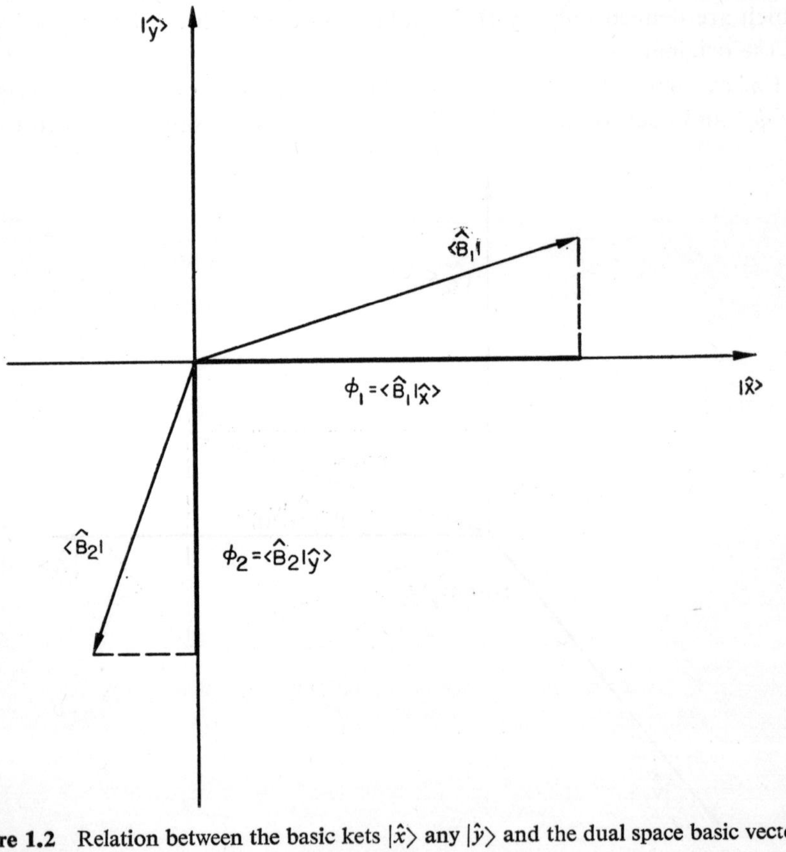

Figure 1.2 Relation between the basic kets $|\hat{x}\rangle$ any $|\hat{y}\rangle$ and the dual space basic vectors $\langle\hat{B}_1|$ and $|\hat{B}_2\rangle$.

In most application, it is convenient to normalize the length of our ket vectors such that

$$\langle A|A\rangle = 1 \tag{1.1.7}$$

1.2 LINEAR OPERATORS

An operator is a quantity that denotes an operation to be performed on a function. An operator α, acting on a ket vector from the left yields a ket that may differ both in magnitude and direction from the original

$$\alpha\,|A_0\rangle = |A\rangle \tag{1.2.1}$$

The operator is always written to the left of the ket.

We will concern ourselves mainly with linear operators. The conditions for linearity may be written

$$\alpha[|A\rangle + |B\rangle] = \alpha\,|A\rangle + \alpha\,|B\rangle \tag{1.2.2}$$

$$\alpha[c\,|A\rangle] = c\alpha\,|A\rangle \tag{1.2.3}$$

where c is a number.

Linear operators may be added

$$[\alpha + \beta]\,|A\rangle = \alpha\,|A\rangle + \beta\,|A\rangle \tag{1.2.4}$$

Multiplication of linear operators e.g. $\alpha\beta$ is carried out by performing first the operation β and then operating on the result with α. The resultant operator $\alpha\beta$ need not be linear.

$$\alpha\beta\,|A\rangle = \alpha[\beta\,|A\rangle] \tag{1.2.5}$$

Let us investigate, for example, the result of the successive operation of the two displacement operators Γ_x and Γ_θ which are defined to produce unit displacement of a vector in the x and θ directions, respectively. As can be seen from Figure 1.3, the order in which the displacements of the vector $\mathbf{R}$ are produced is immaterial, i.e.,

$$\Gamma_x\Gamma_\theta\mathbf{R} = \Gamma_\theta\Gamma_x\mathbf{R} \tag{1.2.6}$$

for the result $\mathbf{R}'$ of the multiplicative operation has the same orientation no matter which order of operation is chosen. However, note that linear operators do not necessarily commute thus their order must be preserved. For example, consider the two independent operations 1. taking a breath and 2. putting your head in a bucket of water. Clearly, performing 2. first and then 1. yields a quite different result then performing the operations in the opposite order.

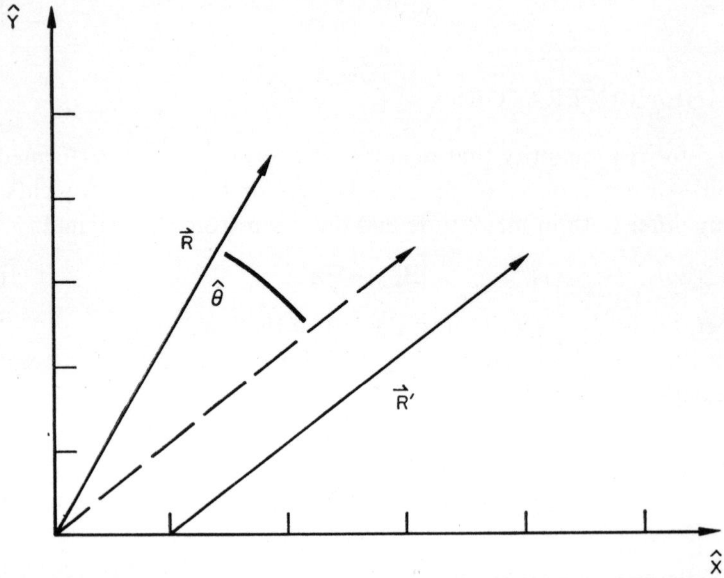

Figure 1.3 The successive operation on **R** with the unit translation operators Γ_x and Γ_θ showing their commutative property.

PROBLEM

(A) Show that the operators

$$\alpha = \frac{\partial}{\partial x} \quad \text{and} \quad \beta = \frac{\partial}{\partial y}$$

commute by operating on the vector

$$\mathbf{A} = \hat{\imath} y^2 z^2 + \hat{\jmath} x^2 z^2 + \hat{k} x^2 y^2$$

(B) Similarly show that

$$\alpha' = x \quad \text{and} \quad \beta' = \partial/\partial x$$

do not commute.

Linear operators may also act on bra vectors and are defined to do so only from the right

$$\overline{[\beta \,|B\rangle]} = \langle B|\, \bar{\beta} \tag{1.2.7}$$

where $\bar{\beta}$ is the adjoint of β.

A linear operator may act on a ket or bra vector in such a way as to produce a vector unchanged in direction but changed only in magnitude

$$\alpha \,|A\rangle = a \,|A\rangle \tag{1.2.8}$$

where a is a constant called the *eigenvalue* of the linear operator α, and the ket $|A\rangle$ is an *eigenket* of α. A similar situation exists for bra vectors.

PROBLEM

Determine if the vector **R**, where

$$\mathbf{R} = \hat{\imath}\,\frac{e^{atx}}{x} + \hat{\jmath}\,\frac{e^{bty}}{y} + \hat{k}\,\frac{e^{ctz}}{z}$$

is an eigenvector of the operator $\partial/\partial t$.

A *real linear operator* is a real operator whose eigenvalues are real numbers. Therefore, as in equation (1.2.8), if α is a real linear operator, a is a real number.

It is easily shown that the eigenvalue a, associated with the eigenket of a real linear operator α is the same as the eigenvalue associated with the eigenbra of the real linear operator $\bar{\alpha}$ by forming the complex imaginary of the eigenvalue equation (1.2.8) which yields

$$\langle A|\,\bar{\alpha} = \langle A|\,a^* \tag{1.2.9}$$

The Orthogonality Theorem

The orthogonality theorem states that two eigenvectors of a real linear operator belonging to different eigenvalues are orthogonal. To prove this we let ξ be a real linear operator, $|A\rangle$ be an eigenket belonging to the eigenvalue a and $|B\rangle$ be an eigenket belonging to the eigenvalue b. Thus we may write the eigenvalue equations

$$\begin{aligned} \xi\,|A\rangle &= a\,|A\rangle \\ \xi\,|B\rangle &= b\,|B\rangle \end{aligned} \tag{1.2.10}$$

Taking the complex imaginary of 1.2.10

$$\langle A|\,\xi = \langle A|\,a \tag{1.2.11}$$

and multiplying by $|B\rangle$ from the right, we find

$$\langle A|\,\xi\,|\beta\rangle = \langle A|\,a\,|B\rangle$$

and

$$\langle A|\,\xi\,|B\rangle = \langle A|\,b\,|B\rangle \tag{1.2.12}$$

Subtracting the equations in (1.2.12) we find

$$\langle A|B\rangle\,(a - b) = 0 \tag{1.2.13}$$

Thus, if $a \neq b$, the scalar product $\langle A|B\rangle$ is equal to zero and the eigenkets

$|A\rangle$ and $|B\rangle$ are orthogonal. When $|A\rangle$ and $|B\rangle$ are orthonormal we can generalize equation (1.2.13) for the case of discrete eigenvalues a and b

$$\langle A|B\rangle = \delta_{a,b} \tag{1.2.14}$$

where $\delta_{a,b}$ is the *Kronecker delta* which has the value 0 for $a \neq b$, and 1 for $a = b$. When $a = b$, $|A\rangle$ and $|B\rangle$ are called *degenerate*.

For the case where we have a continuous range of eigenvalues we may write

$$\langle A'|A''\rangle = \delta(a' - a'') \tag{1.2.15}$$

where $\delta(a' - a'')$ is the *Dirac delta* for continuous eigenvalues and is equal to zero except when $a' = a''$, in which case it has the value 1 and $|A'\rangle$ and $|A''\rangle$ are again degenerate. Note that

$$\int_{-\infty}^{+\infty} \delta_{(a'-a'')} da' = 1 \tag{1.2.16a}$$

and

$$\int_{-\infty}^{+\infty} f_{(x)}\delta_{(x-a)} dx = f_{(a)} \tag{1.2.16b}$$

Observables

Two important assumptions made by Dirac in order to connect his abstract mathematics with experience are that (1) measurable quantities correspond to the eigenvalues of real linear operators. These types of operators are called *observables*. (2) If the eigenvalue of a real linear operator ξ is measurable with the system in a particular state, the act of making a measurement will cause the system to jump into one of a number of allowed states. These states into which the system may jump are assumed to be such that the original state is dependent upon them and thus each must be also an eigenstate of ξ. Thus the original state is dependent on eigenstates of ξ. Since the original state may be any state, we conclude that any state must be dependent on the eigenstates of an observable, and thus an observable must have a complete set of eigenstates, i.e. that any other state be constructable from them. For example, the unit vectors $\hat{\imath}$, $\hat{\jmath}$ and $\hat{k}$ form a complete set in cartesian space since any other cartesian vector may be constructed from a linear combination of these vectors.

Where the eigenvalues of the observable ξ consist of a discrete set of numbers in a certain range such as the energy levels of a harmonic oscillator, the condition that any state $|P\rangle$ is dependent on the complete set of kets $|A_1\rangle, |A_2\rangle \ldots |A_i\rangle \ldots |A_n\rangle$ is written

$$|P\rangle = \sum_{i=1}^{n} c_i |A_i\rangle \tag{1.2.17}$$

where c_i is a number.

If the eigenvalues b' of an observable ξ consist of all of the numbers in a certain range, for example the spatial coordinates of an electron, the condition that any ket $|P\rangle$ is dependent on the complete set of kets $|B_1'\rangle, |B_2'\rangle \ldots |B_i'\rangle \cdots |B_n'\rangle$ is expressed

$$|P\rangle = \int |B'\rangle \, db' \tag{1.2.18}$$

where we integrate over the range of eigenvalues b' necessary to construct $|P\rangle$.

There may exist a state $|P\rangle$ such that it is dependent upon sets of both discrete and continuous states, in which case we write

$$|P\rangle = \sum_{i=1}^{N} c_i |A_i\rangle + \int |B'\rangle \, db' \tag{1.2.19}$$

We shall postulate that if a quantum mechanical system is in an eigenstate $|A\rangle$ of a real linear operator *which* belongs to the eigenvalue a, a measurement of the real linear operator will always give the number a. A real linear operator "which may be measured" is called an ***observable*** and thus any measurement of an observable must yield one of its eigenvalues.

Therefore the eigenstates of an observable form a complete set since any state is dependent on them. Conversely, any real linear operator whose eigenstates form a complete set is an observable.

We can generate an extremely useful linear operator from a complete set of n discrete or n continuous states. Let us form the sum

$$\sum_{i=1}^{n} |A_i\rangle \langle A_i| \tag{1.2.20}$$

and multiply from the right by the ket $|A_J\rangle$

$$\sum_{i=1}^{n} |A_i\rangle \langle A_i|A_J\rangle = \sum_{i=1}^{n} |A_i\rangle \, \delta_{i,J} = |A_J\rangle \tag{1.2.21}$$

Thus the sum (1.2.20) acts as though it were the unit linear operator for discrete kets

$$\sum_{i=1}^{n} |A_i\rangle \langle A_i| = 1 \tag{1.2.22}$$

PROBLEM

Show for the case of continuous eigenvalues

$$\int |B'\rangle\, db' \langle B'| = 1 \tag{1.2.23}$$

Poisson Brackets and Commuting Operators

In classical mechanics it is fundamental that any function which represents the motion of an n dimensional system may be written in terms of a set of $2n$ conjugate coordinates and momenta q_r and p_r, respectively. Of great importance are the *Poisson bracket* relations between these functions. The classical Poisson bracket $[U, V]$ between two functions $U(q_r, p_r)$ and $V(q_r, p_r)$ is given by

$$[U, V] = \sum_{r=1}^{n} \left[\frac{\partial U}{\partial q_r} \frac{\partial V}{\partial p_r} - \frac{\partial U}{\partial p_r} \frac{\partial V}{\partial q_r} \right] \tag{1.2.24}$$

It is easily shown that

$$[U, V] = -[V, U] \tag{1.2.25}$$

and

$$[U, C] = 0 \tag{1.2.26}$$

where C is a constant. If the functions U and V are themselves the coordinate q_i and the momentum p_j, we find

$$[q_i, p_j] = \delta_{ij}$$

$$[p_i, p_j] = 0$$

$$[q_i, q_j] = 0 \tag{1.2.27}$$

Using Hamilton's equations of motion

$$\dot{q}_r = \frac{\partial \mathscr{H}}{\partial p_r} \tag{1.2.28 a}$$

$$-\dot{p}_r = \frac{\partial \mathscr{H}}{\partial q_r} \tag{1.2.28 b}$$

where $\mathscr{H}$ is the Hamiltonian or total energy of the system[2]

$$\mathscr{H} = T + V(r) \tag{1.2.29}$$

we may write, using the Poisson bracket relations

$$[q_r, \mathscr{H}] = \frac{\partial \mathscr{H}}{\partial p_r} = \dot{q}_r \tag{1.2.30}$$

$$[p_r, \mathscr{H}] = -\frac{\partial \mathscr{H}}{\partial q_r} = \dot{p}_r \tag{1.2.31}$$

In general, for any function $G(q_r, p_r)$ we find

$$\frac{\partial G}{\partial t} = \sum_r \left[\frac{\partial G}{\partial q_r}\frac{\partial q_r}{\partial t} + \frac{\partial G}{\partial p_r}\frac{\partial p_r}{\partial t} \right] = [G, \mathscr{H}] \tag{1.2.32}$$

Utilizing the classical definition of Poisson brackets, and demanding that we keep the order of our operators, we may also define Poisson brackets between quantum mechanical operators. The quantum mechanical Poisson bracket between two operators α and β is defined by

$$\left[\frac{\alpha\beta - \beta\alpha}{i\hbar} \right]_{Qm} = [\alpha, \beta]_{Cl} \tag{1.2.33}$$

where $\alpha\beta - \beta\alpha$ is the so-called commutator between α and β.

Problem

Prove

$$q_r q_s - q_s q_r = 0$$

$$p_r p_r - p_s p_r = 0$$

$$q_r p_s - p_s q_r = i\hbar\delta(r, s)$$

where q_r and p_r are considered quantum mechanical operators.

It can be shown that the quantum mechanical Poisson bracket between a time dependent Hamiltonian $\mathscr{H}(t)$ and a time dependent operator $\alpha(t)$ is given by

$$\frac{\partial \alpha(t)}{\partial t} = [\alpha(t), \mathscr{H}(t)] \tag{1.2.34}$$

The commutator

$$\xi\eta - \eta\xi \tag{1.2.35}$$

between two quantum mechanical operators ξ and η is of fundamental importance since, as we shall now prove, when two operators commute, an eigenstate of one may also be an eigenstate of the other.

If a state $|A\rangle$ is simultaneously an eigenstate of two operators ξ and η, then

$$\xi |A\rangle = a_\xi |A\rangle \tag{1.2.36}$$

$$\eta |A\rangle = a_\eta |A\rangle \tag{1.2.37}$$

it follows that

$$\xi\eta |A\rangle = a_\eta a_\xi |A\rangle \tag{1.2.38}$$

and

$$\eta\xi |A\rangle = a_\xi a_\eta |A\rangle \tag{1.2.39}$$

Subtracting, we find

$$(a_\eta a_\xi - a_\xi a_\eta) |A\rangle = 0 \tag{1.2.40}$$

Thus, if the condition exists that the operators ξ and η commute, i.e.,

$$(\xi\eta - \eta\xi) = 0 \tag{1.2.41}$$

the chance that a simultaneous eigenstate exists is most favorable.

It can be shown that there exist so many simultaneous eigenstates for two commuting operators that those eigenstates form a complete set.

PROBLEM

Show that the operators

$$\xi = \frac{\partial^2}{\partial x\, \partial y} \quad \text{and} \quad \eta = \frac{\partial^2}{\partial y\, \partial x}$$

commute by operating on the arbitrary cartesian vector

$$\mathbf{A} = \hat{\imath} g(x, y) + \hat{\jmath} f(x, y)$$

Devise a simultaneous eigenvector of ξ and η.

1.3 REPRESENTATION THEORY

The Representation of Ket Vectors

We have now reviewed enough of Dirac's abstract vector algebra to relate it to the quantum mechanics with which we are familiar.

In general, we may express the state of a quantum mechanical system equally well in any of a number of different representations —in the same way that we can express a point in our more familiar three dimensional space by utilizing, e.g., either the Cartesian, elliptical or spherical representations. In representing the state of a system, however, we are not confined to a three dimensional space but rather we may utilize any of a number of many-dimensional spaces. The basic vectors which define these spaces may be, for example, unit momentum vectors which define the entire momentum space which is allowed to our system. Each basic vector in this momentum representation must be an eigenvector of the momentum operator. We may choose an equivalent *energy* (or Heisenberg) representation in which the basic vectors are eigenvectors of the energy operator and which completely define the allowed energy space.

By far the most familiar description of a quantum mechanical system is given by the Schrödinger wave mechanics. The *Schrödinger representation* is a representation of a quantum mechanical state in a familiar type of coordinate system, i.e., a three dimensional spatial system. By this method small systems are endowed with an easily visualized wave-like character and treated by means of the well-established mathematics of wave motion.

To make practical use of Dirac's powerful abstract formalism, we must first choose the particular representation in which we want to express our state. In practice, as we shall see later, the problem at hand dictates which of the representations we shall choose. At present, however, it will suffice to review the method by which the state of a system is expressed in representation. One first chooses a dual space, i.e., a set of basic bra vectors $\langle B_i|$. In practice it is most useful to choose these basic bras such that they form an orthonormal set. These bra vectors must form a complete set. The scalar product of the state vector $|A\rangle$ is then taken with each basic bra. The resultant numbers $\langle B_i|A\rangle \ldots \langle B_n|A\rangle$ are called the *representative* of the ket $|A\rangle$ in the chosen representation. Thus the representative of a ket $|A\rangle$ can be thought of as corresponding to the coordinates of a point in our chosen space. These coordinates completely define the state $|A\rangle$ in our chosen representation.

In practice we may go from any valid representation to any other. For example, we may represent the velocity of a particle in uniform motion by the vector $|\mathbf{V}\rangle$

$$|\mathbf{V}\rangle = a\,|\hat{V}_x\rangle + b\,|\hat{V}_y\rangle + c\,|\hat{V}_z\rangle \tag{1.30}$$

We may equally well represent the velocity at any time t in a momentum representation in which the basic vectors are $\langle \hat{P}_i|$, $\langle \hat{P}_j|$ and $\langle \hat{P}_k|$. We form the scalar products

$$\begin{aligned} \langle \hat{P}_i|\mathbf{V}\rangle &= a\,\langle \hat{P}_i|\hat{V}_i\rangle = aA \\ \langle \hat{P}_j|\mathbf{V}\rangle &= b\,\langle \hat{P}_j|\hat{V}_j\rangle = bB \\ \langle \hat{P}_k|\mathbf{V}\rangle &= c\,\langle \hat{P}_k|\hat{V}_k\rangle = cC \end{aligned} \tag{1.31}$$

where

$$\langle \hat{P}_i | \hat{V}_i \rangle = A$$
$$\langle \hat{P}_j | \hat{V}_j \rangle = B \tag{1.32}$$
$$\langle \hat{P}_k | \hat{V}_k \rangle = C$$

Thus the velocity of our system is represented in momentum space as the vector

$$|\mathbf{V}\rangle = aA\,|\hat{P}_i\rangle + bB\,|\hat{P}_j\rangle + cC\,|\hat{P}_k\rangle \tag{1.33}$$

as is shown in Figure 1.4.

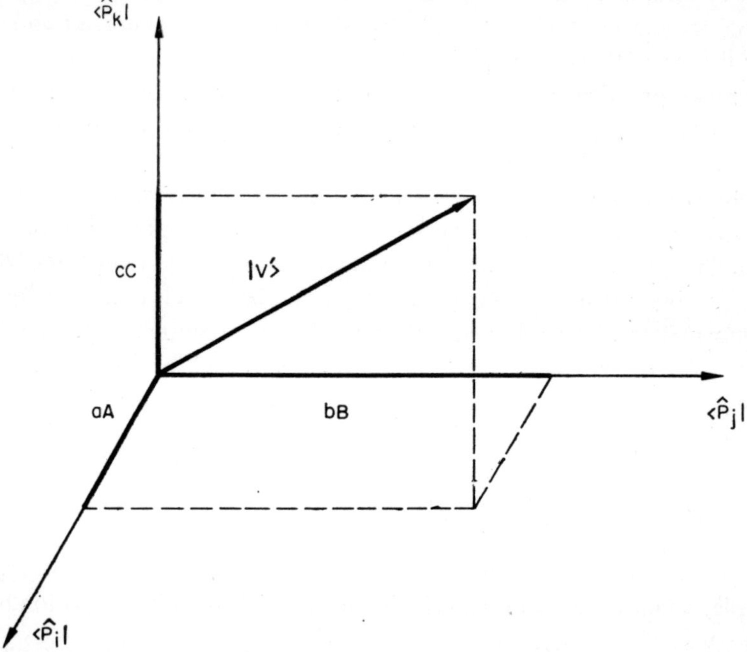

Figure 1.4 The state of the velocity of a system as represented in momentum space.

PROBLEM

Given the vector $|\mathbf{A}\rangle$ where

$$\mathbf{A} = \hat{r}\sin\theta + \theta r^2 + \hat{\phi}\cos\phi$$

find its representative in the Cartesian coordinate system whose basic bras are given by $\hat{i}\hat{j}$, and $\hat{k}$.

It can be shown that the representative of a bra vector is the complex conjugate of the representative of its conjugate ket.

It is easy to see now that any state $|P\rangle$ can be expressed in terms of the conjugate basic kets $|Q_i\rangle$ of any representation, i.e.:

a) Discrete Case

$$|P\rangle = \sum_i |a_i\rangle \langle Q_i|P\rangle \tag{1.33}$$

b) Continuous Case

$$|P\rangle = \int |Q'\rangle \, dq' \, \langle Q'|P\rangle \tag{1.34}$$

where the integration is carried out over the range of eigenvalues q.

The Representation of Operators

To perform calculations in any particular representation, we must also express our operators in that representation. To do this we utilize the same basic bras $\langle Q_i|$ that were used to find the representative of our ket vector. For an n dimensional space, we form the set of n^2 numbers

$$\langle Q_i| \alpha |Q_j\rangle \tag{1.3.5}$$

where the $|Q_j\rangle$ are the conjugate basic ket vectors. This set of numbers is the representative of the linear operator α. It is usually written in matrix form

$$\begin{vmatrix} \langle Q_1| \alpha |Q_1\rangle & \langle Q_1| \alpha Q_2\rangle & \cdots \\ \langle Q_2| \alpha |Q_1\rangle & \cdots & \cdots \\ \vdots & \ddots & \\ \vdots & & \langle Q_n| \alpha |Q_n\rangle \end{vmatrix} \tag{1.3.6}$$

If all of the basic kets of our representation are eigenstates of the linear operator α we find

$$\langle Q_i| \alpha |Q_j\rangle = a_j \langle Q_i|Q_j\rangle = a_j \delta_{ij} \tag{1.3.7}$$

which defines a *diagonal matrix* whose elements are the eigenvalues of the observable. This is a *diagonal representation.* In practice it is, of course, much simpler to determine the eigenvalues of an observable by use of an appropriate diagonal representation.

The diagonal element $\langle Q_j|\alpha|Q_j\rangle$ is the average (or expectation) value of the observable α for the state Q_j. That is, if α is measured a great number of times for the state $|Q_j\rangle$, the average value of all of the measurements will be the number a_j.

The Schrödinger Representation

The Schrödinger representation is one in which the operator q_r (the Cartesian coordinate **r**) is diagonal. The coordinates q_r are assumed to be observables and to have a continuous range of eigenvalues. The canonically conjugate momentum operator p_r is written

$$p_r = -i\hbar\nabla_r \tag{1.3.8}$$

When we want to represent the state $|A\rangle$ of a system in the Schrödinger representation, we use the complete set of basic bras of the form $|r_1\rangle$, $|r_2\rangle \ldots |r_n\rangle$ which are eigenkets of the canonical coordinates q_r. Thus the representation of the state $|A\rangle$ in the Schrödinger description is the set of numbers

$$\langle r_i|A\rangle \tag{1.3.9}$$

As before, the set of numbers (1.3.9) is a unique function $\Psi(\mathbf{r})$. In general, $|A\rangle$ and thus Ψ is changing with time and we may generalize to a function $\Psi(\mathbf{r}, t)$ where t is the time.

$\Psi(\mathbf{r}, t)$ is the function that we utilize to describe the state of a system or a particle in the Schrödinger representation. Note that operators may act on $\Psi(\mathbf{r}, t)$ only from the left.

The equation that describes the motion of $\Psi(\mathbf{r}, t)$ in time, in the Schrödinger description may be shown to be

$$i\hbar\frac{\partial}{\partial t}\Psi(\mathbf{r}, t) = \mathscr{H}\Psi(\mathbf{r}, t) \tag{1.3.10}$$

where $\mathscr{H}(\mathbf{r}, t)$ is the Hamiltonian operator for our system.

For much of our work utilizing the Schrödinger representation we deal with states that are independent of time. These *stationary states* occur when the Hamiltonian which represents our system is time independent. In this case the Hamiltonian may be shown to be equivalent to the total energy

$$\mathscr{H}(\mathbf{r}) = E + V(\mathbf{r}) = \frac{\hbar^2}{2m}\nabla^2 + V(\mathbf{r}) \tag{1.3.11}$$

where E and $V(\mathbf{r})$ are the kinetic and potential energies respectively and m is the mass of our system.

Applying the method of separation of variables we may write

$$\Psi(\mathbf{r}, t) = \psi(\mathbf{r})\,\psi(t) \tag{1.3.12}$$

Substituting (1.3.11) and (1.3.12) into the time dependent Schrödinger equation (1.3.10) yields the time-independent Schrödinger equation

$$\frac{\nabla^2\psi(\mathbf{r})}{\partial r^2} + \frac{2m}{\hbar^2}(E - V(\mathbf{r}))\,\psi(\mathbf{r}) = 0 \tag{1.3.13}$$

whose solutions $\psi(\mathbf{r})$ are stationary in time. $\Psi(\mathbf{r}, t)$ then becomes

$$\Psi(\mathbf{r}, t) = \psi(\mathbf{r})\,e^{\pm i \frac{E}{\hbar} t} \tag{1.3.14}$$

where $E = \hbar\omega$.

Equation (1.3.14) indicates that the function $\Psi(\mathbf{r}, t)$ behaves in an oscillatory manner in time. Thus we can draw an analogy between $\Psi(\mathbf{r}, t)$ and the amplitude of a wave or periodic disturbance, and $\Psi(\mathbf{r}, t)$ is called the wave function. $\Psi(\mathbf{r}, t)$ taken alone has no physical significance. However, we may, by the process of normalizing our original ket $|A\rangle$, generate a quantity which is related to the probability of finding our system in a small region of space. The ket $|A\rangle$ is normalized by utilizing the unit operator (1.2.23)

$$\int |q'_r\rangle\, dq'_r\, \langle q'_r| = 1 \tag{1.3.15}$$

where $|q'_r\rangle$ is a basic ket of the Schrödinger representation. We multiply (1.3.15) from the left by the bra $\langle A|$ and from the right by the ket $|A\rangle$

$$\int_{-\infty}^{+\infty} \langle A|q'_r\rangle\, dq' \langle q'_r|A\rangle = 1 = \langle A|A\rangle \tag{1.3.16}$$

However, $\langle q'_r|A\rangle$ is the representative of the ket $|A\rangle$ in the Schroedinger representation, i.e., $\psi(\mathbf{r})$, thus

$$\int_{-\infty}^{+\infty} \psi^*(\mathbf{r})\,\psi(\mathbf{r})\, d\mathbf{r} = 1 \tag{1.3.17}$$

If we are sure that the system we are attempting to describe exists for certain somewhere in space, we may relate the quantity $\psi^*(\mathbf{r})\psi(\mathbf{r})$ to the probability density $\varrho(\mathbf{r})$ of finding a particle in the elemental volume $d\mathbf{r}$. Equation (1.3.13) imposes the condition (for finite $V(\mathbf{r})$, that $\nabla^2\psi_{(\mathbf{r})}$ be finite everywhere, which leads to the requirement that both $\psi_{(\mathbf{r})}$ and its first derivatives be everywhere continuous. $\psi(\mathbf{r})$ must remain finite, everywhere, and therefore must decrease to zero at infinity and at an infinitely large potential.

$\Psi(\mathbf{r})$, of course, must be single-valued throughout all of space.

PROBLEM

Show by multiplication of (1.3.13) by $\psi(^*\mathbf{r})$ from the left and multiplication of the complex conjugate of 1.3.13 by $\psi(\mathbf{r})$ from the right, that one can define the probability current density by

$$j = \frac{\hbar}{2m}[\psi^*_{(\mathbf{r})}\nabla\psi_{(\mathbf{r})} - \nabla\psi^*_{(\mathbf{r})}\psi_{(\mathbf{r})}] \tag{1.3.18}$$

where $V(r)$ is assumed real.

PROBLEM

Show that the average value of the real linear operator α in the Schrödinger representation is

$$\int_{-\infty}^{+\infty} \psi^*(\mathbf{r})\,\alpha\psi(\mathbf{r})\,d\mathbf{r} \tag{1.3.19}$$

Free Particle

Let us now examine the behavior of a free electron in one dimension. This is most easily visualized by using the Schrödinger representation. For the free electron that experiences a zero or constant potential field V_0, the time independent Schrödinger equation becomes

$$\nabla^2\psi_{(x)} + \frac{2m}{\hbar^2}(E - V_0)\psi_{(\chi)} = 0 \tag{1.3.20}$$

We solve for $\psi_{(\chi)}$ by making the assumption that the solution to (1.3.18) may be of the form

$$\psi_{(x)} = A\,e^{ikx} + B\,e^{-ikx} \tag{1.3.21}$$

where

$$|\mathbf{k}| = \left|\sqrt{\frac{2m}{\hbar^2}(E - V_0)}\right| \tag{1.3.22}$$

is the magnitude of the *wave vector* whose direction is in the direction of propagation. Equation (1.3.21) thus represents the free electron state. If we include the time dependence $e^{-i\omega t}$, it is easily shown that $A\,e^{i[kx-\omega t]}$ represents a plane wave travelling in the positive x direction and $Be^{-i[kx+\omega t]}$ is a wave travelling in the negative x direction. Of course, the general solution to (1.3.20) must include both possibilities.

Taking the derivatives of (1.3.21), we find

$$\frac{d\psi(x)}{dx} = ikA\,e^{ikx} - ikB\,e^{-ikx} \tag{1.3.22}$$

and

$$\frac{d^2\psi(x)}{dx^2} = -k^2 A\, e^{ikx} - k^2 B\, e^{-ikx} = -k^2 \psi_{(x)} \tag{1.3.23}$$

Substitution of (1.3.21) and (1.3.23) into (1.3.20) yields, for the allowed energy of a free particle

$$E = \frac{\hbar^2 k^2}{2m} + V_0 \tag{1.3.24}$$

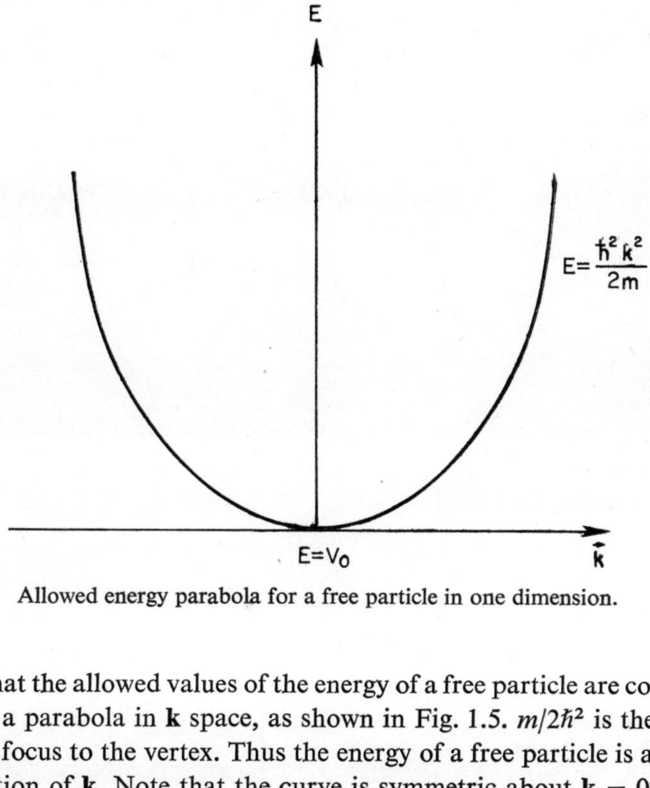

Figure 1.5 Allowed energy parabola for a free particle in one dimension.

tells us that the allowed values of the energy of a free particle are constrained to lie on a parabola in **k** space, as shown in Fig. 1.5. $m/2\hbar^2$ is the distance from the focus to the vertex. Thus the energy of a free particle is a continuous function of **k**. Note that the curve is symmetric about $\mathbf{k} = 0$ and that as the mass of the particle gets larger, the parabola along which its energy is constrained becomes broader, i.e., the mass of the free particle may be defined by the curvature of the allowed energy surface.

$$\frac{1}{m} = \frac{1}{\hbar^2} \frac{\partial^2 E}{\partial k^2} \tag{1.3.25}$$

PROBLEM

Show that the allowed energy levels of a particle trapped in an infinitely deep one-dimensional square potential well of width from $x = -\frac{q}{2}$ to $x = +\frac{q}{2}$ a are given by

$$E_n = \frac{\hbar^2\pi^2 n^2}{2ma^2} \tag{1.3.26}$$

$$n = \pm 1, \pm 2, \pm 3, \ldots$$

The solution (1.2.26) for the allowed energy levels of a particle contained within a box of width a indicates that the imposition of constraints on the free particle has limited the allowed values of $\mathbf{k}$ to the discrete spectrum of numbers $\pm \frac{n\pi}{a}$.

Show that even and odd n correspond to symmetric and antisymmetric wavefunctions of the form

$$\psi_s = \left(\frac{2}{a}\right)^{\frac{1}{2}} \cos\left(\frac{n\pi}{a}x\right) \tag{1.3.27}$$

$$\psi_a = \left(\frac{2}{a}\right)^{\frac{1}{2}} \sin\left(\frac{n\pi}{a}x\right) \tag{1.3.28}$$

respectively.

Tunneling

The requirements for conservation of probability current [equation (1.3.18)] and the continuity of $\Psi_{(\mathbf{r})}$ and its derivatives lead to some interesting and distinctly non-classical results.

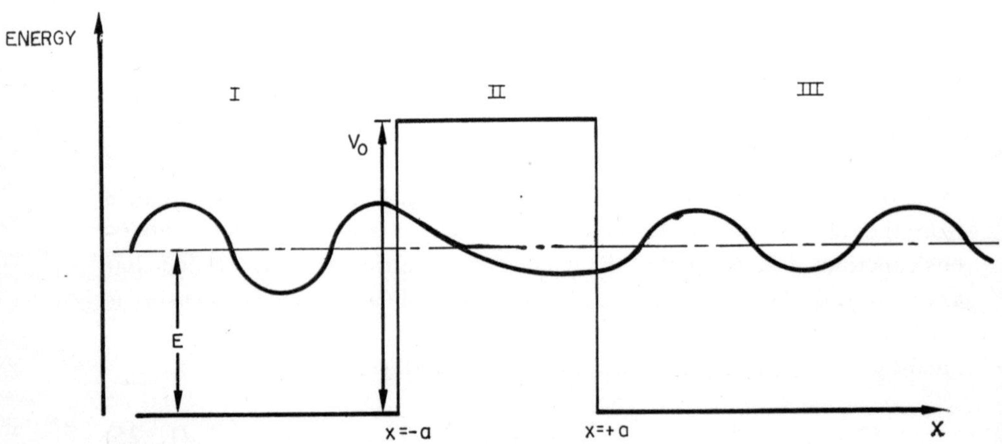

Figure 1.6 Tunneling through a square barrier.

Let us consider the one-dimensional Schrödinger representation of a particle, such as an electron, with energy E, travelling in the x direction and which impinges on a square potential barrier of energy V_0 where $V_0 > E$ as shown in Figure 1.6.

Classically the particle will be repulsed at the barrier. Quantum mechanically, however, the imposition of our boundary conditions at all points of space yields a finite probability that the particle will be found on the far side of the barrier region.

The solutions to the one-dimensional time independent Schrödinger equation (1.3.13) in regions I, II and III are

$$\psi_{\mathrm{I}}(x) = A\, e^{ikx} + B\, e^{-ikx} \tag{1.3.29a}$$

$$\psi_{\mathrm{II}}(x) = C\, e^{\varkappa x} + D\, e^{-\varkappa x} \tag{1.3.29b}$$

$$\psi_{\mathrm{III}}(x) = F\, e^{ikx} + G\, e^{-ikx} \tag{1.3.29c}$$

where

$$\varkappa = \sqrt{\frac{2m}{\hbar^2}(V_0 - E)}$$

$$k = \sqrt{\frac{2mE}{\hbar^2}} \tag{1.3.30}$$

and A, B, C, D, F and G are the amplitudes of the wave functions in the correspondings regions.

Assuming that no wave returns from $x = +\infty$ $(G = 0)$, we now invoke the boundary conditions that $\psi(x)$ and $\partial\psi_{(x)}/\partial x$ be continuous at $x = 0$ and $x = a$ which yield

$$A = F\, e^{+2ika}\left[\cosh(2\varkappa a) - \frac{i}{2}\left[\frac{k}{\varkappa} - \frac{\varkappa}{k}\right]\sinh 2\varkappa a\right] \tag{1.3.31}$$

and

$$B = -\frac{i}{2}F\left[\frac{\varkappa}{k} + \frac{k}{\varkappa}\right]\sinh(2\varkappa a) \tag{1.3.32}$$

Thus we may calculate the relative amplitudes of the transmitted and reflected waves. The transmission coefficient T of a barrier is defined as the ratio of the transmitted probability current density to the incident probability current density

$$T = \frac{[\overset{*}{\psi}_{\mathrm{III}(x)}\, \nabla\psi_{\mathrm{III}(x)} - \nabla\overset{*}{\psi}_{\mathrm{III}(x)}\, \psi_{\mathrm{III}(x)}]}{[\overset{*}{\psi}_{\mathrm{I}(x)}\, \nabla_{\mathrm{I}(x)} - \nabla\overset{*}{\psi}_{\mathrm{I}(x)}\, \psi_{\mathrm{I}(x)}]} \tag{1.3.33}$$

We define the reflection coefficient R of a barrier as the ratio of the reflected probability current density to the incident probability current density. R may be calculated from the requirement for the conservation of probability current density

$$T + R = 1 \tag{1.3.34}$$

The transmission coefficient becomes in the present case

$$T = \frac{|F|^2}{|A|^2} = \frac{1}{\frac{1}{4}\frac{V_0^2}{E(V_0 - E)}\sinh^2(2\varkappa a) + 1} \tag{1.3.35}$$

which reduces in the case of a high, wide barrier to

$$T \approx \frac{16(k\varkappa)^2}{e^{4ka}(k^2 + \varkappa^2)^2} \tag{1.3.36}$$

A graph of the transmission coefficient T as a function of the ratio of particle to barrier energy is shown in Figure 1.7.

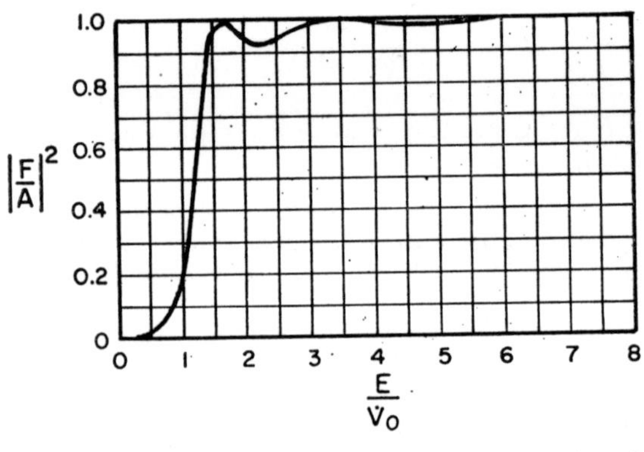

Figure 1.7 Transmission coefficient for a square barrier as a function of energy for $\frac{mV_0}{2}[2a]^2 = 8$ [from L. I. Schiff, *Quantum Mechanics*, used with the permission of McGraw-Hill Book Co., Inc., New York, 1955 Page 95].

Most real potentials, which are not as easily solvable as is the square barrier, require the use of approximation methods. For potential barriers that are much more slowly varying than the discontinuous square barrier, one may make use of the WKB approximation[3].

For a slowly varying one-dimensional barrier of the form $V(x)$ as shown in Figure 1.8, we may assume a solution to (1.3.13) to be of the form

$$\psi = A_{(x)}\, e^{\pm i \int k(x)dx} \tag{1.3.37}$$

where

$$k(x) = \sqrt{\frac{2m}{\hbar^2}(E - V_{(x)})} \tag{1.3.38}$$

Substitution of (1.3.37) into (1.3.13) yields

$$\frac{-i}{k(x)}\frac{d^2A_{(x)}}{dx^2} \pm \left(\frac{2dA_{(x)}}{dx} + \frac{A_{(x)}}{k(x)}\frac{dk(x)}{dx}\right) = 0 \tag{1.3.39}$$

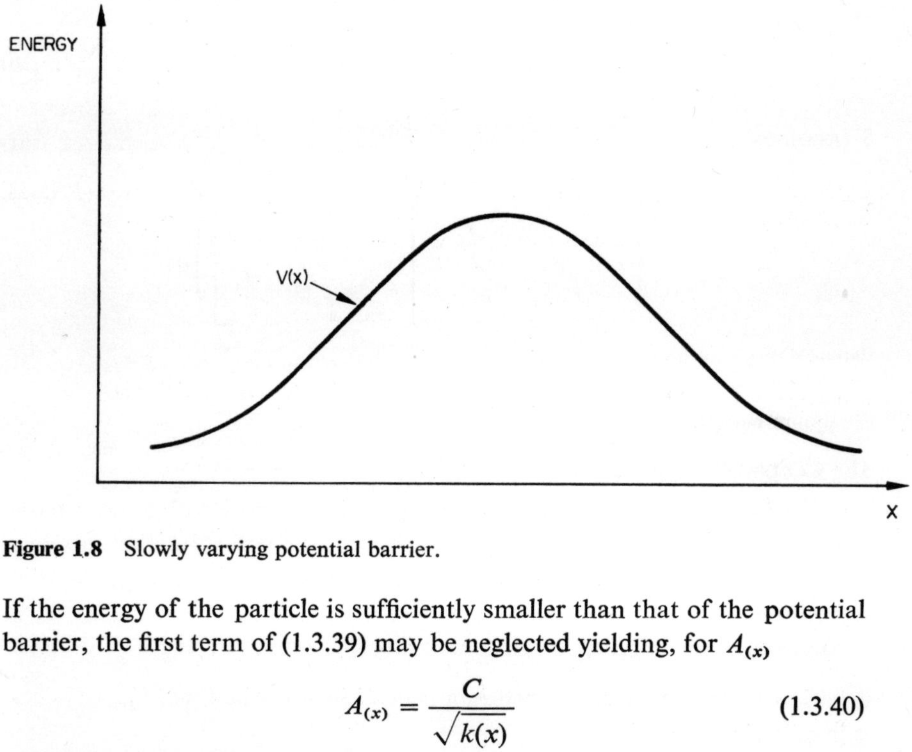

Figure 1.8 Slowly varying potential barrier.

If the energy of the particle is sufficiently smaller than that of the potential barrier, the first term of (1.3.39) may be neglected yielding, for $A_{(x)}$

$$A_{(x)} = \frac{C}{\sqrt{k(x)}} \tag{1.3.40}$$

where C is a constant.

Thus an approximate solution to (1.3.13) is

$$\psi_{(x)} = \frac{C}{\sqrt{k(x)}}\, e^{\pm \int_x k(x)\,dx} \tag{1.3.41}$$

The determination of the transmission coefficient for an arbitrarily shaped barrier is determined as before where

$$\Psi_I = \frac{A}{\sqrt{k(x)}} e^{i\int_x^a k(x)\,dx} + \frac{B}{\sqrt{k(x)}} e^{-i\int_x^a k(x)\,dx}$$

$$\Psi_{II} = \frac{C}{\sqrt{\varkappa(x)}} e^{\int_a^x \varkappa(x)\,dx} + \frac{D}{\sqrt{\varkappa(x)}} e^{-\int_a^x \varkappa(x)\,dx}$$

$$\Psi_{III} = \frac{E}{\sqrt{k(x)}} e^{i\int_b^x k(x)dx} + \frac{F}{\sqrt{k(x)}} e^{-i\int_b^x k(x)\,dx} \qquad (1.3.41)$$

where

$$\varkappa_{(x)} = \sqrt{\frac{2m}{\hbar^2}(V_{(x)} - E)} \qquad (1.3.42)$$

T becomes

$$T = \frac{4}{\left(2\exp\int_a^b \varkappa(x)\,dx + \left[\dfrac{1}{2\exp\int_a^b \varkappa(x)\,dx}\right]\right)^2} \qquad (1.3.43)$$

PROBLEM

Using the WKB approximation determine the approximate transmission coefficient for the square barrier.

The Central Field Problem

The problem of the one-electron hydrogen-like atom is of course a basic foundation for the study of solid state physics. We will solve the one-electron atom and use it as a basis for our study of the allowed energy levels in crystalline solids.[4]

Neglecting the motion of the nucleus of charge $+z$, the Schrödinger equation for an electron trapped in a central field of the form $V_{(r)} = \dfrac{-ze^2}{r}$ is in spherical coordinates

$$\frac{d^2\psi(\mathbf{r})}{dr^2} + \frac{2}{r}\frac{d\psi(\mathbf{r})}{dr} + \frac{1}{r^2}\frac{d^2\psi(\mathbf{r})}{d\theta^2} + \frac{\cot\theta}{r^2}\frac{d\psi(\mathbf{r})}{d\theta}$$

$$+ \frac{1}{r^2\sin^2\theta}\frac{d^2\psi(\mathbf{r})}{d\phi^2} + \frac{2m}{\hbar^2}[E - V_{(r)}]\,\psi(\mathbf{r}) = 0 \qquad (1.3.44)$$

Using the separation of variables technique, we assume that $\psi(\mathbf{r})$ may be written as the product of these independent functions.

$$\psi(\mathbf{r}) = R_{(r)}\,\Theta_{(\theta)}\,\Phi_{(\phi)} \tag{1.3.45}$$

We find, upon substitution of (1.3.45) into (1.3.44)

$$\frac{\sin^2\theta}{R_{(r)}}\frac{d}{dr}\left(r^2\frac{dR_{(r)}}{dr}\right)+\frac{1}{\Phi_{(\phi)}}\frac{d^2\Phi}{d\phi^2}+\frac{\sin\theta}{\Theta_{(\theta)}}\frac{d}{d\theta}\left(\frac{\sin\theta\, d\Theta_{(\theta)}}{d\theta}\right)$$

$$+\frac{2m}{\hbar^2}(E-V_{(r)})\,r^2\,\mathrm{Sin}^2\,\theta = 0 \tag{1.3.46}$$

The second term of (1.3.46) depends only on the coordinate ϕ and thus may be considered as a constant $-m^2$ within the remainder of the equation in which the terms depend only on r and θ. Thus we generate the first of our three separated equations

$$\frac{d^2\Phi}{d\phi^2} = -m^2\Phi \tag{1.3.47}$$

(1.3.46) may then be written

$$\frac{1}{R_{(r)}}\frac{d}{dr}\left(r^2\frac{dR_{(r)}}{dr}\right)+\frac{2mr^2}{\hbar^2}(E-V_{(r)})$$

$$-\frac{m^2}{\sin^2\theta}+\frac{1}{\sin\theta\,\Theta_{(\theta)}}\frac{d}{d\theta}\left(\mathrm{Sin}\,\theta\frac{d\Theta}{d\theta}\right) = 0 \tag{1.3.48}$$

The first two terms of (1.3.48) depend only on the variable r, the last two only on θ. Thus each pair may be set equal to a constant. Setting the first two terms equal to $+\beta$ and the last two terms to $-\beta$ yields the second and third of our separated equations

$$\frac{1}{\sin\theta}\frac{d}{d\theta}\left(\sin\theta\frac{d\Theta_{(\theta)}}{d\theta}\right)-\frac{m^2}{\sin^2\theta}\Theta_{(\theta)}+B\Theta_{(\theta)} = 0 \tag{1.3.49}$$

$$\frac{1}{r^2}\frac{d}{dr}\left(r^2\frac{dR_{(r)}}{dr}\right)-\frac{B}{r^2}R_{(r)}+R_{(r)}\frac{2m}{\hbar^2}(E-V_{(r)}) = 0 \tag{1.3.50}$$

Equation (1.3.47) may be solved to yield, after normalization of $\Phi_m(\phi)$

$$\Phi_m(\phi) = \frac{1}{\sqrt{2\pi}}e^{\pm im\phi} \tag{1.3.50}$$

In order to retain single-valuedness in $\psi(\mathbf{r})$, we must restrict m to the integer values 0, ± 1, ± 2, ... $\pm n$, each of which corresponds to a distinct solution to (1.3.47). m is called the magnetic quantum number.

Equation (1.3.49), Legendre's equation, has solutions only where

$$\beta = l(l+1) \qquad l = 0, 1, 2, \ldots n \tag{1.3.51}$$

where l is the azimuthal quantum number.

The solutions $\Theta_{lm}(\theta)$ are written

$$\theta_{lm}(\theta) = \sqrt{\frac{(2l+1)(l-m)!}{2(l+m)!}}\, P_l^m(\theta) \tag{1.3.52}$$

where $P_l^{|m|}(\theta)$ are the associated Legendre functions

$$P_l^{|m|}(\theta) = \sin^{|m|}(\theta) \frac{d^{|m|}}{d(\cos\theta)^{|m|}} P_l(\cos\theta) \tag{1.3.53}$$

and the $P_l(\cos\theta)$ are the Legendre polynomials defined by

$$P_l(\cos\theta) = \frac{1}{2^l l!} \frac{d^l(\cos^2\theta - 1)^l}{d(\cos\theta)^l} \tag{1.3.54}$$

For problems in which the potential is a function of r alone, it is convenient to represent the angular portion of the wave function in terms of the spherical harmonics

$$Y_{lm}(\theta, d) = (-1)^m \sqrt{\frac{(2l+1)(l-m)!}{4\pi(l+m)!}}\, P_{lm}(\theta)\, e^{imd} \tag{1.3.55}$$

Thus we may write equation (1.3.50) as

$$\frac{1}{r^2}\frac{d}{dr}\left(r^2\frac{dR}{dr}\right) + \frac{2m}{\hbar^2}\left(E + \frac{Ze^2}{r}\right)R_{(r)} - \frac{l(l+1)R_{(r)}}{r^2} = 0 \tag{1.3.56}$$

(1.3.56) is the Laguerre equation whose physically acceptable solutions are written[5]

$$R_{nl}(r) = -\left[\left(\frac{2me^2z}{n\hbar^2}\right)^3 \left(\frac{(n-l-1)!}{\{2n(n+l)!\}^3}\right)\right]^{\frac{1}{2}} e^{-\varrho/2} \varrho^l L_{n+1}^{2l+1}(\varrho) \tag{1.3.57}$$

where n is the principal quantum number and may take on the values 1, 2, 3, ...

$$\varrho = \frac{2mz\,e^2 r}{n\hbar^2} \tag{1.3.58}$$

and the $L_{n+l}^{2l+1}(\varrho)$ are the associated Laguerre functions

$$L_r^s(\varrho) = \frac{d^s}{d\varrho^s} L_r(\varrho) \tag{1.3.59}$$

where $L_r\varrho()$ are the Laguerre polynomials

$$L_r(\varrho) = e^{\varrho} \frac{d^r}{d\varrho^r} (\varrho^r e^{-\varrho}) \tag{1.3.60}$$

The allowed values for the quantum numbers n,l and m

$$n = 1, 2, 3, \ldots$$

$$l = 0, 1, 2, 3, \ldots n - 1$$

$$m = 0, \pm 1, \pm 2, \ldots \pm l$$

may be derived from the condition that the total wave function be everywhere finite.

The states $l = 0, 1, 2, \ldots$ are referred to as s, p, d, f, g, etc. states.

The allowed energy levels are given by

$$E_n = \frac{-me^4z^2}{2h^2\hbar^2} \tag{1.3.61}$$

Graphs of the radial wave functions $R_{nl}(r)$ and the radial electron distribution functions $4\pi r^2[R_{n,l}(r)]^2$ are shown in Figures 1.9 and 1.10 respectively.

The average distance $\langle r_{n,l,m}\rangle$ of an electron from the nucleus is given by

$$\langle r_{nlm}\rangle = \int \psi^*_{n\,lm} r \psi_{nlm} r^2\, dr \sin\theta\, d\theta\, d\varphi = \frac{n^2\hbar^2}{zm\,e^2}\left[1 + \frac{1}{2}\left(1 - \frac{l(l+1)}{n^2}\right)\right] \tag{1.3.62}$$

PROBLEM

Prove that the state wavefunction $\psi_{nlm}(\mathbf{r})$ is spherically symmetric.

Thus the total wave functions allowed for the hydrogen-like atom are

$$\psi_{(\mathbf{r})} = Y_{lm}(\theta, d)\, R_{nl}(r) \tag{1.3.63}$$

where $Y_{lm}(\theta, d)$ and $R_{nl}(r)$ are given by equations (1.3.55) and (1.3.57), respectively. A few of the hydrogen-like wavefunctions are tabulated in Table 1.

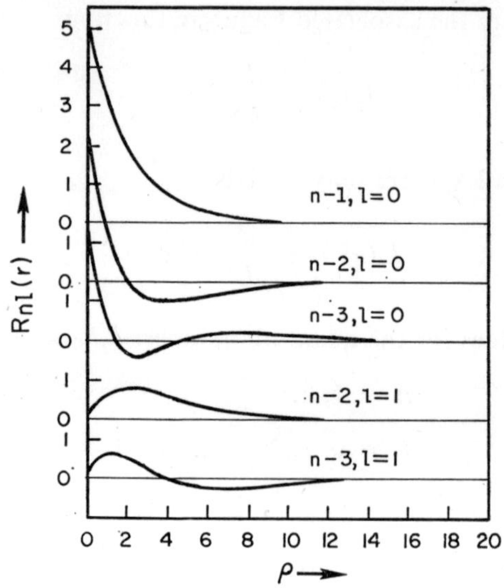

Figure 1.9 Hydrogen-like radial wave-functions $R_{n,l}(r)$ for $n = 1, 2, 3$; $l = 0,1$, [from L. Pauling and E. B. Wilson, *Introduction to Quantum Mechanics*, used with permission of McGraw-Hill Book, Co., Inc., New York, 1935, page 142].

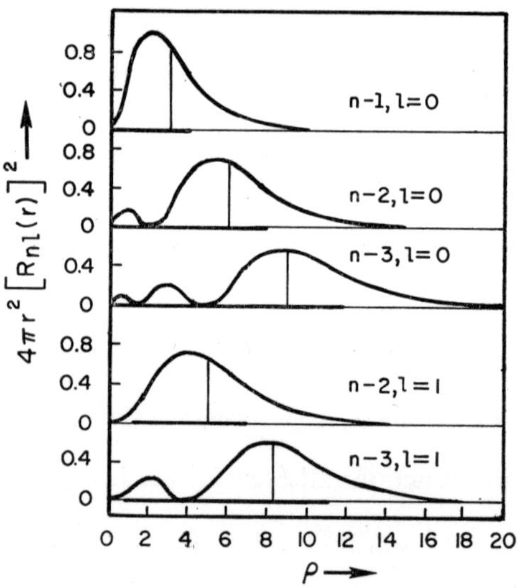

Figure 1.10 Radial electron distribution functions $4\pi r^2[R_{n,l}(r)]^2$ for the hydrogen atom [from L. Pauling and E. B. Wilson, *Introduction to Quantum Mechanics*, used with permission of McGraw-Hill Book Co., Inc., New York, 1935, page 192].

Table 1.1 Wave functions associated with the first four principal quantum numbers [by permission. J. P. McKelvey, *Solid State and Semiconductor Physics*. Harper and Rowe, Pubs., New York, 1966, p. 117]

$n=1$	$l=0$	$m=0$	ψ_{100}	$= N_{100}\, e^{-1/2\varrho}$	$1s$
$n=2$	$l=0$	$m=0$	ψ_{200}	$= N_{200}(2-\varrho)\, e^{-1/2\varrho}$	$2s$
	$l=1$	$m=0$	ψ_{210}	$= N_{210}\varrho\, e^{-1/2\varrho}\cos\theta$	$2p^3$
(4 states)		$m=\pm1$	$\psi_{21\pm1}$	$= N_{211}\varrho\, e^{-1/2\varrho}\sin\theta\, e^{\pm i\phi}$	
$n=3$	$l=0$	$m=0$	ψ_{300}	$= N_{300}(6-6\varrho+\varrho^2)\, e^{-1/2\varrho}$	$3s$
	$l=1$	$m=0$	ψ_{310}	$= N_{310}\varrho(4-\varrho)\, e^{-1/2\varrho}\cos\theta$	$3p^3$
		$m=\pm1$	$\psi_{31\pm1}$	$= N_{311}\varrho(4-\varrho)\, e^{-1/2\varrho}\sin\theta\, e^{\pm i\phi}$	
	$l=2$	$m=0$	ψ_{310}	$= N_{320}\varrho^2\, e^{-1/2\varrho}(3\cos^2\theta-1)$	$3d^5$
		$m=\pm1$	$\psi_{32\pm1}$	$= N_{321}\varrho^2\, e^{-1/2\varrho}\sin\theta\cos\theta\, e^{\pm i\phi}$	
(9 states)		$m=\pm2$	$\psi_{32\pm2}$	$= N_{322}\varrho^2\, e^{-1/2\varrho}\sin\theta\, e^{\pm i\varphi}$	
$n=4$	$l=0$	$m=0$	ψ_{400}	$= N_{400}(24-36\varrho+12\varrho^2-\varrho^3)\, e^{-1/2\varrho}$	$4s$
	$l=1$	$m=0$	ψ_{410}	$= N_{410}\varrho\, e^{-1/2\varrho}(20-10\varrho+\varrho^2)\cos\theta$	$4p^3$
		$m=\pm1$	$\psi_{4\pm11}$	$= N_{411}\varrho\, e^{-1/2\varrho}(20-10\varrho+\varrho^2)\sin\theta\, e^{\pm i\phi}$	
	$l=2$	$m=0$	ψ_{420}	$= N_{420}\varrho^2(6-\varrho)\, e^{-1/2\varrho}(3\cos^2\theta-1)$	$4d^5$
		$m=\pm1$	$\psi_{42\pm1}$	$= N_{421}\varrho^2(6-\varrho)\, e^{-1/2\varrho}\sin\theta\cos\theta\, e^{\pm i\phi}$	
		$m=\pm2$	$\psi_{42\pm2}$	$= N_{422}\varrho^2(6-\varrho)\, e^{-1/2\varrho}\sin^2\theta\, e^{\pm 2i\phi}$	
	$l=3$	$m=0$	ψ_{430}	$= N_{430}\varrho^3\, e^{-1/2\varrho}(5/3\cos^3\theta-\cos\theta)$	$4f^7$
		$m=\pm1$	$\psi_{43\pm1}$	$= N_{431}\varrho^3\, e^{-1/2\varrho}(5\cos^2\theta-1)\sin\theta\, e^{\pm i\phi}$	
(16 states)		$m=\pm2$	$\psi_{43\pm2}$	$= N_{432}\varrho^3\, e^{-1/2\varrho}\sin^2\theta\cos\theta\, e^{\pm 2i\phi}$	
		$m=\pm3$	$\psi_{43\pm3}$	$= N_{433}\varrho^3\, e^{-1/2\varrho}\sin^3\theta\, e^{\pm 3i\phi}$	

$$N_{nlm} = -\left[\left(\frac{2m_0 e^2}{n\hbar^2}\right)^3 \frac{(n-l-1)!\,(l-m)!\,(2l+1)}{4\pi(2n[n+l]!)^3\,(l+m)!}\right]^{1/2}$$

The Momentum Representation

It is frequently convenient in our study of solids to work in the representation in which the basic vectors are eigenvectors of the momentum operator $\boldsymbol{p}$. We will find it profitable to use the momentum related variable $\mathbf{k}$ where

$$\mathbf{k} = \frac{2\pi}{\lambda} = \frac{2\pi}{h}P = \frac{\mathbf{P}}{\hbar} \tag{1.3.64}$$

The wave functions $\psi(\mathbf{k})$ in this momentum space are defined by the Fourier transform of the wave function in Cartesian space

$$\psi(\mathbf{k}) = \left\{\frac{1}{\sqrt{2\pi}}\right\}^3 \int d\mathbf{r}\psi(\mathbf{r})\, e^{-i\mathbf{kr}} \tag{1.3.65}$$

$$\psi(\mathbf{r}) = \left(\frac{1}{\sqrt{2\pi}}\right)^3 \int d\mathbf{k}\psi(\mathbf{k})\, e^{i\mathbf{kr}} \tag{1.3.66}$$

PROBLEM

Show that the Fourier transform of the Schrödinger equation leads to the integral equation of motion

$$\left(\frac{\hbar^2 k^2}{2m} - E\right)\psi(\mathbf{k}) = -\int d\mathbf{k}\psi(\mathbf{k}')\, V'(\mathbf{k}-\mathbf{k}')\left[\frac{1}{\sqrt{2\pi}}\right]^3 \tag{1.3.67}$$

where

$$V'(\mathbf{k}) = \left\{\frac{1}{\sqrt{2\pi}}\right\}^3 \int d\mathbf{r}\, e^{-i\mathbf{k}\cdot\mathbf{r}} V(\mathbf{r}) \tag{1.3.68}$$

The Heisenberg Representation

In our generalized theory of quantum mechanical systems we represent a state by a ket vector which may well be a function of time t. These ket vectors are operated on by time-independent operators. In the Schrödinger representation we saw that the so-called wave function which represents our system retains the time dependent character of the state.

In the Heisenberg representation the basic bras and kets must be eigenfunctions of the Hamiltonian operator which is therefore diagonal. The Heisenberg picture differs from that of the other representations in that a linear time-dependent transformation is performed on our time-dependent state vectors rendering these states time-independent. The inverse of this transformation is then applied to the relevant operators transforming them into time-dependent operators. The application, then, of these time dependent operators on the time independent states produces the equations of motion of our quantum mechanical system in the Heisenberg representation.

If our quantum mechanical state is time-independent to begin with, our operators in the Heisenberg representation will of course remain time-independent. It is easily shown that (1.2.34) holds for time-dependent Heisenberg operators. For time-independent Heisenberg operators it becomes

$$\frac{d\alpha}{dt} = i\hbar[\alpha, \mathscr{H}] \tag{1.3.69}$$

The Harmonic Oscillator

In this section we shall, by means of a simple example, demonstrate the utility of Dirac's generalized quantum mechanical formalism. This section will serve also as an introduction to the concepts necessary for future studies in which we shall take up the subject of electron-phonon interactions.

We will derive the allowed vibrational energy levels for a simple one-dimensional quantum mechanical harmonic oscillator whose Hamiltonian may be written

$$\mathscr{H} = \frac{P^2}{2m} + \frac{m\omega^2}{2} r^2 \tag{1.3.70}$$

where $\omega = \sqrt{\frac{k}{m}}$, m is the mass, K is the effective spring constant, p is the momentum, and r is the deviation of the oscillating particle from its equilibrium position.

Using equations (1.2.33) and (1.3.69), we may write

$$i\hbar\dot{r} = r\mathscr{H} - \mathscr{H}r \tag{1.3.71}$$

and

$$i\hbar\dot{p} = p\mathscr{H} - \mathscr{H}p \tag{1.3.72}$$

$$\dot{r} = p/m \tag{1.3.73}$$

and

$$\dot{p} = -m\omega^2 r \tag{1.3.74}$$

We shall define the following operator

$$\beta = (2m\hbar\omega)^{-\frac{1}{2}}\,(m\omega r + ip) \tag{1.3.75}$$

and form its complex conjugate

$$\beta^* = (2m\hbar\omega)^{-\frac{1}{2}}\,(m\omega r - ip) \tag{1.3.76}$$

Note that β and β^* are not real linear operators, and thus are not observables.

It is easily shown that

$$\beta\beta^* = \frac{\mathscr{H}}{\hbar\omega} + \frac{1}{2} \tag{1.3.77}$$

and

$$\beta^*\beta = \frac{\mathscr{H}}{\hbar\omega} - \frac{1}{2} \tag{1.3.78}$$

from which we find the commutation relation between β and β^*

$$\beta\beta^* - \beta^*\beta = 1 \tag{1.3.79}$$

Using (1.3.77), (1.3.78) and (1.3.79) it is now possible to find the commutation relations between $\mathscr{H}$, β and β^*

$$\beta\mathscr{H} - \mathscr{H}\beta = \hbar\omega\beta \tag{1.3.80}$$

$$\mathscr{H}\beta^* - \beta^*\mathscr{H} = \hbar\omega\beta^* \tag{1.3.81}$$

Let us assume that $|H'\rangle$ is an eigenket of the Hamiltonian operator and has the eigenvalue E' From (1.3.80) we find

$$\mathscr{H}\beta\,|H'\rangle = (E' - \hbar\omega)\,\beta\,|H'\rangle \tag{1.3.82}$$

Equation (1.3.82) indicates that the new ket vector $|H''\rangle$ generated by the operation of β on $|H'\rangle$ i.e.,

$$\beta\,|H'\rangle = |H''\rangle$$

is also an eigenket of the Hamiltonian, and has the eigenvalue $(E' - \hbar\omega)$. The application of $\beta\, n$ times will produce an eigenket

$$(\beta)^n\,|H'\rangle \tag{1.3.83}$$

whose eigenvalue is $(E' - n\hbar\omega)$.

Thus it seems that the application of β on a particular eigenstate yields the next lower eigenstate. The repeated application of β will in time generate negative eigenvalues such that $E' < n\hbar\omega$. Since we are not interested in negative energy levels, let us attempt to determine the lowest allowable positive energy of our harmonic oscillator. From equation (1.3.78) we find

$$\hbar\omega\beta^*\beta = \mathscr{H} - \tfrac{1}{2}\hbar\omega \tag{1.3.84}$$

Multiplying (1.3.84) from the left and from the right by $\langle H'|$ and $|H'\rangle$, respectively, we find

$$\hbar\omega(\langle H'|\,\beta^*)\,(\beta|H'\rangle) = \left(E' - \frac{\hbar\omega}{2}\right) \tag{1.3.85}$$

The left hand side of (1.3.85) must be equal to or greater than zero, since the quantity in brackets is the "length" of the ket $\beta|H'\rangle$. Therefore $E' - \dfrac{\hbar\omega}{2}$ must also always be greater than or equal to zero. Thus we may repeatedly apply β^* to an eigenstate of the harmonic oscillator only until we reach the lowest allowed energy which is $\tfrac{1}{2}\hbar\omega$.

In the same manner, using equation (3.2.81) we find

$$\mathscr{H}(\beta^*\,|H'\rangle) = (E' + \hbar\omega)\,(\beta^*\,|H'\rangle) \tag{1.3.86}$$

and by the same arguments as before we may state that $E' + n\hbar\omega$ is again an eigenvalue of the Hamiltonian. Thus we have found the allowed energy eigenvalues for the harmonic oscillator

$$(E' - n\hbar\omega),\ (E' - (n-1)\,\hbar\omega)\ \ldots\ E'(E' + \hbar\omega)\ \ldots\ (E' + n\hbar\omega) \tag{1.3.87}$$

or

$$(n + \tfrac{1}{2})\,\hbar\omega \tag{1.3.88}$$

The operators β^* and β are called *creation* and *annihilation* operators, respectively. Multiplication of (1.3.79) from the right by β^* yields

$$\beta\beta^{*2} - \beta^*\beta\beta^* = \beta^* \tag{1.3.89}$$

Using (1.3.79) we find also

$$\beta\beta^{*2} - \beta^{*2}\beta = 2\beta^* \tag{1.3.90}$$

This procedure may be repeated many times to yield

$$\beta\beta^{*n} - \beta^{*n}\beta = n\beta^{*(n-1)} \tag{1.3.91}$$

Letting $|H_0\rangle$ be our lowest energy state, we may write

$$\beta\beta^{*n}|H_0\rangle - \beta^{*n}\beta|H_0\rangle = n\beta^{*(n-1)}|H_0\rangle \tag{1.3.92}$$

Thus, since $\beta H_0\rangle = 0$,

$$\beta[\beta^{*n}|H_0\rangle] = n\beta^{*(n-1)}|H_0\rangle \tag{1.3.93}$$

If we let $|H_n\rangle$ represent the ket produced by successive applications of β^* on the zero state, i.e.

$$\beta^{*n}|H_0\rangle = |Hn\rangle \tag{1.3.94}$$

we can find the result of operating with β

$$\frac{\beta|Hn\rangle}{(n!)^{\frac{1}{2}}} = \frac{\beta\beta^{*n}|H_0\rangle}{(n!)^{\frac{1}{2}}} \tag{1.3.95}$$

where $(n!)^{\frac{1}{2}}$ is introduced for normalization,

Using (1.3.93) we may write

$$\frac{\beta^n\beta^{*n}|H_0\rangle}{(n!)^{\frac{1}{2}}} = \frac{n\beta^{*(n-1)}|H_0\rangle}{(n!)^{\frac{1}{2}}} \tag{1.3.96}$$

which yields

$$\frac{\beta|Hn\rangle}{(n!)^{\frac{1}{2}}} = \frac{n^{\frac{1}{2}}|H_{(n-1)}\rangle}{[(n-1)!]^{\frac{1}{2}}} \tag{1.3.97}$$

Thus the application of β on an eigenket of $\mathscr{H}$ yields the next lower energy eigenket. The name *demotion* or *annihilation* operator is given to β

In a similar manner, by considering the action of β^* on an eigenket $|Hn\rangle$ we find

$$\frac{\beta^*|Hn\rangle}{(n!)^{\frac{1}{2}}} = \frac{(n+1)^{\frac{1}{2}}(|H_{(n-1)}\rangle)}{[(n+1)!]^{\frac{1}{2}}} \tag{1.3.98}$$

Thus β^* acts as a *promotion* or *creation* operator. When it acts on an eigenstate of the Hamiltonian it promotes that state by one quantum ($\hbar\omega$) of energy.

The ground state wave function ψ_0 is easily found by operating on $|H_0\rangle$ with β. Utilizing equations (1.3.8) and (1.3.76)

$$\beta\,|H_0\rangle = (2m\hbar\omega)^{-\frac{1}{2}}\,[m\omega r + \hbar\,\nabla_r]\,|H_0\rangle = 0 \tag{1.3.99}$$

Multiplication of (1.3.99) from the left by the basic bra vectors $\langle r'|$ of the Schrödinger representation gives

$$(2m\hbar\omega)^{-\frac{1}{2}}\,(m\omega r\,\langle r'|H_0\rangle + \hbar\nabla_r\langle r'|H_0\rangle = 0 \tag{1.3.100}$$

which yields the normalized ground state wave function

$$\langle r'|H_0\rangle = \psi_0 = \left(\frac{m\omega}{\pi\hbar}\right)^{\frac{1}{4}} e^{-m\omega r^2/2\hbar} \tag{1.3.101}$$

PROBLEM

Prove equation (1.3.101)

1.4 PERTURBATION THEORY

In the simplest of quantum mechanical problems, what we look for are the allowed energy eigenvalues for the system. The easiest way to go about this is to formulate a representation in which the Hamiltonian is a diagonal operator, for then we need only look at the diagonal elements. Unfortunately, not all quantum mechanical systems have Hamiltonians for which a diagonal representation is as easy to find as it was for the harmonic oscillator. Thus this standard method for finding the eigenvalues of our system is not always feasible. When this happens we utilize perturbation theory.

To use the method of perturbation theory, the intractable Hamiltonian $\mathscr{H}$ must be separable into two portions:

$$\mathscr{H} = \mathscr{H}_0 + \mathscr{H}_1 \tag{1.4.1}$$

The first portion $\mathscr{H}_0$ is a Hamiltonian easily solvable by the usual methods. The second part $\mathscr{H}_1$ need only satisfy the condition that it produce a small effect relative to $\mathscr{H}_0$.

When both portions of the total Hamiltonian are not explicitly functions of time, perturbation theory becomes very simple.

We form the eigenvalue equation

$$\mathscr{H}\,|H\rangle = E'\,|H\rangle \tag{1.4.2}$$

which yields

$$(E' - \mathscr{H}_0)\,|H\rangle = \mathscr{H}_1\,|H\rangle \tag{1.4.3}$$

where $|H\rangle$ is assumed to be an eigenstate of the total Hamiltonian. We assume that the eigenket $|H\rangle$ of the total Hamiltonian can be expressed in series form as the sum of many eigenkets denoted by $|0\rangle$, $|1\rangle$, $|2\rangle$, ... $|n\rangle$ in which the first ket $|0\rangle$ of the series is assumed to be the eigenket of the unperturbed portion $\mathscr{H}_0$ of the Hamiltonian. We may write

$$|H\rangle = |0\rangle + |1\rangle + |2\rangle + \dots |n\rangle \tag{1.4.4}$$

and the eigenvalues may be written

$$H' = E_0 + E_1 + E_2 + \dots + E_n \tag{1.4.5}$$

where E_1 is the first order correction to the energy eigenvalue.

Substituting (1.4.4) and (1.4.5) into (1.4.3) yields

$$\left[\sum_{j=0}^{n} E_j - \mathscr{H}_0\right] \sum_{i=0}^{n} |i\rangle = \mathscr{H}_1 \sum_{i=0}^{n} |i\rangle \tag{1.4.6}$$

Separation of terms according to order gives

$$(E_0 - \mathscr{H}_0)\,|0\rangle = 0 \tag{1.4.7}$$

$$(E_0 - \mathscr{H}_0)\,|1\rangle + E_1\,|0\rangle = \mathscr{H}_1\,|0\rangle \tag{1.4.8}$$

$$(E_0 - \mathscr{H}_0)\,|2\rangle + E_1\,|1\rangle + E_2\,|2\rangle = \mathscr{H}_1\,|1\rangle \tag{1.4.9}$$

Equation (1.4.7) tells us that H_0 is the energy eigenvalue of the unperturbed system. The first order correction to the unperturbed system is found from equation (1.4.8). First well rewrite (1.4.8) in the Heisenberg (energy) representation in which $\mathscr{H}_0$ is diagonal and where the basic bras are given by $\langle E_{j,l}|$

$$\langle E_{j,l}|\,(E_0 - \mathscr{H}_0)\,|1\rangle + \langle E_{j,l}|\,E_1\,|0\rangle = \langle E_{j,l}|\,\mathscr{H}_1\,|0\rangle \tag{1.4.10}$$

where we introduce the index l to generalize to the case where more than one eigenket belongs to the eigenvalue E_j.

Using the unitary operator

$$\sum_{l'} |E_{j,l'}\rangle\,\langle E_{j,l'}| = 1 \tag{1.4.11}$$

we find

$$\langle E_{j,l}|\,(E_0 - \mathscr{H}_0)\,|1\rangle + E_1\langle E_{j,l}\,|0\rangle = \sum_{l'} \langle E_{j,l}|\mathscr{H}_1\,|E_{j,l'}\rangle\,\langle E_{j,l'}\,|0\rangle \tag{1.4.12}$$

Setting

$$\langle E_{j,l}\,|0\rangle = \delta_{j,0}G(l) \tag{1.4.13}$$

where $G(l)$ is a function of l, we find, for the case where the perturbation produces only a very small effect

$$E_1\,\delta_{j,0}G(l) - \sum_{l'} \langle E_{j,l}|\,\mathscr{H}_1|E_{j,l'}\rangle\,\delta_{l',0}G_{(l')} = 0 \qquad (1.4.14)$$

where the unperturbed state is non-degenerate we find

$$E_1 = \langle E_0|\,\mathscr{H}\,|E_0\rangle \qquad (1.4.15)$$

It way be readily shown that the second order correction E_2 to the energy is, for the non-degenerate case

$$E_2 = \sum_{E_j \neq E_0} \frac{\langle 0|\,\mathscr{H}_0\,|E_j\rangle\,\langle E_j|\,\mathscr{H}_0\,|0\rangle}{E_0 - E_j} \qquad (1.4.16)$$

PROBLEM

Show that for a time-independent perturbation $\mathscr{H}_1$, the first order correstion E_1 to the energy in the Schrödinger representation is given by

$$E_1 = \int_{-\infty}^{+\infty} \psi^*_{(\mathbf{r})}\,\mathscr{H}_,\,\psi_{(\mathbf{r})}\,d\mathbf{r}$$

PROBLEM

For a particle trapped in the potential well shown in Figure 1.11, determine to first order the perturbed energy of the symmetric state

$$\psi_s = \left(\frac{2}{a}\right)^{\frac{1}{2}} \cos\left(\frac{n\pi}{a}\right)\psi$$

for $n = 2$

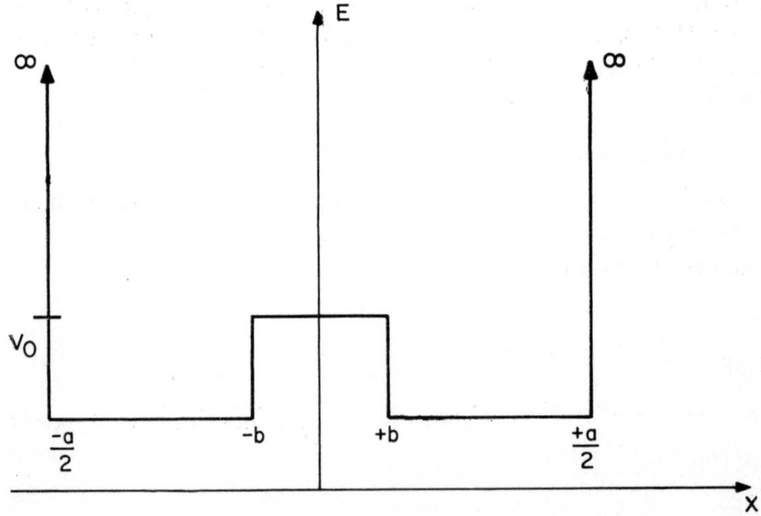

Figure 1.11 Infinite potential well with perturbation extending from $-b \leqq x \leqq +b$.

It is sufficient simply to state the results for the case where the perturbing portions of the total Hamiltonian is time-dependent.

A time-dependent perturbation is usually thought of as causing a transition from the unperturbed to a new perturbed state, the transition probability $P_{A,B}$ is given by the matrix element for the transition

$$P_{A,B} = \frac{1}{\hbar^2}\left| \int_{t_0}^{t} dt \langle A| e^{-\frac{iH_0(t'-t_0)}{\hbar}} \mathscr{H}^*_{1(t)} e^{\frac{iH_0(t'-t_0)}{\hbar}} |B\rangle \right| \qquad (1.4.17)$$

where $\langle A|$ and $|B\rangle$ signify the initial and final states respectively.

One finds that the transition probability from a state $|A\rangle$ to a final state $|C\rangle$ through an intermediate state $|B\rangle$ is, to second order, given by

$$P_{A,C} = \frac{1}{\hbar^2}\left| \int_{t_0}^{t} \langle C| \mathscr{H}_0 |A\rangle \, dt - \frac{i}{\hbar} \sum_{\substack{|B\rangle \neq \\ |A\rangle, |C\rangle}} \int_{t_0}^{t} \langle C| \mathscr{H}_{1(t)} |A\rangle \, dt \right| \qquad (1.4.18)$$

The Variational Method

There are many complicated quantum mechanical systems for which the descriptive Hamiltonian is not easily separable into two portions as we assumed in the last section. Some systems, although they are separable as above, are not such that the perturbing portion $\mathscr{H}_1$ of the Hamiltonian yields only *a small correction* to the energy determined for the unperturbed state. In cases such as these there are other approximate methods to which one can turn in order to calculate the energy of the different states of the system. The method which we will treat here is due to Eckart[6] and is called the variational method. This method makes use of the fact that the quantity

$$E = \langle A| \mathscr{H} |A\rangle \qquad (1.4.19)$$

where $\mathscr{H}$ is the total Hamiltonian of the system and $|A\rangle$ is an approximation to the true, as yet unknown, state of the system, is the upper limit to the true energy E_0. The state $|A\rangle$ is called the variational state. It is obvious that if $|A\rangle$ were the *exact* state of our system and E_0 its true energy, (1.4.19) would be exact.

Expansion in Orthogonal Functions

The proof that (1.4.19) is an upper limit to the true energy of the system is a simple one and follows quite simply from an expansion of our approximate state $|A\rangle$ in terms of a series of the n excited orthogonal eigenstates $|B_i\rangle$ of $\mathscr{H}$ (if they exist) which satisfy the conditions

$$E^i = \langle B_i| \mathscr{H} |B_i\rangle \qquad (1.4.20)$$

where $E^i < E^{i+1}$. We construct

$$|A\rangle = \sum_{i=0}^{n} c_i |B_i\rangle \tag{1.4.21}$$

where the C_i's are constant coefficients which are constrained by the relation

$$\sum_i c_i^* c_i = 1 \tag{1.4.22}$$

Substitution into (1.4.20) yields

$$E = \sum_{i=0}^{n} \sum_{j=0}^{n} c_i^* c_j \langle B_i | \mathscr{H} | B_j \rangle \tag{1.4.23}$$

Since the states $|B_i\rangle$ and $|B_j\rangle$ are orthogonal

$$E^i \delta_{ij} = \langle B_i | \mathscr{H} | B_j \rangle \tag{1.4.24}$$

and

$$E = \sum_{k=0}^{n} c_k^* c_k E^k \tag{1.4.25}$$

Subtracting E_0 from equation (1.4.24) yields

$$E - E_0 = \sum_{k=0}^{n} c_k^* c_k (E^k - E_0) \tag{1.4.26}$$

Since we postulated that the E^k are the eigenvalues of the excited orthogonal eigenstates of $\mathscr{H}$ we have

$$E^k - E_0 \geqq 0 \tag{1.4.27}$$

Thus E is always greater than or equal to the true E_0.

Offtimes, for the first approximation, a simple state will be chosen for the function $|A\rangle$. In order to determine whether, if we had chosen a slightly different state, we would have gotten a value of E smaller than we did in our first approximation, we change only an arbitrary parameter (say z) in our first approximation. If we are fairly certain that the approximate state we have chosen is a good approximation to the true state, it is a simple matter to determine the particular value z_0 of the parameter z which will yield the lowest value for the energy. To do this we set $\dfrac{\partial E}{\partial z} = 0$ and utilize the resulting z_0 in a second approximate system state.

Expansion in non-orthogonal functions

In many cases we must choose the expansion (1.4.21) in terms of non-orthogonal states. In this case we perform the expansion of the approximate system state $|A\rangle$ in terms of a series of non-orthogonal eigenstates $|D_i\rangle$ with constant coefficients a_i which must be determined such that they minimize the energy.

We have

$$|A\rangle = \sum_{i=0}^{n} a_i \, |D_i\rangle \tag{1.4.28}$$

In this case equation (1.4.19) becomes

$$E = \frac{\sum_{i=0}^{n} \sum_{j=0}^{n} a_i^* a_j \, \langle D_i | \mathscr{H} | D_j \rangle}{\sum_{i=0}^{n} \sum_{j=0}^{n} \langle D_i | D_j \rangle \, a_i^* a_j} \tag{1.4.29}$$

Taking the derivative of (1.4.29) with respect to a_l^* yields the condition for a minimum of E

$$0 = \frac{\partial}{\partial a_l^*} \left[\sum_i \sum_j a_i^* a_j \, \langle D_i | \mathscr{H} | D_j \rangle - E \sum_i \sum_j \langle D_i | D_j \rangle \, a_i^* a_j \right] \tag{1.4.30}$$

or

$$\sum_j^n a_j [H_{l,j} - E\Delta_{l,j}] = 0 \tag{1.4.31}$$

where we have used the common notation

$$H_{lj} = \langle D_l | \mathscr{H} | D_j \rangle \tag{1.4.32}$$

$$\Delta_{lj} = \langle D_l | D_j \rangle \tag{1.4.33}$$

It should be noted that Δ_{lj} is a measure of the lack of the orthogonality between the states $|D_l\rangle$ and $|D_j\rangle$ and is called the overlap between these states.

Equations (1.4.31) are a set of linear, homogeneous equations in a_l and as such the determinant of the coefficients of a_l must be equal to zero in order for a nontrivial solution to exist. Thus (1.4.31) yields the matrix condition

$$0 = \begin{vmatrix} H_{11} - \Delta_{11}E & \cdots & H_{1n} - \Delta_{1n}E \\ \vdots & \ddots & \vdots \\ \vdots & & H_{nn} - \Delta_{nn}E \end{vmatrix} \tag{1.4.34}$$

It can be shown that the lowest value of E obtained by the solution of (1.4.34) is greater than or equal to the energy eigenvalue E_0 of the exact state which we are trying to approximate by our series. The next higher root of equation (1.4.34) can be shown to be greater than or equal to the exact energy of the next excited state of our system, and so on for all roots of equation (1.4.34)[7].

1.5 MANY-PARTICLE SYSTEMS

Much of the research in solids requires the consideration of all of the electrons in a particular system. This problem requires the utilization of theoretical methods by which we can, at least in an elementary fashion, determine the combined effect of many portions of a system. It has recently become fashionable to use many-body techniques in the solution of such problems. Later we shall give a review of some of the more elementary of the techniques, but for now, we shall be content with a description of some many-body states.

If we are given a system of n particles, each satisfying an eigenvalue equation of the form

$$\mathscr{H}_i\,|A_i\rangle = E_i\,|A_i\rangle \tag{1.5.1}$$

a first approximation to the correct treatment of the collective system is to assume that the state $|A_T\rangle$ corresponding to the total system of electrons is the resultant product of each independent electron state

$$|A_T\rangle = \prod_i^n |A_i\rangle \tag{1.5.2}$$

and that the Hamiltonian operator $\mathscr{H}_T$ for the system is the sum of all of the individual $\mathscr{H}_i$

$$\mathscr{H}_T = \sum_i^n \mathscr{H}_i \tag{1.5.3}$$

Thus the eigenvalue equation which, to a first approximation, represents the motion in time of a many particle system is

$$\sum_i^n \mathscr{H}_i \prod_k^n |A_k\rangle = E \prod_k^n |A_k\rangle \tag{1.5.4}$$

It is a simple matter to show that the total system energy E is the resultant sum of the energies of the individual states. We multiply (1.5.4) from the left by $\prod_j^n \langle A_j|$ yielding

$$\prod_j^n \langle A_j| \sum_i^n \mathscr{H}_i \prod_k^n |A_k\rangle = \prod_i^n \prod_j^n \sum_i \langle A_j| \mathscr{H}_i |A_k\rangle \tag{1.5.5}$$

where, of course, only $|A_1\rangle$ is an eigenfunction of $\mathscr{H}_1$ etc. Thus, since $\mathscr{H}_i$ commutes with $|A_j\rangle$ for $i \neq j$, we will only get a contribution where $i = k = j$. Thus (1.5.5) becomes

$$\sum_i^n E_i \prod_j^n \prod_k^n \langle A_j|A_k\rangle \tag{1.5.6}$$

Since $|A_j\rangle$ is normalized, we find, finally, that

$$\sum_i^n \mathscr{H}_i \prod_k^n |A_k\rangle = \sum_l^n E_l \prod_k^n |A_k\rangle \tag{1.5.7}$$

and thus

$$E = \sum_l^n E_l \tag{1.5.8}$$

Symmetric and Antisymmetric States

In practice it is always found that particles or systems are in either symmetric or antisymmetric states. A symmetric many-body state, S, is a state which remains unchanged if we permute the particles in the system.

$$P\,|S\rangle = +\,|S\rangle \tag{1.5.9}$$

where we have used the so-called single permutation operator P which permutes the order by which the single particle states are multiplied to produce the many-body state. An antisymmetric state is one which changes its sign when operated on an odd number of times by the permutation operator.

$$P_{\text{odd}}\,|A\rangle = -|A\rangle \tag{1.5.10}$$

However, since the probability amplitude $\langle A|A\rangle$ or $\langle S|S\rangle$ represents the state for the whole system, it must remain the same when we exchange particle coordinates for particles that are indistinguishable.

We are aware that an electron must obey the Pauli exclusion principle which prohibits more than one fermion from occupying a particular state. It is quite easy to show that a normalized antisymmetric wave function which will satisfying the Pauli exclusion principle may be written

$$|A\rangle = \frac{1}{\sqrt{n!}} \begin{vmatrix} |a_{11}\rangle & |a_{12}\rangle & \cdots & \cdots \\ |a_{21}\rangle & \cdots & \cdots & \cdots \\ \vdots & \ddots & & \\ \vdots & & \ddots & \\ \vdots & & & |a_{nn}\rangle \end{vmatrix} \tag{1.5.11}$$

This determinantal state was first considered by Slater[8] for the case of both orbital and spin states. The right hand portion of (1.5.11) is called a Slater determinant, and states of this type are such that their eigenvalue E is again the sum of the single state energies E_i. The interchange of either two rows or two columns, which will exchange either quantum numbers or particle coordinates respectively, will change the sign of the determinant, and thus (1.5.11) represents an antisymmetric wave function.

PROBLEM

Show by considering an unallowed determinantal state in which a singleton state is duplicated at least once, that the determinantal wave function goes to zero and thus satisfies the Pauli exclusion principle.

References

1. See P. A. M. Dirac, *The Principles of Quantum Mechanics*, Oxford at the Clarendon Press, Fourth Ed., 1957.
2. In the present work we will deal only with those conservative classical and quantum mechanical systems in which the Hamiltonian $\mathscr{H}$ is the total energy of the system

   $$\mathscr{H} = E + V$$

   where E and V are the kinetic and potential energies of the system, respectively.
3. G. Wentzel, *Z. Physik*, **38**, 518 (1926). H. A. Kramers, *Z. Physik*, **39**, 828 (1926). L. Brillouin, *C. R. Acad. Sci., Paris*, **183**, 24 (1926).
4. For a more detailed treatment, the student is referred to R. M. Eisberg, *Fundamentals of Modern Physics*, p. 293, John Wiley and Sons, Inc., New York, 1961.
5. In general the solutions to (1.3.56) are

   $$R_{n,l}(r) = e^{-\varrho/2}\varrho^l F(l + 1 - n, 2l + 2), \varrho) \tag{1.9.1.}$$

   where F is the congruent hypergeometric series

   $$F(a, b, c) = 1 + \frac{ac}{b} + \frac{a(a+1)c^2}{b(b+1)2!} + \ldots \tag{1.9.2}$$

   when n is can integer, in which case F reverts to the laguerre polynomial. See, for instance, A. Messiah, Quantum Mechanics, Vol. I, Interscience Publishers, Inc. New York, 1961, Page 414.
6. C. Eckart, Phys. Rev. **36,** 878 (1930).
7. E. MacDonald, Phys. Rev. **43,** 830 (1933).
8. J. C. Slater, **34,** 1293 (1929).

CHAPTER 2

Thermodynamics and Statistical Mechanics

2.1 THERMODYNAMICS

As was the case in numerous other branches of physics, a detailed study of the thermodynamic properties of solids, liquids and gases was begun before the theoretical tools were available to interpret the experimental data in terms of microscopic phenomena. As a consequence a large body of data was built up concerning relations between macroscopic parameters such as pressure, temperature, volume, etc. and described only in terms of simple macroscopic models. It was not until the 19th century, when the study of the statistics of aggregates of many individual particles was begun, that the macroscopic observations could be adequately explained in terms of behaviour on an atomic scale.

For this reason, students of thermodynamics, who must necessarily study the results of macroscopic observations, often view it as a closed subject. It is not until the study of the statistical mechanics of many particles is begun that one begins to see the open-endedness of the field of statistical-thermodynamics, in much the same manner as do students of classical mechanics before their introduction to quantum theory. In this short review of thermodynamics, the student must bear in mind that he is preparing the macroscopic foundations for an atomistic theory of thermodynamics in which a correspondence principle must be served, i.e., that the atomistic picture, averaged over a large number of atoms, must yield in turn our macroscopic observations.

The First Law

We will limit our review to the thermodynamics of reversible processes in which system changes are performed infinitesimally slowly in order that the system never leave the equilibrium state.

An equilibrium state, in turn, is defined as the state that a thermodynamic system will spontaneously "settle down" to in the absence of any outside influences over a long period of time. The equilibrium state of a system is characterized by *state variables* such as the internal energy U, the volume V, and the temperature T (of course many other state variables exist as we shall see). Each time the system returns to a particular equilibrium state, each state variable will assume its unique value for that state. It is measurability of many of these state variables, along with their specific values for each equilibrium state that forced the realization that a science could be built up to describe the thermodynamic properties of materials. The first to quantitatively show the relation between heat and mechanical energy was James P. Joule who, in 1843 reported to the British Association that approximately 800 foot-pounds of mechanical work was necessary to raise the temperature of one pound of water 1° Fahrenheit. It was not until 1847, however, that this concept began winning acceptance.[1] The early work of Benjamin Thompson (Count Rumford) (1798), in which he equated the mechanical work necessary to bore a cannon with the by-product heat that raised the temperature of the muzzle, was the first to equate heat with a form of energy. As we now know, a quantity of heat, ΔQ is energy which flows by virtue of a temperature difference ΔT. It flows from the higher to the lower temperature.

The experiences of Rumford and Joule eventually led to the formulation of the first law of thermodynamics

$$dU = đQ + đW \tag{2.1.1}$$

which states that the change in total internal energy dU of a thermodynamic system is equal to the work energy transferred to the system because of work done on the system (taken as a positive quantity) plus the heat energy transferred to the system by virtue of a temperature difference (also a positive quantity). We note that in an adiabatic process no heat is transferred to or from a system. The quantities $đQ$ and $đw$ are called imperfect differentials. The result of integration of quantities of this type depends on the path of integration as well as the limits.

Intensive and Extensive Parameters, Entropy

In general, thermodynamic parameters may be classed as being either *intensive* or *extensive* quantities. Extensive parameters, such as the volume V, the mole number N, and the internal energy U, are quantities which change as the number of particles or mass of a system changes.[2] Parameters which remain constant although the mass of a system is changed, for example the temperature T or pressure P, are called *intensive* parameters.

As a consequence of the properties of state variables, we may assume that at equilibrium, the *extensive state* variable U is expressible as a function of m other extensive state variables which we shall designate as L_i

$$U = U[L_1, L_2 \ldots L_i \ldots L_m] \tag{2.1.2}$$

We may differentiate (2.1.2) with respect to the extensive variables L_i

$$dU = \sum_{i=0}^{m} \left(\frac{\partial U}{\partial L_i}\right)_{Lj} dL_i \tag{2.1.3}$$

where the subscript j indicates that all other extensive variables be kept constant in the ith differentiation. It can be shown that the terms $\left(\dfrac{\partial U}{\partial L_i}\right)_{Lj}$

are intensive varables, and thus the change in the internal energy of a thermodynamic system is the sum of products of intensive and extensive quantities.

We can gain insight into the meaning of (2.1.3) if we relate it to the Hamilton equations of motion. Equation (1.2.28 b) may be written, in the case of only one generalized coordinate

$$d\mathscr{H} = -\dot{p}\, dq = -\mathbf{F} \cdot \mathbf{dq} \tag{2.1.4}$$

Thus we see that the energy lost $d\mathscr{H}$ by a conservative system may be expressed as the negative of the product of a generalized force produced by the system with the generalized "distance" dq through which that force acts (or, alternately, the work $đw$ that the system performs).

Comparison between (2.1.4) and (2.1.3) reveals that the extensive variables play the role of generalized coordinates, and the intensive variables play the role of generalized forces, e.g.,

$$\left(\frac{\partial U}{\partial L_i}\right)_{Lj} = -F_i \tag{2.1.5}$$

The term $đW$ may include the energy from many types of work which may be done on a system, e.g., work done electrically $\mathbf{E} \cdot đ\mathbf{D}$ or magnetically, $\mathbf{H} \cdot d\mathrm{M}$, where **E** and **H** are the electric and magnetic fields, respectively, **D** is the electric displacement vector and **M** is the magnetization.

PROBLEM

Show that the mechanical work done on a gas enclosed in a container by compressing the gas with a piston is $-P\,dV$, where dV is the change in volume of the system.

We now postulate that an integrating factor for the imperfect differential $đQ$ exists, and is of the form $\frac{1}{T}$ where T is the temperature. The quantity $\frac{đQ}{T}$ is thus a perfect differential, which we shall call dS, the change in the entropy of a system. When an amount $đQ$ of heat is added to a system the entropy is increased by an amount

$$dS = \frac{đQ}{T} \tag{2.1.6}$$

The entropy change that a system undergoes depends only on the limits of the integral

$$S_B - S_A = \int_A^B \frac{đQ}{T} \tag{2.1.7}$$

and not on the path. Thus S is also a state variable. We shall postulate that S is an extensive quantity and that the entropy of a system in equilibrium is a maximum. We shall further define the entropy of a system in the following way; (1) the change in entropy due to a reversible thermodynamic process is equal to zero, (2) the entropy change involved in any irreversible process is always positive and may be thought of as a measure of the irreversibility of that process. It follows that the entropy change involved in a reversible, adiabatic process ($đQ = 0$) is zero.

Since all physical processes are irreversible to some degree, the total entropy of the universe is continually increasing. Thus an amount of energy $dU = T\,dS$ has been irreversibly changed from a useful to useless form. Thus, an equilibrium state of a system is reached when its internal energy, or the total amount of useful energy still left in the system, reaches a minimum. We see from Equation (2.1.7) that we can compute only the entropy difference between two states.

We may write (2.1.3) as

$$dU = T\,dS - P\,dV + \sum_{j=1}^{r} \mu_j\,dN_j + \sum_{i=1}^{s} X_i dL_i \tag{2.1.8}$$

where we have defined the corresponding thermodynamic forces (intensive parameters) as follows

$$\begin{aligned} X_1 &= T = \left(\frac{\partial U}{\partial S}\right)_{L_k} \\ X_2 &= -P = \left(\frac{\partial U}{\partial V}\right)_{L_i} \\ X_j &= \mu_j = \left(\frac{\partial U}{\partial N_j}\right)_{L_n} \\ X_i &= \left(\frac{\partial U}{\partial L_i}\right)_{L_n} \end{aligned} \tag{2.1.9}$$

where X_i is any generalized force, μ_j is the electrochemical potential of the jth component with mole number N_j, and n is the number of types of component molecules comprising the system.

Specific Heats and Other Useful Derivatives

In this section we list some useful derivatives of some of our extensive parameters with respect to a number of intensive parameters.

1. The Coefficient of Thermal Expansion:

$$\varkappa_0 = \frac{1}{V}\left(\frac{\partial V}{\partial T}\right)_{L_j} \tag{2.1.10}$$

This quantity is the relative change in volume (per unit volume) which takes place in a system which undergoes a change in temperature.

2. Coefficient of Compressibility (Isothermal Bulk Modulus) β

$$\beta = -\frac{1}{V}\left(\frac{\partial V}{\partial P}\right)_{L_j} \tag{2.1.11}$$

The isothermal bulk modulus gives the normalized decrease in volume per unit increase in pressure at constant temperature and mole number.

3. Specific Heat at Constant Volume c_V

$$c_V = \frac{T}{N}\left(\frac{\partial S}{\partial T}\right)_V = \frac{1}{N}\left(\frac{đQ}{dT}\right)_V \tag{2.1.12}$$

The specific heat at constant volume is the relative change in entropy per unit temperature per mole at a particular temperature T of a system constrained at constant volume.

4. Specific Heat at Constant Pressure c_p

$$c_p = \frac{T}{N}\left(\frac{\partial S}{\partial T}\right)_p = \frac{1}{N}\left(\frac{đQ}{dT}\right)_p \tag{2.1.13}$$

The specific heat at constant pressure is the relative increase in entropy per unit increase in temperature, per mole, at a fixed temperature T for a system constrained at constant pressure.

PROBLEM

For an ideal gas the state variable U may be expressed as $U = \frac{3}{2}\mathrm{NRT}$ $R = 1.986 \frac{\mathrm{Cal}}{\mathrm{Mole\text{-}^\circ K}}$. Show that

$$c_v = \frac{3}{2}R \qquad \varkappa_0 = \frac{1}{T}$$

$$c_p = \frac{5}{2}R \qquad \beta = \frac{1}{P}$$

Other Thermodynamic Representations

The Legendre Transform

We have found that we can represent a small change in the internal energy of a thermodynamic system in terms of the variation in the extensive parameters. It is more convenient in certain experimental situations to change to another valid thermodynamic representation. We will replace the description of a thermodynamic system in which the extensive parameters are the independent variables with a description in which one or more of the independent extensive variables have been replaced by their corresponding intensive variables.

The mathematical method by which a functional term $x\,dy$ in which y is the independent variable is replaced by $y\,dx$ in which x is the independent variable is called a *Legendre Transform* and is a simple but powerful tool. It is illustrated below.

A situation may exist where one has a thermodynamic function U of m extensive variables L_i and wishes to replace a number r of the independent extensive variables with their corresponding intensive variables. To perform this transformation one forms the function R by subtracting from U the product of each unwanted extensive variable with its intensive conjugate

$$R = U - \sum_{j=1}^{r} X_j L_j \tag{2.1.14}$$

The subscript j denotes the extensive variables which are to be replaced by their intensive counterparts. Differentiating R we get

$$dR = dU - \sum_{j=1}^{r} (X_j\,dL_j + L_j\,dX_j) \tag{2.1.15}$$

Recalling that

$$dU = \sum_{k=1}^{m} X_k\,dL_k \tag{2.1.16}$$

we find by substitution

$$dR = \sum_{k=r+1}^{m} X_k\,dL_k - \sum_{j=1}^{r} L_j\,dX_j \tag{2.1.17}$$

Thus we have created a function R whose independent variable include a chosen set of intensive variables X_j which replace their corresponding extensive variables.

Helmholtz Free Energy F

The Helmholtz free energy F is the thermodynamic state function which results from the replacement of the entropy by its corresponding intensive parameter the temperature T.

$$F = U - TS \tag{2.1.18}$$

Differentiating F and replacing dU, we get the differential Helmholtz free energy

$$dF = -S\,dT - P\,dV + \sum_l \mu_l\,dN_l + \sum_j X_j\,dL_j \tag{2.1.19}$$

Thus,

$$F = F(T, V, N_l, L_j) \tag{2.1.20}$$

The Gibbs Free Energy G

The Gibbs free energy (or Gibbs function) is the Legendre transform of the internal energy derived by the simultaneous replacement of both the volume and the entropy

$$G = U - TS + PV \tag{2.1.21}$$

Again, substitution for U yields

$$dG = -S\,dT + V\,dP + \sum_l \mu_l\,dN_l + \sum_j X_j\,dL_j \tag{2.1.22}$$

Thus

$$G = G(T, P, N_l, L_j) \tag{2.1.23}$$

The Enthalpy H

When one replaces the volume as an independent variable by its corresponding intensive parameter pressure, one obtains the enthalpy H, the hear function of a system

$$H = U + PV \tag{2.1.24}$$

Thus the variation of H is given by

$$dH = T\,dS + V\,dP + \sum_l \mu_l\,dN_l + \sum_j X_j\,dL_j \tag{2.1.25}$$

Thus

$$H = H(S, P, N_1, L_j)$$

PROBLEM

Show, by a negative Legendre transform that one may obtain equation (1.2.28b) from (1.2.28a).

Conditions for Equlibrium

The Lagrange Transform

In this section we will develop a general method by which we will be able to determine the necessary conditions for thermodynamic equilibrium. We have postulated that, at equilibrium, the internal energy U reaches a minimum and the entropy S is maximized. Thus, if we express the state of our system by either of the state variables U or S, we can take advantage of their equilibrium extremum values to determine the conditions for equilibrium.

The calculus of variations provides us with a powerful tool for the determination of the necessary conditions for the existence of an extremum of a given mathematical function. Let us take a function F of a number n of mutually independent variables X_i. We define an extremum to be the point at which a small variation δF in the function F, resulting from small variations in the independent parameters X_i, is equal to zero. Thus

$$\delta F = \sum_{\substack{i=1\\i\neq j}}^{n} \left(\frac{\partial F}{\partial X_j}\right)_{X_j} dX_i = 0 \tag{2.1.27}$$

which yields

$$\left(\frac{\partial F}{\partial X_i}\right)_j = 0 \quad \text{(for each } X_i\text{)} \tag{2.1.28}$$

which determines the conditions for an extremum.

If, however, the variables X_i are not all mutually independent, but coupled, one to another, by a set of m conditions or constraint equations of the form

$$g_k(X_1 \dots X_i \dots X_n) = 0 \tag{2.1.29}$$

only a number $(n - m)$ of the X_i are independent. Thus (2.1.28) follows from (2.1.27) for only $(n - m)$ of the variables.

A method exists, however, for the determination of the conditions for an extremum in this case also. We multiply each of the constraint equations (2.1.29) by a parameter λ_k, where λ_k is called the ***Lagrange multiplier*** for the kth constraint equation. Differentiating (2.1.29) we get m equations of the form

$$\lambda_k \sum_{i=1}^{n} \left(\frac{\partial g_k}{\partial X_i}\right) dX_i = 0 \tag{2.1.30}$$

Addition of the m equations (2.1.30) to (2.1.27) yields

$$\sum_{i=1}^{n} \left\{\left(\frac{\partial F}{\partial X_i}\right)_{X_j} + \sum_{k=1}^{m} \lambda_k \left(\frac{\partial g_k}{\partial X_i}\right)_{X_j}\right\} dX_i = 0 \tag{2.1.31}$$

The m parameters λ_k may be chosen to make m of the terms in curly brackets vanish, i.e.,

$$\left(\frac{\partial F}{\partial X_i}\right)_{X_j} + \sum_{k=1}^{m} \lambda_k \left(\frac{\partial g_k}{\partial X_i}\right)_{X_j} = 0 \quad i = 1, 2, \dots m \tag{2.1.32}$$

Thus the remainder of (2.1.31)

$$\sum_{i=m+1}^{n} \left\{ \left(\frac{\partial F}{\partial X_i} \right)_{X_j} + \sum_{k=1}^{m} \lambda_k \left(\frac{\partial g_k}{\partial X_i} \right)_{X_j} \right\} dX_i = 0 \tag{2.1.33}$$

contains only $n - m$ variables each independent of the other. Thus the conditions for an extremum obtains from

$$\left(\frac{\partial F}{\partial X_i} \right)_{X_j} + \sum_{k=1}^{m} \lambda_k \left(\frac{\partial g_k}{\partial X_i} \right)_{X_j} = 0 \quad i = 1, 2 \ldots k \tag{2.1.34}$$

PROBLEM

Using Lagrangian multipliers, determine the the area of the largest isosceles triangle that can be inscribed in a square of side *a*.

We are now in a position to determine the conditions necessary for equilibrium between closed thermodynamic systems. Let us consider two similar simple systems each having m different types of particles separated by a diathermal, semipermeable piston which allows for the flow of some types of matter (J-type particles), but not others, from one system to another (see Figure 2.1). We can represent a virtual displacement from equilibrium for each system as

(System 1)

$$\delta U_1 = T_1\, dS_1 - P_1\, dV_1 + \sum_{i=1}^{m} \mu_{i_1}\, dN_{i_1} \tag{2.1.35}$$

(System 2)

$$\delta U_2 = T_2\, dS_2 - P_2\, dV_2 + \sum_{k=1}^{m} \mu_{k_2}\, dN_{k_2} \tag{2.1.36}$$

where each system is defined by the same number of parameters.

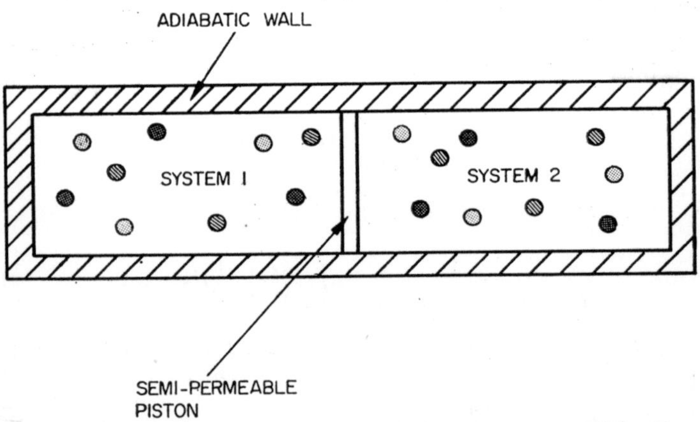

Figure 2.1 Two systems separated by a semipermeable piston which allows for the exchange of particles.

Since our systems are closed, except to one another, $dV \equiv dV_1 = -dV_2$, $dN_{j_1} = -dN_{j_2} \equiv dN_i$, and $\delta U = -\delta U$. We assume a reversible process, thus $dS_1 = -dS_2 \equiv dS$. Adding (2.1.36) to (2.1.35) we find

$$(T_1 - T_2)\,dS - (P_1 - P_2)\,dV + \sum_{j=1}^{m} (\mu_{j_1} - \mu_{j_2})\,dN_j = 0 \qquad (2.1.37)$$

Thus our conditions for equilibrium are

$$\begin{aligned} T_1 &= T_2 \\ P_1 &= P_2 \\ \mu_{j_1} &= \mu_{j_2} \end{aligned} \qquad (2.1.38)$$

Thus for a system made up of two simple systems in which there is allowed exchange of heat energy, pressure and particles, the conditions for equilibrium are that the temperature and pressure be equal in each of the separate subsystems. In addition, we require that the electrochemical potential μ_j of those components of each system allowed to penetrate the membrane be equal.

PROBLEM

By the variational technique, show for a system in contact with both pressure and temperature reservoirs, that the equilibrium condition is $dG = 0$.

The Maxwell Relations

By formal manipulation of the total differentials of our state functions, we can generate a set of quite useful relations between the extensive and intensive parameters of a system in equilibrium. These relations are known as the *Maxwell relations*. They are derived by making use of the quality of mixed partial derivatives, as in the following example. From (2.1.19) we find

$$\left(\frac{\partial F}{\partial T}\right)_{V,N_l} = -S \qquad (2.1.39\text{a})$$

$$\left(\frac{\partial F}{\partial V}\right)_{T,N_l} = -P \qquad (2.1.39\text{b})$$

$$\left(\frac{\partial^2 F}{\partial V\,\partial T}\right)_{N_l} = -\left(\frac{\partial S}{\partial V}\right)_{T,N_l} \qquad (2.1.39\text{c})$$

$$\left(\frac{\partial^2 F}{\partial T\,\partial V}\right)_{N_l} = -\left(\frac{\partial P}{\partial T}\right)_{V,N_l} \qquad (2.1.39\text{d})$$

Thus we have

$$\left(\frac{\partial S}{\partial V}\right)_{T,N_l} = = \left(\frac{\partial P}{\partial T}\right)_{V,N_l} \tag{2.1.40}$$

which is a differential relation between the intensive and extensive parameters which describe our system. Many more relations of the form (2.1.40) may be generated from the state functions we have defined. In fact, for a system whose state function is describable by n variables, one can generate $\frac{n(n-1)}{2}$ Maxwell relations.

In certain situations where the change in one thermodynamic variable with respect to another is known, it is convenient to be able to quickly replace this change by an equivalent expression involving other thermodynamic parameters. A simple method by which this is done is the ***method of conjugate variables***. Given a relation between two thermodynamic parameters such as

$$-\left(\frac{\partial P}{\partial N}\right)_{T,V} \tag{2.1.41}$$

the conjugate Maxwell relation is determined as follows: The numerator of the conjugate relation is the differential of the variable conjugate to the denominator of (2.1.41) and the denominator is the differential of the variable conjugate to the numerator of (2.1.41). For example

$$-\left(\frac{\partial P}{\partial N}\right)_{T,V} \Leftrightarrow \left(\frac{\partial \mu}{\partial V}\right)_{T,N} \tag{2.1.42}$$

where the constant quantities T and N are determined by first noting which of the constant quantities in (2.1.41) is *not* a conjugate variable of either the numerator or denominator (T in this case). This parameter is kept constant in the conjugate Maxwell relation. The second constant quantity is the parameter conjugate to the numerator of our derived relation. If either relation is a ratio of only intensive or extensive parameters the sign of that ratio is changed.

Given one half of a Maxwell relation, it is quite simple to determine from which state function it derives by forming a general function A of all of the parameters in the given expression. For instance, given the left hand side of (2.1.42)

$$A = A(N, V, P, T) \tag{2.1.43}$$

The only intensive variable in A which occurs without its conjugate variable is T. Since the internal energy U, from which all of our state functions have been derived, is a function of the extensive variables S, N and V, A must have been produced by a Legendre Transform which replaced the extensive variable S by its intensive conjugate T. Thus (2.1.42) was derived from the Helmholtz Free energy F.

PROBLEM

Find the Maxwell relations conjugate to

$$\left(\frac{\partial \mu}{\partial P}\right)_{T,N}$$

and

$$\left(\frac{\partial V}{\partial T}\right)_{P,N}$$

From which state functions were they derived?

PROBLEM

Calculate the change in entropy of a solid which under a compression of 10^5 dynes/cm^2 changes its volume by 8.3×10^{-5} cc per degree Kelvin.

2.2 STATISTICAL MECHANICS

We shall review statistical mechanics in an effort to understand on an atomic level the phenomena which govern the macroscopic observations of thermodynamics, and to predict these observations from a knowledge of the microscopic dynamics.

If we encounter even a simple system such as one mole of a gas, we immediately see the horrendous problem which we face. One mole of a substance contains Avogadro's number of atoms or molecules. It would be impossible in a lifetime to write the equations governing the dynamics of each of these molecules. An attempt at solution of these equations would be all but impossible. It is necessary then to find a simpler method by which we can predict the macroscopically observable parameters by a knowledge of the dynamics of a representative member or members of our system.

The Mathematics of Probability

The mathematics of probability can predict for us the probable outcome of an event when that event is repeated many times. If we consider that the equations which describe the dynamics of each of our particles at a time t describe an event, we may apply probability theory to many of these events (many particles) and successfully predict macroscopically observable parameters.

When a particular experiment or trial results in an event A, we say that $P(A)$ is the probability of the occurrence of event A. If our experiment is such that no event other than A is possible, and A must occur for every trial, the probability $P(A)$ is equal to 1. If the event A is an impossible event in a particular trial, the probability $P(A)$ is equal to zero for that trial. In general, the probability that an event A will not occur in a particular trial is $(1 - P(A))$. Let us consider a particular experiment in which any of the independent, mutually exclusive events A, B, C, D, E and F are possible and, for each trial, one and only one of these events must occur. We postulate that the separate probabilities of events A, B, C, D, E and F are additive, and since a result is demanded for each trial, the total probability for each trial must be 1, i.e., for each trial

$$P(A) + P(B) + P(C) + P(D) + P(E) + P(F) = 1 \tag{2.2.1}$$

In this experiment, the probability that either of the independent mutually exclusive events A or E will occur is written

$$P(A) + P(E) = P(A \text{ or } E) \tag{2.2.2}$$

From (2.2.1) we find the probability that the independent, exclusive event A will not occur in a single experiment is

$$1 - P(A) = P(B) + P(C) + P(D) + P(E) + P(F) \tag{2.2.3}$$

Given a set of independent events a, b, c, d, e and f not all mutually exclusive, it may be possible that a number of these events may occur in a single trial. For instance, it may be possible that both events a and b may occur simultaneously. The probability $P(ab)$ of this is then written

$$P(ab) = P(a)\,P(b) \tag{2.2.4}$$

It follows that the probability of either event a or event b or both occurring in this circumstance is

$$P(a \text{ or } b) = P(a) + P(b) - P(ab) \tag{2.2.5}$$

PROBLEM

What is the probability that two consecutive throws of a single die will yield a total of 6? Is this the same as the probability that a single throw of a pair of dice will yield a total of 6?

A sum of the form

$$\sum_{\substack{i=\text{all}\\ \text{possible}\\ \text{events}}} P(i) = 1 \tag{2.2.6}$$

where $P(i)$ may include the simultaneous probability of a number of events or the mutually exclusive probability of a number of events, is a statement that one expects a result for each trial. The left hand side of (2.2.6) is called the probability distribution for a particular trial. It lists the distribution of probability amongst each of the possible outcomes of our trial. It is clear that the sum of (2.2.6) over many trials is equal to the number of trials. It follows from (2.2.6) that for each trial

$$P(A) = 1 - \sum_{i \neq A} P(i) \tag{2.2.7}$$

In many instances the probability of an event A, although it is constant for each trial, apparently depends on the total number of trials. As a very simple example let us consider the hypothetical experiment which consists of flipping a coin. The probability for obtaining a "head" is $\frac{1}{2}$ for each trial, as is the probability of a "tail". If, however, we perform many trials we may tabulate our results as shown in Table 2.1. Thus, as the number of trials increases, we see that the apparent probability of obtaining a head not only changes, but tends toward the actual value of $\frac{1}{2}$. The function which describes how the apparent probability of an event varies with the number of trials is called a distribution function.

Table 2.1 Apparent probability of getting head as a function of a number of trials

Trials	Heads	Apparent Probability of Heads
1	1	1
10	3	0.3
100	43	0.43
1000	411	0.411
10,000	4920	0.49

It is of great importance in the theory of probability to be able to determine the total number of possible outcomes of a given trial. A little thought will show that the number of ways that N objects may be permuted amongst themselves in $N!$, since the first may be placed into any of N positions, the second into any of the remaining $(N - 1)$, etc.

If not all of the objects are distinct, but a number r of them are indistinguishable one from another, we must reduce the number of distinct permutations by $r!$ Thus, for example, 5 objects, 3 of which are indistinguishable, may be permuted into

$$\frac{N!}{r!} = \frac{5!}{3!} \tag{2.2.8}$$

distinct ways.

If two independent events are mutually exclusive, and event A may occur in m ways while event B may occur in r ways, then either event A or event B can occur in $m + r$ different ways.

Entropy and Probability

In our review of thermodynamics, we introduced, in a mysterious fashion, the extensive parameter entropy, and postulated that this parameter would be maximum for a system in equilibrium. We will now investigate the relation between the entropy and the probability distribution of the allowed microscopic energy states within a thermodynamic system.

Let us take, for example, a simple thermodynamic system consisting of one mole of an ideal gas of A-type molecules. At equilibrium, the state of the entire system may be specified, as before, in terms of its total entropy S.

We shall assume that each of the 6.023×10^{23} molecules has available to it m distinct microscopic entropy states, and that each of these entropy states is equally probable for each molecule. Thus, the m states may be permuted amongst the 6.023×10^{23} molecules in

$$m^{(6.023 \times 10^{23})} \tag{2.2.9}$$

ways. Each of these permutations produces an allowed entropy state for the total system. Not all of these permutations, however, will lead to the equilibrium value S for the entropy. The system may utilize only one of a few of these permutations to reach equilibrium.

We shall define the "probability" P_s of the state of a system as the number of distinct ways the system can rearrange itself to yield the state S. We assume that each permutation is independent, mutually exclusive and equally probable. Thus the probability that the system is in one of the set of equilibrium permutations which leads to an equilibrium value S of the entropy may be written

$$P_s = \prod_{i=1}^{6.023 \times 10^{23}} P_{ij} \tag{2.2.10}$$

where P_{ij} is the probability that the ith molecule is in the jth microscopic entropy state.

It is clear that the probability of the occurrence of an equilibrium entropy state P_s depends on the multiplicative probability of each of the microscopic entropy states, whereas the entropy itself is an extensive parameter. The relation between the entropy of a system and probability of the occurrence of an equilibrium entropy state was devised by Boltzmann, who suggested that the entropy of a system is related to the logarithm of the probability of the occurrence of an equilibrium entropy state

$$S = k_0 \ln P_s \tag{2.2.11}$$

(Boltzmann's Principle)

where k_0 is Boltzmann's constant ($k_0 = R/6.023 \times 10^{23} = 1.38 \times 10^{-16}$ ergs-°K^{-1}). As we saw before, the entropy is a maximum at equilibrium. Thus the equilibrium state of a system is the state of maximum probability distribution. If we assume that the entropy states of the individual molecules are independent of one another, we can show that relation (2.2.11) holds for the case of the ideal gas.

If we double the number of molecules, the entropy S_2 of the resultant state becomes

$$S_2 = k_0 \ln \left(\prod_{i}^{2\times 6.023\times 10^{23}} P_{ij} \right) = k_0 \sum_{j}^{2\times 6.023\times 10^{23}} \ln P_{ij} \tag{2.2.12}$$

Since each molecular entropy state is equally probable,

$$S_2 = 1.2 \times 10^{24} k_0 \ln P_{ij} \tag{2.2.13}$$

whereas it is easily shown that the entropy S of the original system before doubling is given by

$$S = 6.023 \times 10^{23} k_0 \ln P_{ij} \tag{2.2.14}$$

Thus the entropy defined by (2.2.11) behaves in the proper extensive manner.

Classical Statistics

The Maxwell-Boltzmann Distribution Function

We are now in a position to calculate the probability distribution functions of individual noninteracting particles amongst the allowed thermodynamic states of a system. To do this, we will follow the method we have used before in the calculation of equilibrium conditions between macroscopic thermodynamic parameters. We shall organize our thermodynamic system into a number of identical smaller subsystems each in a particular quantum state, and we shall calculate the conditions for equilibrium between these subsystems. Our macroscopic system will now be an ensemble of many similar subsystems. We shall first concern ourselves with the socalled *classical approximation*, in which a particular quantum substate may be occupied by any number of distinguishable particles.

We can separate our macroscopic system of N distinguishable particles into m subsystems, each assumed to be in the same quantum substate j determined by an allowed set of quantum numbers. The number of ways that these particles may be permuted amongst the m subsystems is

$$\frac{N!}{\prod_{j=1}^{m} n_j!} \tag{2.2.15}$$

since there are $n_j!$ permutations of the particles in the j_{th} subsystem which leave the state of that subsystem unchanged. We must, however, modify equation (2.2.15) to take into account the number of ways that the allowed quantum numbers for each particle may be permuted to yield the same particle state. This degeneracy l_k of the state of each particle produces a number

$$\prod_{k=1}^{n_j} L_k \tag{2.2.16}$$

of allowed permutations for each permutation we have already calculated in (2.2.15). If we assume that this degeneracy is the same for each particle in a subsystem, (2.2.16) becomes $(L_j)^{n_j}$.

Thus the total number of distinct ways that our identifiable particles may be permuted amongst the m subsystems (the probability of our state) including degeneracies is

$$P_S = \frac{N! \prod_{j=1}^{m} L_j^{n_j}}{\prod_{j=1}^{m} n_j!} \tag{2.2.17}$$

Thus we have the entropy of our classical system

$$S = k_0 \ln \left(\frac{N! \prod_{j=1}^{m} L_j^{n_j}}{\prod_{j=1}^{m} n_j!} \right) \tag{2.2.18}$$

We shall take advantage of the condition that entropy is maximum at equilibrium in order to determine the probability distribution $f(E_j)$ for the number of particles in each energy state E_j. As before

$$\sum_j f(E_j) = 1 \tag{2.2.19}$$

It is clear that in the present case

$$f(E_j) = \frac{n_j}{L_k} \tag{2.2.20}$$

Using the method of Lagrange multipliers, the condition for a maximum in entropy becomes

$$\frac{\partial}{\partial n_l}\left\{k_0 \ln\left[\frac{N!\prod_{j=1}^{m} L_j^{n_j}}{\prod_{j=1}^{m} n_j!}\right]\right\} + \lambda_1 \frac{\partial}{\partial n_l}\left(U - \sum_{j=1}^{m} n_j E_j\right)$$

$$+ \lambda_2 \frac{\partial}{\partial n_l}\left(N - \sum_{j=1}^{m} n_j\right) = 0 \tag{2.2.21}$$

where we have used the constraints that our closed system is adiabatically isolated from the outside world

$$\sum_{j=1}^{m} n_j = N \tag{2.2.22A}$$

$$\sum_{j=1}^{m} n_j E_j = U \tag{2.2.22B}$$

E_j is the internal energy of the jth quantum state, and U is the internal energy of our macroscopic system. We find, noting that $\frac{\partial U}{\partial n_l}$ denotes merely a rearrangement of particles during which the total internal energy must remain constant

$$k_0 \frac{\partial}{\partial n_l} \ln N! + k_0 \frac{\partial}{\partial n_l} \ln \prod_{j=1}^{m} L_j^{n_j} - k_0 \frac{\partial}{\partial n_l} \ln \prod_{j=1}^{m} n_j!$$

$$- \lambda_1 \sum_{j=1}^{m} \frac{\partial}{\partial n_l}(n_j E_j) - \lambda_2 \sum_{j=1}^{m} \frac{\partial}{\partial n_l}(n_j) = 0 \tag{2.2.23}$$

Using Stirling's approximation, valid for large n

$$n! \approx (2\pi n)^{\frac{1}{2}} n^n e^{-n} \tag{2.2.24}$$

or

$$\ln(n!) \approx n \ln n - n \tag{2.2.25}$$

we find

$$k_0 \frac{\partial}{\partial n_l} \sum_{j=1}^{m} n_j \ln L_j - k_0 \frac{\partial}{\partial n_l} \sum_{j=1}^{m} (n_j \ln n_j - n_j)$$

$$-\lambda_1 \sum_{j=1}^{m} \frac{\partial}{\partial n_l} (n_j E_j) - \lambda_2 \sum_{j=1}^{m} \frac{\partial}{\partial n_l} (n_j) = 0 \qquad (2.2.26)$$

Equation (2.2.26) may be written

$$k_0 \ln\left(\frac{n_l}{L_l}\right) + \lambda_1 E_l + \lambda_2 = 0 \qquad (2.2.27)$$

using (2.2.20) we find the distribution function $f(E)$ for the number of classical particles in the energy state E_l

$$f(E_l) = e^{-\frac{(\lambda_1 E_l + \lambda_2)}{k_0}} \qquad (2.2.28)$$

We can find the values of the Lagrange multipliers by remembering that the Maxwell-Boltzmann system is an approximation to a classical system in which the particles are indistinguishable. Thus we must divide (2.2.17) by $N!$. Using Stirling's approximation and (2.2.22) we find, as before

$$S = k_0 \sum_{j=1}^{m} n_j \ln f^{-1}(E_j) \qquad (2.2.29)$$

Substitution of equation (2.2.28) for the distribution function yields

$$S = \lambda_2 \sum_{j=1}^{m} n_j + \lambda_1 \sum_{j=1}^{m} n_j E_j \qquad (2.2.30)$$

or

$$U = \frac{S}{\lambda_1} - \frac{\lambda_2}{\lambda_1} N \qquad (2.3.31)$$

We know, however, that

$$\left(\frac{\partial U}{\partial S}\right)_{Lj} = T \qquad \left(\frac{\partial U}{\partial N}\right)_{Lj} = \mu \qquad (2.2.32)$$

Thus

$$\lambda_1 = \frac{1}{T} \qquad \lambda_2 = \frac{-\mu}{T} \qquad (2.2.33)$$

and

$$f(E_l) = e^{+\frac{\mu - E_l}{k_0 T}} \qquad (2.2.34)$$

However

$$\sum_j f(E_j) = 1 \tag{2.2.35}$$

Thus

$$e^{+\frac{\mu}{k_0 T}} = \frac{1}{\sum_j e^{-E_j/k_0 T}} \tag{2.2.36}$$

It follows that

$$f(E_l)_{MB} = \frac{e^{-E_l/k_0 T}}{\sum_l e^{-E_l/k_0 T}} = e^{-\frac{E_l+\mu}{k_0 T}} \tag{2.2.37}$$

or

$$f(E)_{MB} = \frac{e^{-E/k_0 T}}{Z} \tag{2.2.38}$$

where

$$Z = \sum_j e^{-E_l/k_0 T} \tag{2.2.39}$$

is called the *partition* function which can be shown to be

$$Z = e^{-F/k_0 T} \tag{2.2.40}$$

where F is the Helmholtz free energy.

PROBLEM

Show that for a system at constant temperature and volume the Fermi level μ is the Helmholtz free energy per particle.

Equation (2.2.37) is the *Maxwell-Boltzmann* distribution function. It enumerates the most probable number of classical particles in each allowed energy state of a system. A plot of the Maxwell-Boltzmann distribution function vs. energy for three different temperatures is shown in Figure 2.2.

Our partition function Z requires that we sum over all of the available quantum states E_l. For convenience we can, however, assume that each of our substates is a single particle and that the total energy of the state E_l is the sum of a number of independent substates, e.g., $E_1 = E_{\text{rotational}}$, $E_2 = E_{\text{vibrational}}$ or $E_3 = E_{\text{translational}}$. In this case we may write the partition function Z for a system of N similar particles each of which has p degrees of freedom as

$$Z = \left[\sum_{i=1}^{p} e^{-E_i/k_0 T}\right]^N \tag{2.2.41}$$

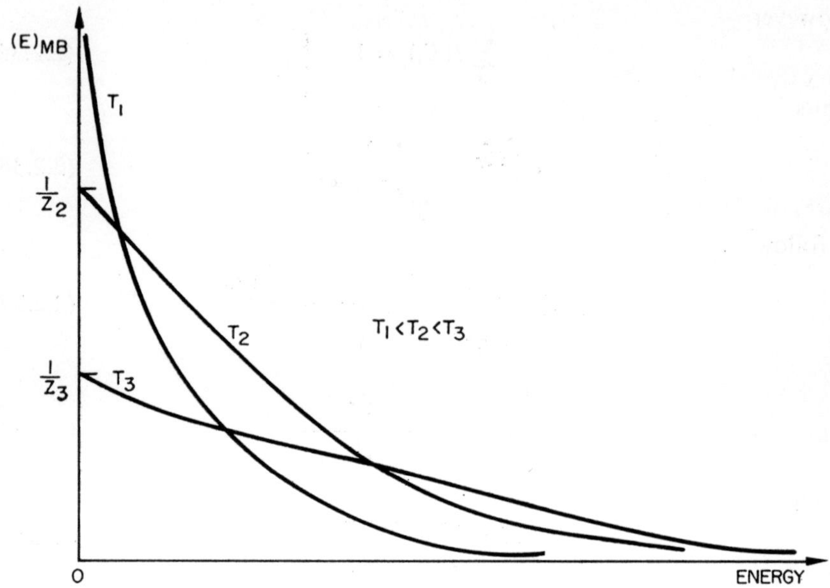

Figure 2.2 The Maxwell-Boltzmann distribution at three temperatures.

Thus Z may be factored into a number of single particle distribution functions z_i

$$Z = \prod_i z_i \tag{2.2.42}$$

One should remember that our derivation assumed that the Maxwell-Boltzmann particles are distinguishable. In order to correct for the indistinguishability of particles, we shall assume that the ratio of the number of particles N to the number of available states is small. We may then make the approximation

$$Z_{\text{ind}} = \frac{Z_{\text{d ist}}}{N!} \tag{2.2.43}$$

Thus the Maxwell-Boltzmann distribution function becomes

$$f(E_l)_{MB} = \frac{N!\, e^{-E_l/k_0T}}{(z_i)^N} \tag{2.2.44}$$

Let us calculate the partition function for a number of noninteracting Maxwell-Boltzmann type particles which are constrained within the volume V enclosed by infinitely high potential walls located at $\pm\frac{lx}{2}$, $\pm\frac{ly}{2}$, and $\pm\frac{lz}{2}$.

We will consider the case where the particles have only translational energy. The allowed energy levels for this system are given by an obvious extension of equation (1.3.26) to three dimensions

$$E_n(k) = \frac{\hbar^2 k^2}{2m} = \frac{\hbar^2}{2m}[k_x{}^2 + k_y{}^2 + k_z{}^2] \tag{2.2.45}$$

where, as before, the k's may take on the values

$$k_x = \frac{\pi n_x}{l_x}, \quad k_y = \frac{\pi n_y}{l_y}, \quad k_z = \frac{\pi k_z}{l_z} \tag{2.2.46}$$

where n_x, n_y and n_z may take on the values 0, ± 1, ± 2, etc.

We shall assume that the allowed energy levels are dense enough to allow us to replace the summation in (2.2.39) by an integral. Thus the single particle partition function becomes

$$z_i = \frac{1}{h^3} \iiint_{-\infty}^{\infty} \exp\left[\frac{-\hbar^2 k^2}{2mk_0 T}\right] d^3k\, d^3r \tag{2.2.47}$$

and thus

$$Z_{\text{Ind.}} = \frac{z_i^N}{N!} = \left(\frac{1}{N!}\right) V^N \left(\frac{2\pi m k_0 T}{h^4}\right)^{3/2N} \tag{2.2.48}$$

Quantum Statistics

The Bose-Einstein Distribution Function

As we know, quantum mechanical particles may have either symmetric or antisymmetric wave functions and those with antisymmetric wave functions must obey the Pauli Exclusion Principle. Systems which have symmetric wave functions, however, obey no exclusion principle and may pack as may particles as possible into each allowed quantum level. In addition, quantum theory demands that similar particles be indistinguishable one from another. The first type of quantum statistics we shall study is that valid for indistinguishable particles with symmetric wave functions.

Typical particles with symmetric wave functions are photons and atoms composed of an even number of subparticles with antisymmetric wave functions. We shall determine the probability distribution for these Bose-Einstein type particles amongst a set of allowed energy levels.

Let us again divide our macroscopic system into a set of m subsystems. This time, however, each subsystem will be made up of n_j particles each of which may be in any of L quantum states allowed in the subsystem. We shall allow, also, the free movement of particles between each subsystem.

Considering the jth subsystem, we can permute its n_j indistinguishable particles amongst its L_j allowed quantum states in

$$\frac{[n_j + L_j - 1]!}{n_j!\,(L_j - 1)!} \tag{2.2.49}$$

ways. The probability for our entire macroscopic system-state becomes

$$P = \prod_{j=1}^{m} \frac{(L_j + n_j - 1)!}{n_j!\,(L_j - 1)!} \tag{2.2.50}$$

And

$$S = k_0 \ln \prod_{j=1}^{m} \frac{(L_j + n_j - 1)!}{n_j!\,(L_j - 1)!} \tag{2.2.51}$$

Again maximizing S

$$\frac{\partial}{\partial n_l}\left[k_0 \ln \prod_{j=1}^{m} (n_j + L_j - 1)! - k_0 \ln \prod_{j=1}^{m} n_j!(L_j - 1)!\right]$$

$$+ \lambda_1 \frac{\partial}{\partial n_l}\left(\sum_{j=1}^{m} n_j - N\right) + \lambda_2 \frac{\partial}{\partial n_l}\left(\sum_j n_j E_j - U\right) = 0 \tag{2.2.52}$$

Using the constraints (2.2.22) and Stirling's approximation, we find

$$k_0 \ln\left(\frac{n_l + L_l - 1}{n_l}\right) + \lambda_1 + \lambda_2 E_l = 0 \tag{2.2.53}$$

Assuming that $\frac{1}{n_l} \ll 1$, i.e., that the number of particles in the lth energy state is large, we find

$$\ln\,[1 + f^{-1}(E_l)_{BE}] + \frac{\lambda_1}{k_0} + \frac{\lambda_2 E_l}{k_0} = 0 \tag{2.2.54}$$

where $f(E_l)_{BE} = \frac{n_l}{L_l}$ is the probability distribution of particles within each similar subsystem and thus of the whole system. Thus the Bose-Einstein distribution function may be written

$$f_{(E_l)BE} = \frac{1}{e^{-\frac{\lambda_1 - \lambda_2 E_l}{k_0}} - 1} \tag{2.2.55}$$

To find the values of the Lagrange multipliers λ_1 and λ_2 we note that (2.2.51) may be expressed as

$$S = k_0 \sum_j \left\{ n_j \ln\left(1 + \frac{L_j}{n_j}\right) + (L_j) \ln\left(\frac{n_j}{L_j} + 1\right)\right\} \tag{2.2.56}$$

where we have again used Stirling's approximation and assumed that $L_j \gg 1$ and $n_j \gg 1$. We find

$$U = -\frac{S}{\lambda_2} - \frac{\lambda_1 N}{\lambda_2} - \frac{k_0}{\lambda_2} \sum_j L_j \ln\left(1 + e^{\frac{\lambda_1 + \lambda_2 E_j}{k_0}}\right) \tag{2.2.57}$$

Utilizing

$$\left(\frac{\partial U}{\partial S}\right)_{Lj} = T$$

$$\left(\frac{\partial U}{\partial N_j}\right)_{Lk} = \mu_j \tag{2.2.58}$$

we find

$$\lambda_2 = -\frac{1}{T} \qquad \lambda_1 = \frac{\mu_j}{T} \tag{2.2.59}$$

where μ_j is the electrochemical potential.

And we get the Bose-Einstein distribution function

$$f(E_l)_{BE} = \frac{1}{e^{-\frac{\mu + E_l}{k_0 T}} - 1} \tag{2.2.60}$$

The Fermi-Dirac Distribution Function

The second type of quantum statistics which we will investigate is that valid for particles with antisymmetric wave functions which obey the Pauli Exclusion Principle (which limits the number of particles which may occupy any quantum state to 1). Again we shall assume that Fermi-Dirac particles, such as the electron and neutron are indistinguishable one from another.

As before, to determine the distribution function, we separate our macroscopic system in m subsystems each having n particles distributed amongst L allowed quantum states. Of course the number of particles N may not exceed L_j.

We must take into account, however, the degeneracy g_j associated with each jth quantum state, e.g., the spin multiplicity of the $1S$ level of a hydrogen atom. The ionized state of a hydrogen atom may accept an electron in either one of two available states, spin up or spin down. Once an electron is captured in one of these spin states, however, Coulombic repulsion makes it highly improbable that the state with the reverse spin orientation will be occupied.

It is easily shown that the number of ways n_j indistinguishable particles may be permuted amongst L_j quantum states, one particle to a state, is

$$\frac{L_j!}{n_j!\,(L_j - n_j)!} \tag{2.2.61}$$

We find that $g_j^{n_j}$ new allowed permutations are produced by the inclusion of the multiplicity g_j of each state. Thus, from Boltzmann's theorem, we find

$$S = k_0 \ln \prod_j^m \frac{L_j!\, g_j^{\,nj}}{n_j(L_j - n_j)!} \tag{2.2.62}$$

Proceeding as before, we maximize the entropy to determine the Fermi-Dirac distribution function

$$f(E_j)_{FD} = \frac{1}{1 + g_j^{-1}\, e^{\frac{E_j - \mu}{k_0 T}}} \tag{2.2.63}$$

PROBLEM

Prove Equation (2.2.54).

The electrochemical potential μ usually is called the Fermi Energy. When dealing with large numbers of similar particles and energy states, it is usually (but not always, as we shall see) the case that the state multiplicity g_j may be deliberately ignored (set equal to one) and the effects of state multiplicity taken care of by other approximate means, such as multiplying the total number of available states by the factor g_j.

The Fermi-Dirac distribution function with multiplicity 1 becomes

$$f(E_l)_{FD} = \frac{1}{e^{\frac{E_l - \mu}{k_0 T}} + 1} \tag{2.2.64}$$

A graph of the Fermi-Dirac Distribution Function at various temperatures is shown in Figure 2.3.

Thus, at $T = 0°K$, the Fermi-Dirac distribution function is 1 and all energy states below the energy μ_0 are filled, and all above μ_0 are empty. As the temperature is increased, the distribution begins to "smear out" around the Fermi level, becoming smaller below it and correspondingly greater above it.

PROBLEM

Show that the condition $E - \mu \gg kT$ leads to

$$f_{\text{F.D.}}(E) = f_{\text{B.E.}}(E) = f_{\text{M.B.}}(E)$$

It is useful to contrast the different ways that a given number of allowed energy levels are filled by particles which obey the three types of statistics we have discussed. A simple but generally useful group of states are those energy levels available to a particle constrained within the boundaries l_x, l_y and l_z of the three dimensional potential well with infinitely high walls. The allowed energy levels E_n for this system are given by equation (1.3.26).

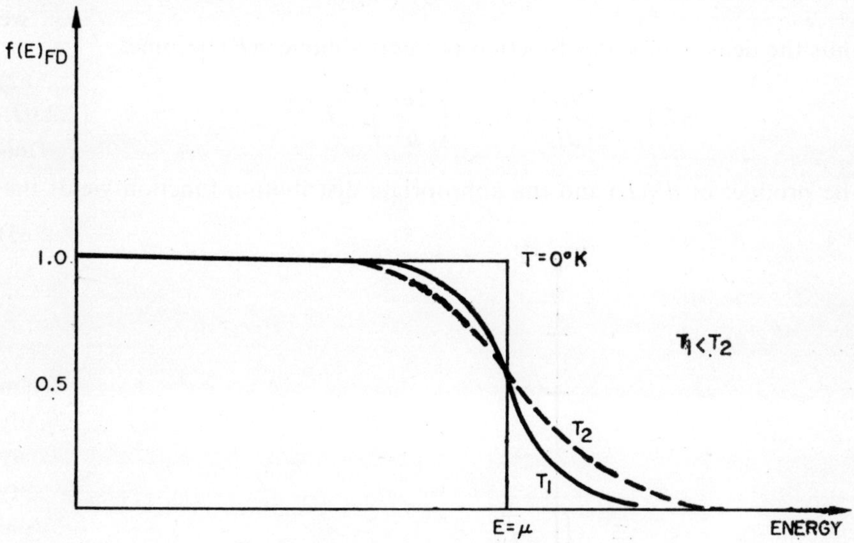

Figure 2.3 The Fermi-Dirac distribution at three different temperatures.

We will determine the density of allowed energy states $g(E)$ as a function of energy for this system. We may construct a momentum space with basic vectors $\hat{k}_x$, $\hat{k}_y$, and $\hat{k}_z$ such that each point in this space represents one allowed value of the momentum k of our system (see Figure 2.4). It is evident that the density $\varrho(\mathbf{k})$ of allowed momentum values is constant everywhere in our space, and that there is one allowed momentum state per unit volume in this space.

In general, the number of allowed energy levels per unit energy is dependent on the shape of the chosen energy surface. To determine a general expression for the density of states function $g(E)$, we imagine that the function $E_1(\mathbf{k})$ represents a surface in **k** space. A similar, larger surface $E_2(\mathbf{k})$ is formed as shown in Figure 2.5 by displacing each point in $E_2(\mathbf{k})$ outwards a distance dk in a direction normal to $E_1(\mathbf{k})$. The number of states N enclosed between $E_1(\mathbf{k})$ and $E_2(\mathbf{k})$ is

$$N = \varrho(\mathbf{k})\, dV = \varrho(\mathbf{k}) \int E_1(\mathbf{k})\, d\mathbf{k} \qquad (2.2.65)$$

Let us take for our purposes a spherical constant energy surface. In this case equation (2.2.65) yields for the number of distinct states

$$N(E) = \frac{\pi k^3}{6} - \frac{4\pi}{3}\left(\frac{2m}{h^2}\right)^{3/2} E^{3/2} \qquad (2.2.66)$$

Thus the density of states function per unit volume $g(E)$ becomes

$$g(E) = \frac{dN(E)}{dE} = 2\pi\left(\frac{2m}{h^2}\right)^{3/2} E^{1/2} \qquad (2.2.67)$$

The product of $dN(E)$ and the appropriate distribution function yields the

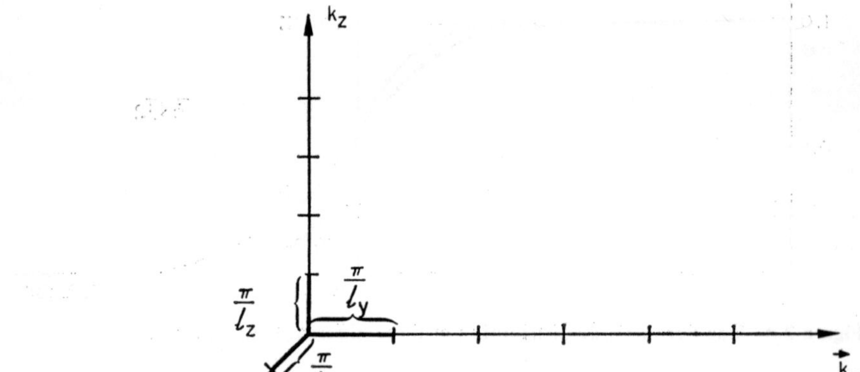

Figure 2.4 Allowed momentum space for a particle constrained within the walls of a infinitely deep potential cube of sides l_x, l_y and l_z.

number of particles per unit energy. Thus we have

$$\frac{dN(E)}{dE_{FD}} = 2\pi\left(\frac{2m}{h^2}\right)^{3/2} \frac{E^{1/2}}{e^{\frac{E-\mu}{k_0T}} + 1}$$
$$\frac{dN(E)}{dE_{BE}} = 2\pi\left(\frac{2m}{h^2}\right)^{3/2} \frac{E^{1/2}}{e^{\frac{E-\mu}{k_0T}} - 1} \qquad (2.2.68)$$
$$\frac{dN(E)}{dE_{MB}} = 2\pi\left(\frac{2m}{h^2}\right)^{3/2} \frac{E^{1/2}}{e^{\frac{E-\mu}{k_0T}}}$$

Figures 2.6, a and b, contrast the filling of our simple collection of states by an ideal gas of Fermi-Dirac, Bose-Einstein and Maxwell-Boltzmann particles for the cases where $\mu/k_0T = 0.1$ and $\mu/k_0T = 10$. For both cases all three statistics predict roughly the same density of particles only in the classical region, i.e., where $E \gg \mu$. In this region the density of particles per

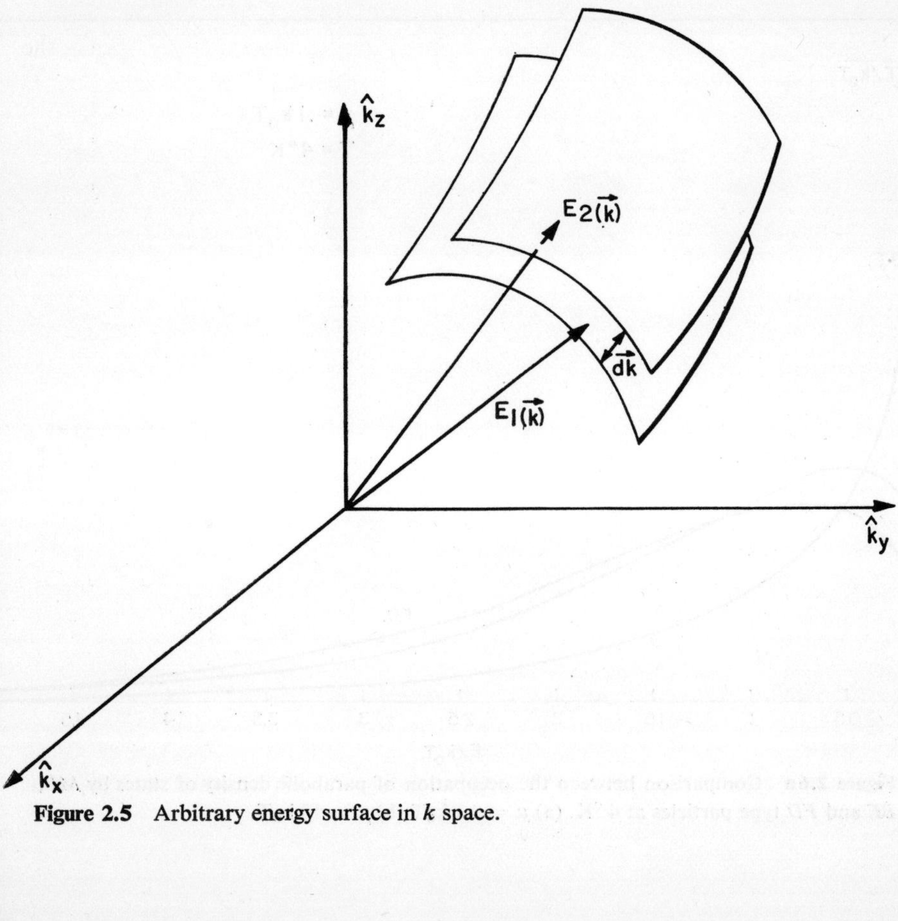

Figure 2.5 Arbitrary energy surface in k space.

unit energy is small. Thus each particle sees so many unoccupied allowed states that the differences in occupation probability produce negligible effects. The differences in the filling of our density of states $g(E)$ arise in the region $E < \mu$. For the Fermi-Dirac particles, most of the states below μ are filled and most above μ are empty. Many of our Bose-Einstein particles, on the other hand, have condensed into the $E = 0$ state and the number of particles per unit energy decreases drastically with energy. Note that both the Bose-Einstein and the Maxwel-Boltzmann statistics allow for more than one particle per state and thus both of these curves may, and indeed do, lie above the curve of $g(E)$. The Fermi-Dirac curve of course lies below $g(E)$ at all points. Note that all three curves have been normalized such that the total number of particles in each system is equal to N and

$$\int_0^{\mu} g(E)\, dE = N \tag{2.2.69}$$

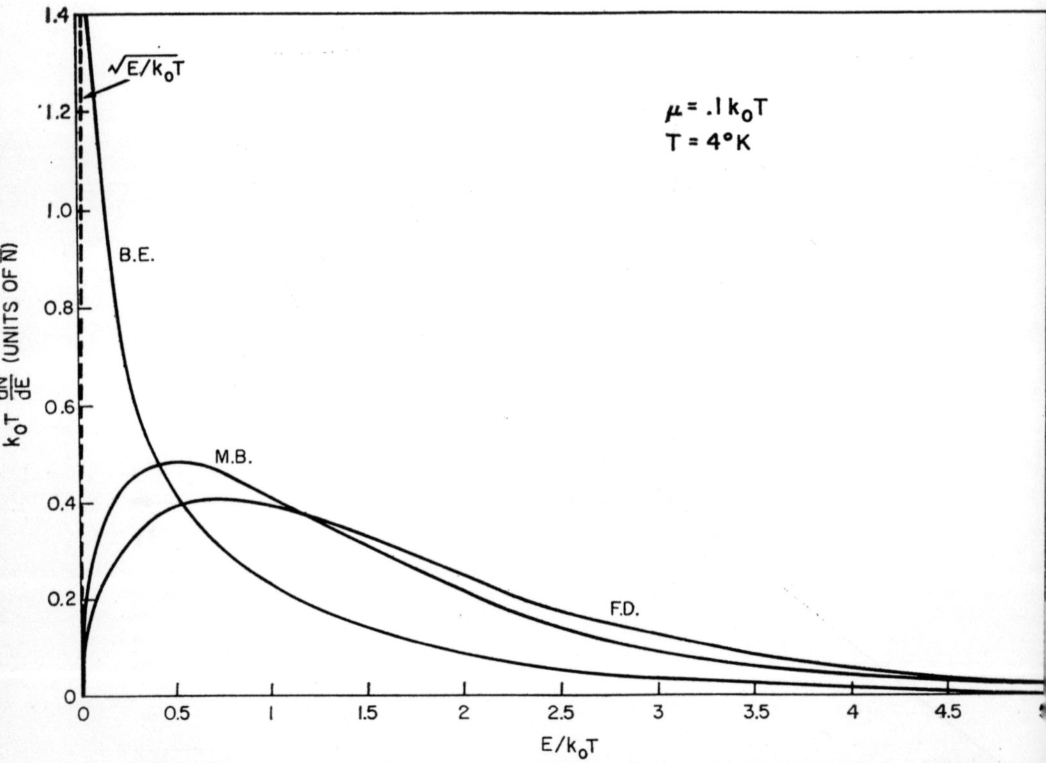

Figure 2.6a Comparison between the occupation of parabolic density of states by *MB*, *BE* and *FD* type particles at 4 °K. (a) $\mu = 0.1\ k_0T$ (b) $\mu = 10\ k_0T$.

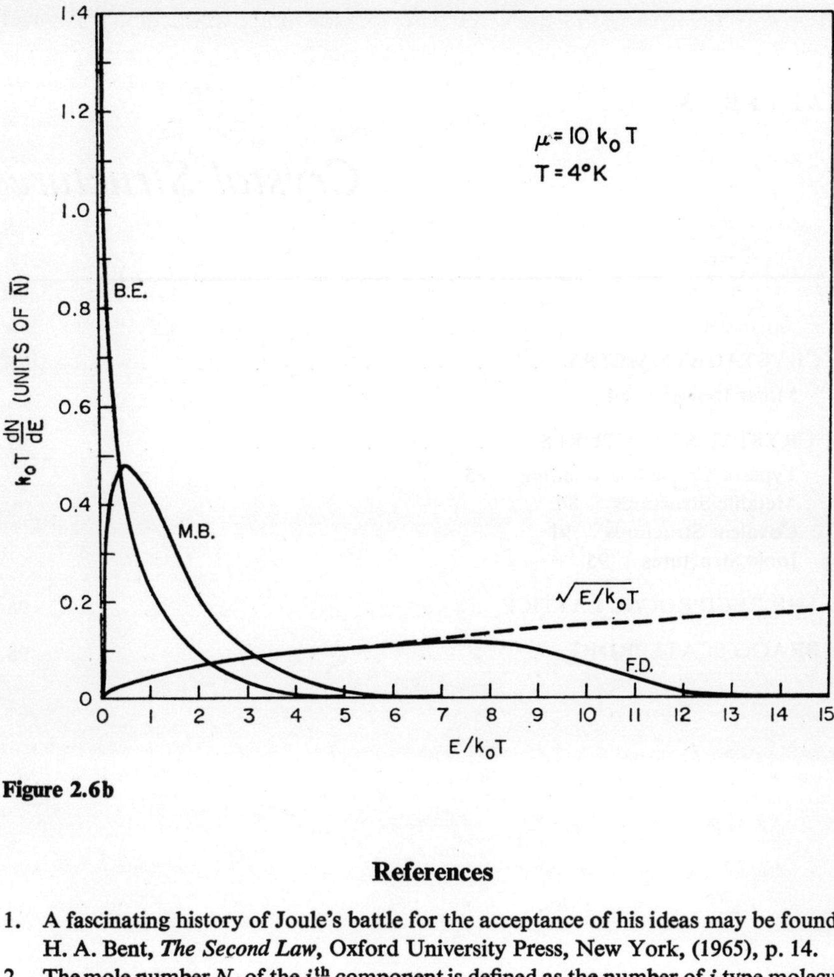

Figure 2.6b

References

1. A fascinating history of Joule's battle for the acceptance of his ideas may be found in H. A. Bent, *The Second Law*, Oxford University Press, New York, (1965), p. 14.
2. The mole number N_j of the jth component is defined as the number of j-type molecules divided by Avogadro's number (6.024×10^{23}).

CHAPTER 3

Crystal Structure

3.1 CRYSTAL SYMMETRY

The theory of solids was developed to its present state primarily by making use of the regularity of the electronic potential in single crystal materials. In this chapter we will learn a little about the types of crystalline order that occur naturally so that we may utilize this order in our study of the theory of electron motion in solids.

A lattice is defined as a regular array of points in which any point may be generated from any other by one of a set of lattice vectors **R**

$$\mathbf{R} = n_i\hat{a}_i + n_j\hat{a}_j + n_k\hat{a}_k \qquad (3.1.1)$$

in which n_i, n_j and n_k are integers and $\hat{a}_i$, $\hat{a}_j$ and $\hat{a}_k$ are the basic vectors of the lattice. These basic vectors may be thought of as unit vectors by which the entire lattice space may be spanned. They are usually chosen in the following way: A lattice site is chosen and the first basic vector is directed from that lattice site to its nearest neighbor. The second basic vector spans the distance to the closest atom in a direction not collinear with the first. The third basic vector is directed to the closest neighbor not coplanar with the first two. The magnitudes a_i, a_j and a_k are called the *lattice constants*, and the unit vectors $\hat{a}_i/a_i$, $\hat{a}_j/a_j$ and $\hat{a}_k/a_k$ define the *principal directions* of the lattice.

The angles between the basic lattice vectors are given by

$$\begin{aligned} \hat{a}_i \cdot \hat{a}_j &= a_i a_j \cos\gamma \\ \hat{a}_i \cdot \hat{a}_k &= a_i a_k \cos\beta \\ \hat{a}_j \cdot \hat{a}_k &= a_j a_k \cos\alpha \end{aligned} \qquad (3.1.2)$$

Examples of two dimensional rectangular and hexagonal lattices are shown in Figures 3.1 a and 3.1 b respectively. The two basic vectors which generate the two dimensional square lattice are equal in magnitude and are orthogonal. The basic vectors which generate the two dimensional hexagonal lattice are also equal in magnitude and make an angle of 60° with one another.

A lattice described by lattice vectors in which every lattice point has the same kind of atom is called a simple or Bravais lattice. In the field of crystallography, a lattice refers to a regular array of atoms or molecules forming a three dimensional periodic structure. A lattice has the property that the entire structure may be generated by successive unit translations of a *unit cell*. In the simplest types of lattices, such as the simple cubic lattice shown in

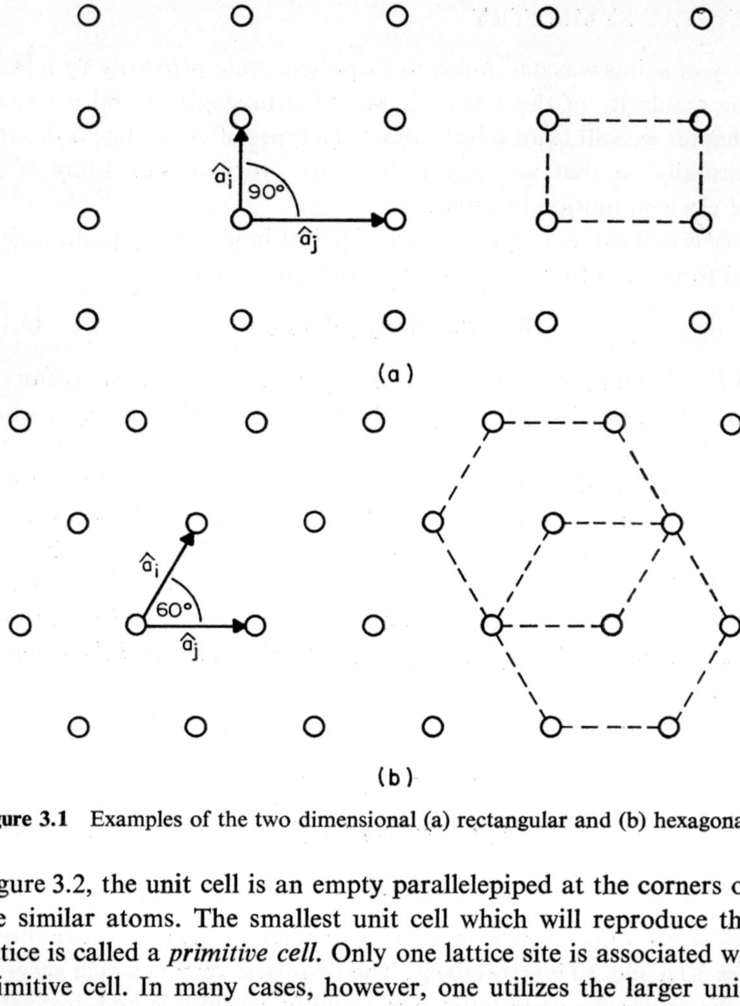

Figure 3.1 Examples of the two dimensional (a) rectangular and (b) hexagonal lattices.

Figure 3.2, the unit cell is an empty parallelepiped at the corners of which are similar atoms. The smallest unit cell which will reproduce the entire lattice is called a *primitive cell.* Only one lattice site is associated with each primitive cell. In many cases, however, one utilizes the larger unit cell in order to more clearly demonstrate the symmetry properties of the lattice. Figure 3.3 shows for example both the rhombohedral primitive cell and the cubic unit cell for the face-centered cubic lattice (a simple cubic lattice with one atom in the center of each face.)

In the three dimensional lattice, the unit cell is the parallepiped whose sides are formed from parallel translations of the three basic lattice vectors $\hat{a}_i$, $\hat{a}_j$ and $\hat{a}_k$. It is a simple exercise, left to the student to show that the volume V of a unit cell is given by

$$V = \hat{a}_i \cdot \hat{a}_j \times \hat{a}_k \tag{3.1.3}$$

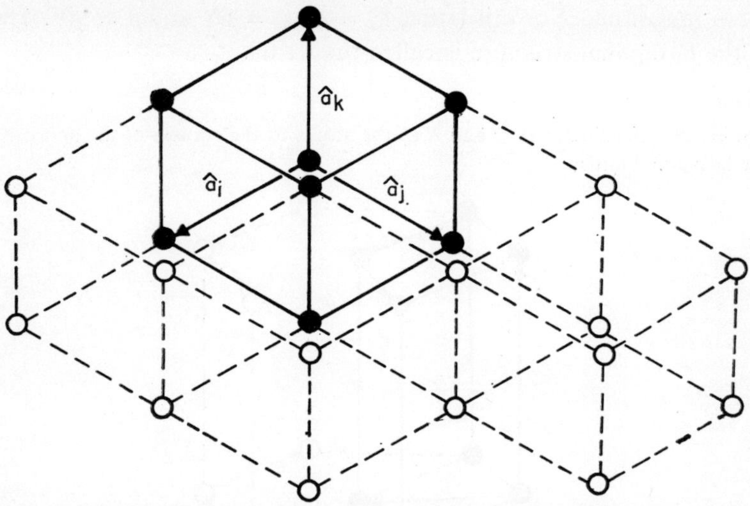

Figure 3.2 The three dimensional simple cubic lattice showing the cubic unit cell.

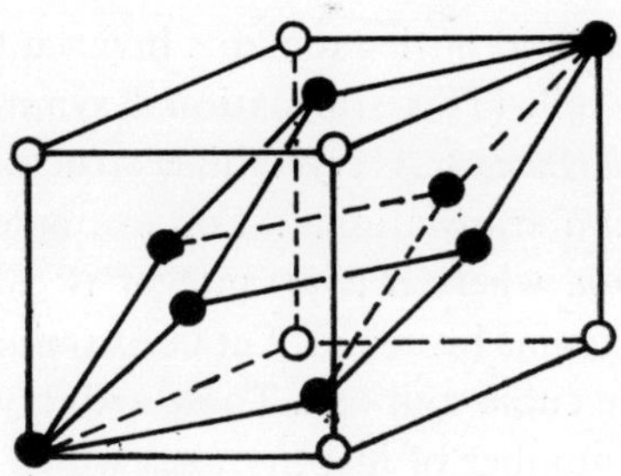

Figure 3.3 The cubic unit cell of the face-centered cubic lattice with the rhombohedral primitive cell enclosed.

The positions of the atoms in a unit cell are designated by their distances *P*, *Q* and *R* in unit lattice translations from a corner of the unit cell. Thus the atoms at the corners of the unit cell of the three dimensional simple cubic lattice (Figure 3.2) have the positions 100, 010, 001, 011, 110, 101, 111, and 000. The unit cell of the simple cubic lattice contains 1 atom, 1/8 contribution from each atom at the corners of the unit cell.

In Figure 3.4 the unit cell of the three dimensional simple hexagonal lattice is shown outlined in heavy lines. The lattice vectors which generate the simple hexagonal lattice are not mutually perpendicular, nor are they equal in magnitude. For this lattice $a_i = a_j$, $\gamma = 60°$ and $\beta = 90°$. The axis $\hat{a}_k$ in the hexagonal structure is called the c axis.

PROBLEM

Determine the coordinates P, Q and R of the atoms at the corners of the unit cell of the simple hexagonal lattice.

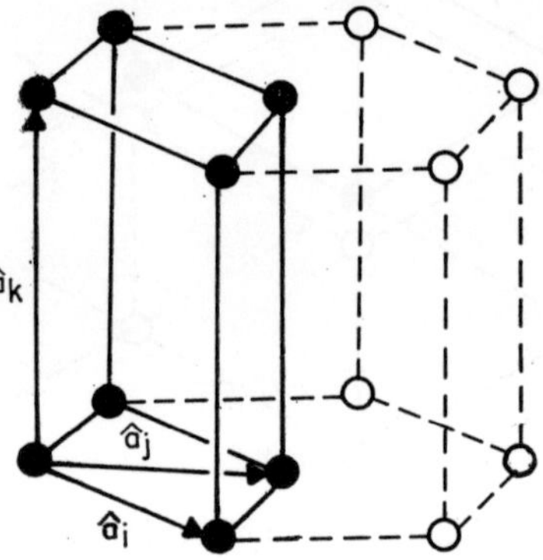

Figure 3.4 The simple hexagonal lattice. The unit cell is outlined in black.

As we have seen, a crystal lattice remains invariant under successive unit translations of a unit cell. (This translational symmetry of course breaks down at the crystal surfaces.) A crystalline structure also displays types of symmetry other than translational. For example, an axis of rotation through an angle $(2\pi)/n$, where n is an integer is called a rotation axis of order n if the crystal remains invariant. Let us examine the axes of rotational symmetry of the simple cubic unit cell. These are shown in Figure 3.5. (Not shown are the infinite number of identity axes which require a full rotation of 360° to bring the cell into coincidence with itself.) There are six diad axes of symmetry which require a rotation of 180° for coincidence, four triad axes which require a rotation of 120° and three tetrad axes which require a 90° rotation. It can be shown quite generally that symmetry operations of order higher than that of the hexad axis (60° rotation) are not allowed in a three dimensional system. The requirement for translational symmetry in crystalline structure makes pentagonal symmetry impossible.

A crystal may remain invariant after reflection through a plane. For example, as shown in Figure 3.6, there are three such planes of symmetry parallel to the faces of our cubic unit cell, and six diagonal planes of symmetry. Reflection symmetry is symbolized by the letter m. A unit cell consains a center of symmetry if any point on its face may be reflected to an identical point on an opposite face.

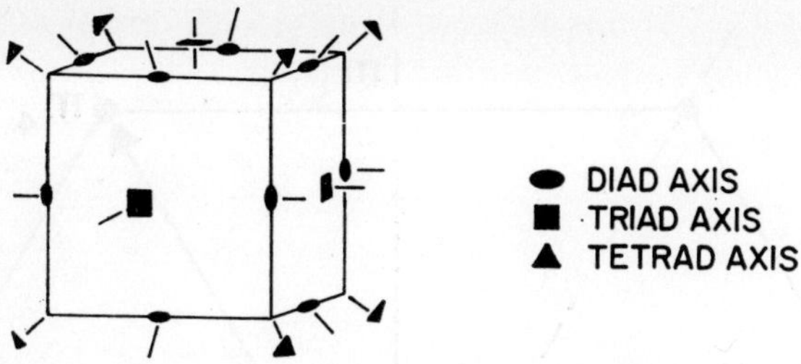

Figure 3.5 The thirteen axes of rotational symmetry of the cubic unit cell (from *An Introduction to Crystallography*, by F. C. Phillips, Longmans, Green and Co Inc., London 1963, p. 6).

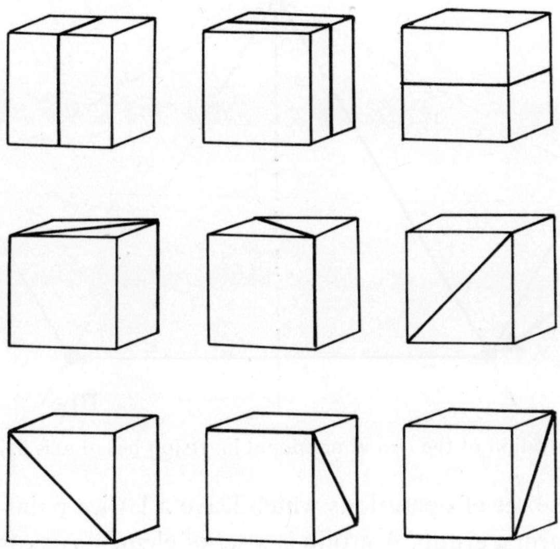

Figure 3.6 The nine mirror planes of symmetry of the simple cubic unit cell (Note that the unit cell of the three dimensional simple rectangular lattice has no mirror planes along the diagonals.)

Combinations of any of the above operations may also produce invariance. For example, a rotation-inversion axis will leave a point invariant after the compound operation (1) rotation about a given angle and (2) inversion through the axis of rotation. A rotation-inversion axis is symbolized by the order of the rotation topped by a bar, e.g., $\bar{1}$, $\bar{2}$, $\bar{3}$, $\bar{4}$, or $\bar{6}$. Figure 3.7 shows the inversion hexad axis $\bar{6}$.

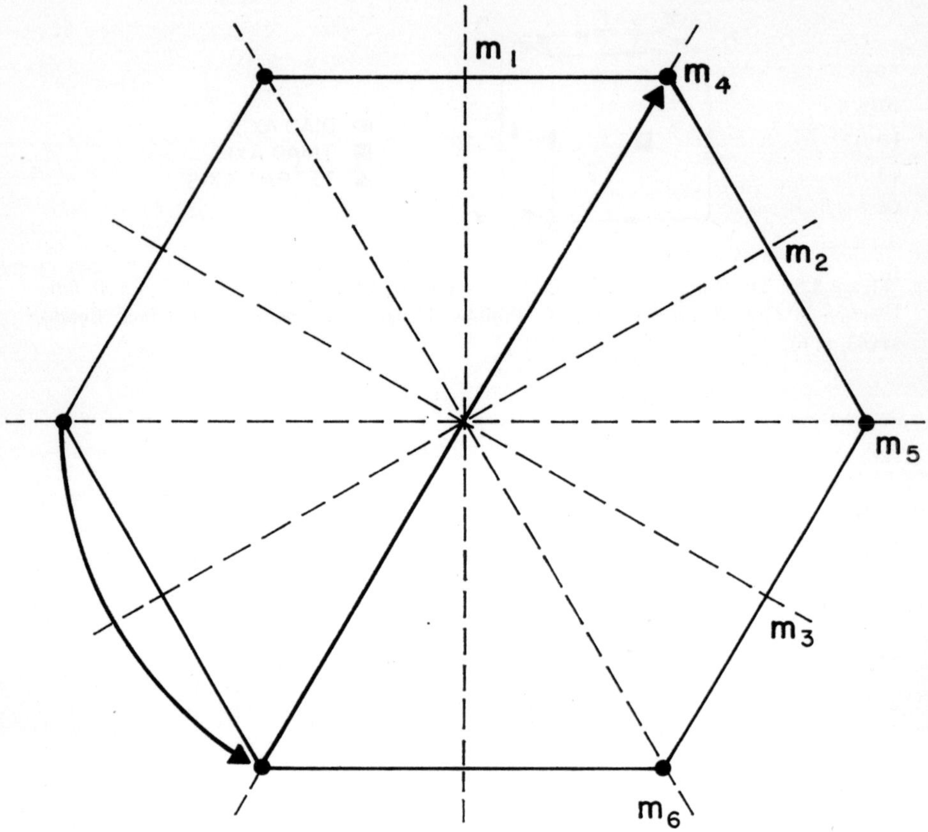

Figure 3.7 Illustration of the two dimensional inversion hexad axis $\bar{6}$.

The complete set of operations which leave a lattice point invariant may be shown to form a group. A group is a set of elements or operations which obey the following rules:

(1) One member of group is the identity operator E which commutes with all of the other members of the group.

(2) Each member A of the group has an inverse A^{-1} defined by $AA^{-1} = A^{-1}A = E$.

(3) The successive operation of two members of the group results in a third member of the group.

(4) Multiplication is associative: $A(BC) = (AB)C$. A group is called Abelian if each member commutes with every other member.

The symmetry operations which leave a two dimensional hexagon invariant (see Figure 3.7) are E, m_1, m_2, m_3, m_4, m_5, m_6, $6Z$, $3Z$, $2Z$, $3Z^2$ and $6Z^5$ where m denotes a minor reflection and z is the axis of rotation (out of the page). These operations may be written in the form of a multiplication table as shown in Table 3.1. Note that the operations do not necessarily commute. For example, application of a $3Z$ rotation followed by a reflection m_6 is equivalent to an m_4 reflection. However, penforming m_6 first followed by $3Z$ yields m_5. Thus we must adhere to the convention that the operations in the top row are applied first and those in the left column second.

Table 3.1 Multiplication table for two dimensional hexagonal symmetry

Apply first → / Apply second ↓	E	m_1	m_2	m_3	m_4	m_5	m_6	$6Z$	$3Z$	$2Z$	$3Z^2$	$6Z^5$
E	E	m_1	m_2	m_3	m_4	m_5	m_6	$6Z$	$3Z$	$2Z$	$3Z^2$	$6Z^5$
m_1	m_1	E	$3Z$	$3Z^2$	$6Z$	$2Z$	$6Z^5$	m_4	m_2	m_5	m_3	m_6
m_2	m_2	$3Z^2$	E	$3Z$	$6Z^5$	$6Z$	$2Z$	m_5	m_3	m_6	m_1	m_4
m_3	m_3	$3Z$	$3Z^2$	E	$2Z$	$6Z^5$	$6Z$	m_6	m_1	m_4	m_2	m_5
m_4	m_4	$6Z^5$	$6Z$	$2Z$	E	$3Z$	$3Z^2$	m_2	m_5	m_3	m_6	m_1
m_5	m_5	$2Z$	$6Z^5$	$6Z$	$3Z^2$	E	$3Z$	m_3	m_6	m_1	m_4	m_2
m_6	m_6	$6Z$	$2Z$	$6Z^5$	$3Z$	$3Z^2$	E	m_1	m_4	m_2	m_5	m_3
$6Z$	$6Z$	m_6	m_4	m_5	m_1	m_2	m_3	$3Z$	$2Z$	$3Z^2$	$6Z^5$	E
$3Z$	$3Z$	m_3	m_1	m_2	m_6	m_4	m_5	$2Z$	$3Z^2$	$6Z^5$	E	$6Z$
$2Z$	$2Z$	m_5	m_6	m_4	m_3	m_1	m_2	$3Z^2$	$6Z^5$	E	$6Z$	$3Z$
$3Z^2$	$3Z^2$	m_2	m_3	m_1	m_5	m_6	m_4	$6Z^5$	E	$6Z$	$3Z$	$2Z$
$6Z^5$	$6Z^5$	m_4	m_5	m_6	m_2	m_3	m_1	E	$6Z$	$3Z$	$2Z$	$3Z^2$

In a three dimensional crystal, the requirement for translational periodicity yields thirty two allowed point groups i.e., groups of operations which leave the 'center of mass' of a unit cell unchanged in position. Seven of these groups make up the seven crystal systems from which we get fourteen allowed Bravais lattices (see Table 3.2 and Figure 3.8). The remaining twenty five point groups come about by associating a number of atoms with each lattice site.

Most of the complicated crystal structures occurring in nature are not simple Bravais lattices. However many can be described by the addition of a number of *basis* vectors to the basic vectors of Bravais lattices. Using this

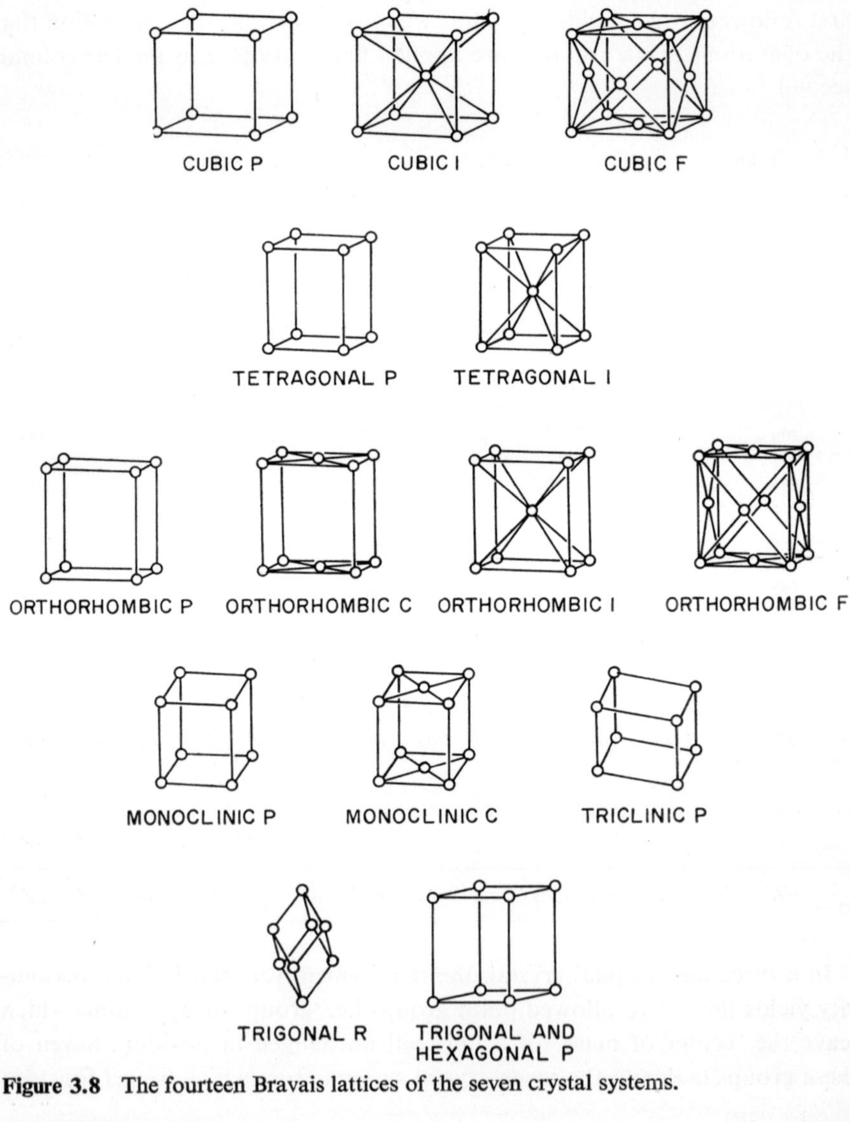

Figure 3.8 The fourteen Bravais lattices of the seven crystal systems.

Table 3.2 The fourteen Bravais lattices of the seven crystal systems

System	Number of Lattices in System	Nature of Unit Cell Axes and Angles	Symmetry
Triclinic	1	$\hat{a}_i \neq \hat{a}_j = \hat{a}_k$ $\alpha \neq \beta \neq \gamma \neq 90°$	none
Monoclinic	2	$\hat{a}_i \neq \hat{a}_j \neq \hat{a}_k$ $\alpha = \beta = 90° \neq \gamma$	one two fold rotation axis
Orthorhombic	4	$\hat{a}_i \neq \hat{a}_j \neq \hat{a}_k$ $\alpha = \beta = \gamma = 90°$	three mutually perpendicular two-fold rotation axes
Tetragonal	2	$\hat{a}_i = \hat{a}_j \neq \hat{a}_k$ $\alpha = \beta = \gamma = 90°$	one 4-fold rotation axis or a 4-fold rotation-inversion axis
Cubic	3	$\hat{a}_i = \hat{a}_j = \hat{a}_k$ $\alpha = \beta = \gamma = 90°$	Four 3-fold rotation axes
Trigonal (Rhombohedral)	1	$\hat{a}_i = \hat{a}_j = \hat{a}_k$ $\alpha = \beta = \gamma \neq 90°$	One 3-fold rotation axis
Hexagonal	1	$\hat{a}_i = \hat{a}_j \neq \hat{a}_k$ $\gamma = \beta = 90°$ $\alpha = 120°$	One 6-fold rotation axis

procedure, one can construct most, but not all, lattice structures. A basis vector designates the position of an extra atom in the unit cell of a Bravais lattice. An example of a two dimensional Bravais lattice with a basis can be seen in Figure 3.9. This structure, characteristic of the composition AB_2, is described as a simple square lattice of atoms A with basic vectors $\hat{a}_i$ and $\hat{a}_j$, with a set of basis vectors $\hat{b}_i = \frac{1}{2}\hat{a}_i$ and $\hat{b}_j = \frac{1}{2}\hat{a}_j$ which designate the positions of atom B.

It should be pointed out that the representation of a lattice in terms of a Bravais lattice with a basis is useful for purposes of explanation, but it may not be the most fundamental description, and, therefore, the unit cell determined by this Bravais-basis description is, in most cases, not the most fundamental unit cell for the structure.

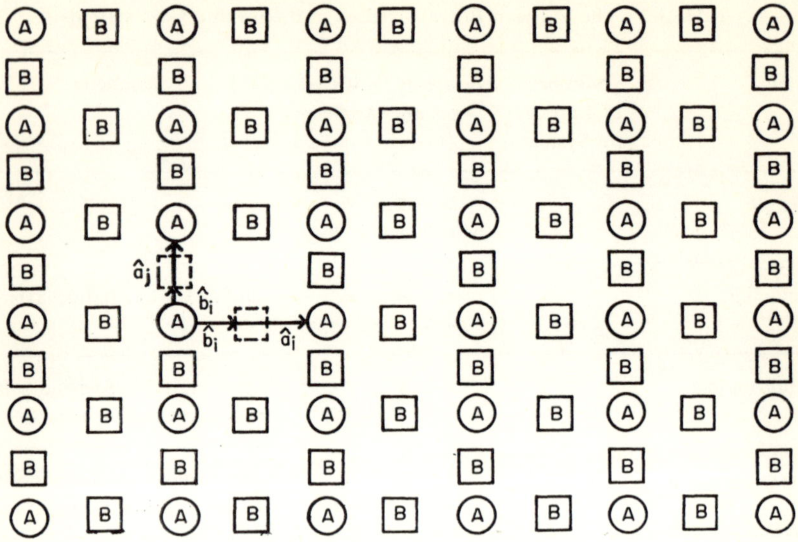

Figure 3.9 The two dimensional AB_2 lattice lilustrated as a simple lattice with basic vectors $\hat{a}_i$ and $\hat{a}_j$ and with basis vectors $\hat{b}_i$ and $\hat{b}_j$.

Miller Indices:

It is often useful to define a concise notation by which we may designate specific planes or sets of crystalline planes and directions. For this purpose, one usually chooses Miller indices. To find the Miller indices for a particular plane, one makes use of three noncollinear atoms, each of which lies in one of the principal directions and all of which lie in the particular crystal plane. The Miller indices are found as follows: the coordinates of each atom in units of lattice distances are written yielding, for example $(P, 0, 0)$, $(0, Q, 0)$, $(0, 0, R)$ where P, Q and R are integers. The inverse of P, Q and R are then computed. The smallest set of integers h, k, and l, which are in the same ratio to each other as are $1/P$, $1/Q$ and $1/R$ are called the Miller indices and are written (h, k, l).

PROBLEM

The law of rational indices states that the Miller indices which designate a crystal face are rational numbers. Prove this for the simple cubic and simple hexagonal lattices.

To specify a particular direction in a crystal, one chooses a vector in that direction, the scalar components of which are in units of lattice distances. A crystal direction is written $[P, Q, R]$ and a set of identical directions is written $\langle PQR \rangle$. A set of identical planes is written $\{h, k, l\}$. A negative integer is designated with a bar, e.g., $(h, \bar{k}, l)$.

PROBLEM

Show vectorially that a set $\{h_0, k_0, l_0\}$ of the Miller indices may indicate many parallel planes.

PROBLEM

Find the vector directions normal to the $\{111\}$ and $\{1, 0, 1\}$ planes in a simple cubic crystal.

3.2 CRYSTAL STRUCTURES

Types of Crystalline Bonding

We shall briefly discuss the major types of atomic bonding which result from atomic interaction at short distances and the types of crystalline structure which are to be expected as a result of these binding forces. One should remember that a crystalline structure need not be the result of only one of these types of bonding but may be due to a combination of two or more. In all cases, however, although the structure is usually a result of more than one type of bonding force, one will be dominant and we shall be able to determine from the crystalline structure, the type of binding force which is dominant. We can distinguish four major types of electrostatic (Coulombic) bonding forces, each of which produces characteristic crystalline structures:

The electrons in metallic crystals move about quite freely and behave very much like a gas of free electrons. The *metallic* bond is envisaged as a bond between the metallic ions and the electron gas in which they are immersed. Examples of crystals bound metallically are Li, Mg, K and Rb.

The *covalent* bond arises from the penetration of the electronic charge clouds of two atoms. This penetration results in the sharing of these electrons and usually in highly directional crystalline bonding. A covalent bond may exist between like atoms, in which case it is called a *homopolar* bond. Examples of homopolar bonding are crystalline Ge and Si. When a covalent bond is produced between two unlike atoms, it is called a *heteropolar* bond. Examples of heteropolar bonding are crystalline InP, GaAs and InAs.

The *ionic* bond (an extreme case of heteropolar bonding) is produced by the electrostatic attraction of highly electropositive and electronegative ions. In this case one atom is assumed to donate all of its valence electrons to its neighboring ion. It is often the case that ionic bonding may coexist with a covalent bond. For example the ionic character of the II-VI compounds such as ZnSe and CdS is usually stronger than that of the III-V compounds such as GaAs and GaP. When ionic and covalent bonds coexist, the resultant bond is usually further strengthened. It is typical that ionic crystals show excellent ionic conductivity at high temperatures. Examples of crystals ionically bound are the Alkali Halides.

The *Van der Waal's* bond is due to the induced dipole moment in one atom due to the constantly fluctuating dipole moment of an adjacent atom.[1] This process of induction leads to a very short ranged attractive force between the two atoms. The Van der Waal's force is usually found to be of the analytical form $F = A/(r^6)$ where A is a constant, and is much weaker than any of the other binding forces we have discussed, thus yielding solids with low melting points. The Van der Waal's bond is typical of molecular crystals in which, although each individual molecule may be a collection of covalently bound atoms, the entire crystal structure is bound together by a Van der Waal's bond.

In the case of solidified noble gas elements in which none of the three previous examples of binding forces are present, the binding force is of the Van der Waal's type.

Metallic Structures

The metallic bond is envisioned as a bond between the metallic ions at the lattice sites and the free electron gas in which they are immersed. We would expect that, in a crystal metallically bound, the resulting crystal structure would come about when equilibrium is reached between the ion-ion repulsive force and the ion-electron gas attractive force. In a first approximation in which we assume that the potential due to the metallic ions is spherical, the crystal structure will be determined by how closely the ions can be packed when equilibrium is reached.[2]

There are two methods by which spheres may be packed so as to minimize the unoccupied volume. Both of these methods of close-packing begin with the close-packing of the spheres in a plane in which each sphere is placed in contact with six others, as is shown in Figure 3.10. Spheres are then placed on the top of this layer in the depressions formed between any three contiguous spheres of the first layer. There are two ways to arrange the third layer to create the close-packing arrangements:

Hexagonal Close-Packing:

When the third layer is added such that every atom in the third layer is directly above the atoms in the first layer, hexagonal symmetry arises. This vertical stacking is repeated with the subsequent layers to build the lattice. The stacking for the close-packed hexagonal structure is shown in cross

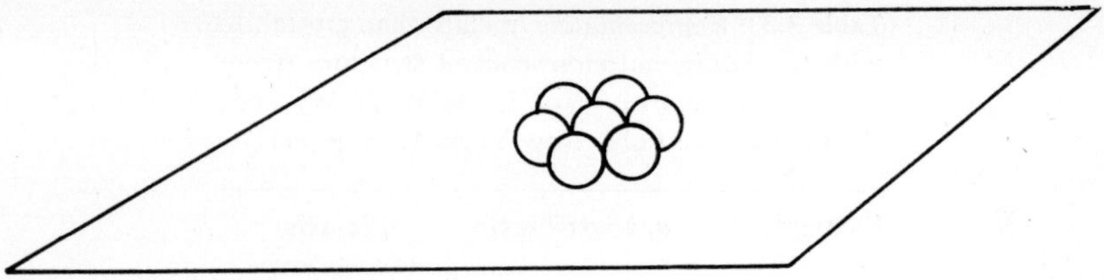

Figure 3.10 Close-packed spheres in a plane.

section in Figure 3.11. Figure 3.12 shows the hexagonal close-packed structure which arises from this type of packing. The close-packed hexagonal structure may be described as a simple hexagonal lattice with a single basis vector

$$\mathbf{b} = \frac{\hat{a}_i}{3} + \frac{2\hat{a}_j}{3} + \frac{\hat{a}_k}{2} \tag{3.2.1}$$

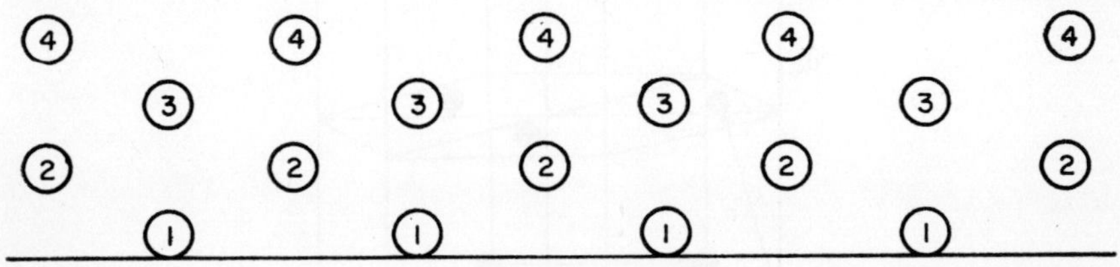

Figure 3.11 Cross-section of the stacking that results in the close-packed hexagonal structure.

The close-packed hexagonal lattice may not be described as a simple lattice. Its fundamental representation is that of a lattice with a basis. Each atom has six nearest neighbors at a distance a_j and three others at a distance, $\sqrt{\frac{a_j^2}{3} + \frac{a_k^2}{1}}$, three above and three below the basal plane. The volume of a unit cell is $\sqrt{3/2}a_j^2 a_k$.

Representative metals with hexagonal close-packed structure and their corresponding interatomic distances are shown in Table 3.3.

Cubic Close-Packing (Face-Centered Cubic):

When each atom of the third layer is added in the depressions of the second layer, the resultant structure obtains cubic symmetry in which each atom has six nearest neighbors. Figure 3.3 shows the unit cell of the cubic close-packed structure.

Table 3.3 Representative metals that crystallize with the hexagonal close-packed structure (from *Crystal Structures*, Vol. I, R. W. G. Wyckoff, Interscience Pubs. New York, 1965, p. 11)

Element	a_i ångströms)	a_k (*c*-axis) (ångströms)
Be	2.286	3.583
Cr	2.722	4.427
Li	3.111	5.093
Mg	3.209	5.210
Na	3.767	6.154

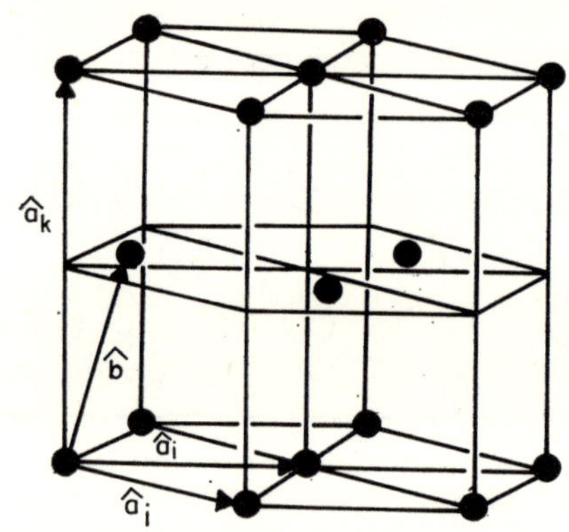

Figure 3.12 The unit cell of the close-packed hexagonal structure.

The description of the face-centered cubic lattice as a simple lattice is made via the vectors

$$\hat{a}_i = \frac{\hat{A}_y + \hat{A}_z}{2}$$

$$\hat{a}_j = \frac{\hat{A}_x + \hat{A}_y}{2}$$

$$\hat{a}_k = \frac{\hat{A}_x + \hat{A}_z}{2} \tag{3.2.2}$$

where $\hat{A}_x$, $\hat{A}_y$ and $\hat{A}_z$ refer to the orthogonal axes of the cubic unit cell.

The cubic close-packed structure may also be described as a simple cubic lattice with three basis vectors

$$\hat{b}_1 = \tfrac{1}{2}(\hat{A}_x + \hat{A}_y)$$
$$\hat{b}_2 = \tfrac{1}{2}(\hat{A}_x + \hat{A}_z)$$
$$\hat{b}_3 = \tfrac{1}{2}(\hat{A}_j + \hat{A}_z) \tag{3.2.3}$$

Table 3.4 shows some metals which commonly crystallize with the cubic close-packed structure.

Table 3.4 Representative metals that crystallize with the face-centered cubic sructure (from Wyckoff p. 10.)

Element	A_l (ångströms)
Ag	4.086
Al	4.049
Au	4.078
Fe	3.591
Cu	3.615

PROBLEM

What are the volume of the cubic unit and primitive cells of the face-centered cubic structure?

PROBLEM

Show by using Lagrangian multipliers that the interstitial volume is $\approx 35.9\%$ of the unit cell for both hexagonal close-packed and for cubic close-packed structures.

The Body-Centered Cubic Structure:

Another crystal structure commonly found for the metallic elements is the body-centered cubic structure. This lattice is built up in the following way: A square lattice of spheres is first laid out on a horizontal plane. A second layer is then built up by placing spheres in the depressions formed between the spheres in the first layer. The atoms of the third layer are placed vertically above the atoms in the first layer and so on. The square lattice arrangement of the first plane of atoms forms the foundation for the body-centered cubic arrangement, the unit cell of which is shown in Figure 3.13.

The body-centered cubic structure may be described as a cubic Bravais lattice with a single basis vector

$$\hat{b}_1 = \tfrac{1}{2}(\hat{A}_x + \hat{A}_y + \hat{A}_z), \tag{3.2.4}$$

where $\hat{A}_x$, $\hat{A}_y$ and $\hat{A}_k$ are the lattice vectors describing the edges of the unit cube. This structure may alternatively be thought of as two interpenetrating simple cubic lattices displaced from one another by one-half of the body diagonal.

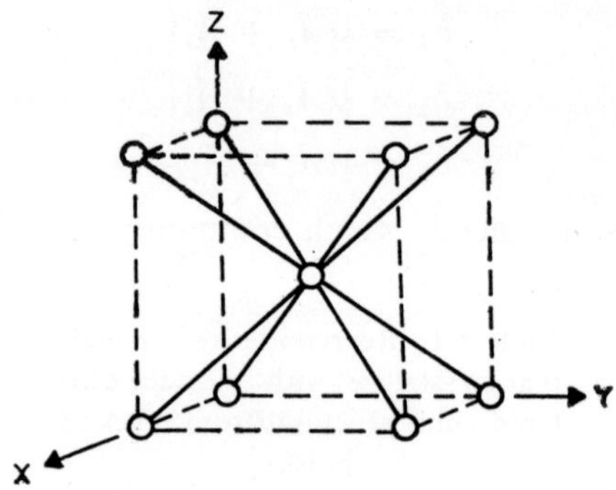

Figure 3.13 The body-centered cubic unit cell.

The representation of the body-centered cubic structure as a simple lattice is made by use of the three lattice vectors

$$\hat{a}_i = \frac{\hat{A}_x}{2} + \frac{\hat{A}_y}{2} + \frac{\hat{A}_z}{2}$$

$$\hat{a}_j = \frac{\hat{A}_x}{2} - \frac{\hat{A}_y}{2} + \frac{\hat{A}_z}{2}$$

$$\hat{a}_k = \frac{\hat{A}_x}{2} + \frac{\hat{A}_y}{2} - \frac{\hat{A}_z}{2} \tag{3.2.5}$$

The cubic close-packing arrangement leaves an unoccupied volume of 29.9% of the unit cell, and thus the atoms of this structure are quite closely packed. Table 3.5 lists some representative metals which crystallize in the body-centered cubic structure.

Table 3.5 Representative metals that crystallize with the body centered cubic structure (from Wyckoff, p. 16)

Element	*a* (ångströms)
Ba	5.025
K	5.225 (5°K)
	5.247 (78°K)
Cs	6.045 (5°K)
	6.067 (78°K)
Mo	3.147
Rb	5.585 (5°K)
	5.605 (78°K)

Covalent Structures:

The Diamond Structure (Cubic):

The diamond structure is shown in Figure 3.14a. The diamond lattice may be thought of as being made up of two face-centered cubic lattices displaced from one another by one-quarter of a body diagonal. Each atom has four nearest neighbors at a distance $A\frac{\sqrt{3}}{4}$ arranged at the corners of a regular tetrahedron. The positions of the atoms in the cubic unit cell are shown in Figure 3.14b.

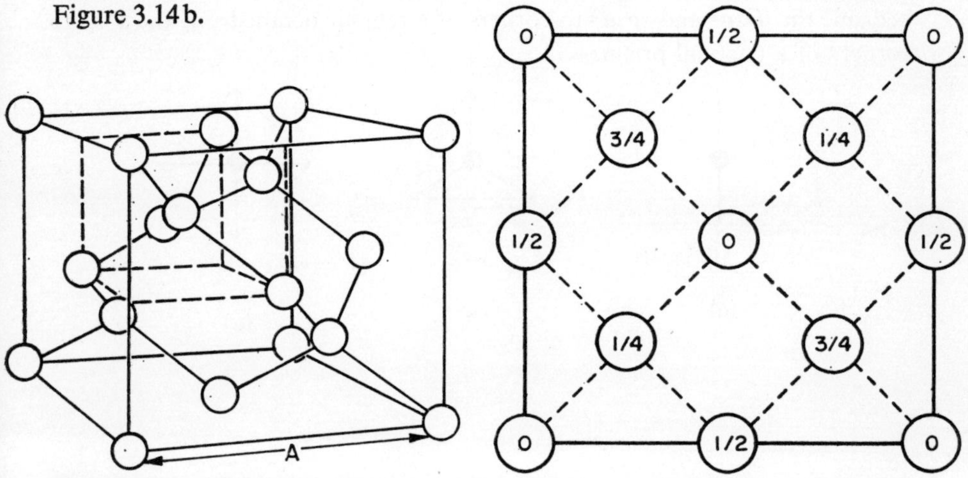

Figure 3.14 (a) The diamond structure. (b) Coordinates of the atoms in the diamond structure (in units of A).

The diamond structure is described as a simple cubic lattice with basis vectors

$$\hat{b}_1 = \tfrac{1}{2}(\hat{A}_x + \hat{A}_y),$$
$$\hat{b}_2 = \tfrac{1}{2}(\hat{A}_x + \hat{A}_z),$$
$$\hat{b}_3 = \tfrac{1}{2}(\hat{A}_y + \hat{A}_z), \qquad (3.2.6)$$

which designate points on the faces of the unit cube and

$$\hat{b}_4 = \tfrac{1}{4}(\hat{A}_x + \hat{A}_y + \hat{A}_z),$$
$$\hat{b}_5 = \tfrac{1}{4}(3\hat{A}_x + 3\hat{A}_y + \hat{A}_z),$$
$$\hat{b}_6 = \tfrac{1}{4}(3\hat{A}_x + \hat{A}_y + 3\hat{A}_z),$$
$$\hat{b}_6 = \tfrac{1}{4}(\hat{A}_x + 3\hat{A}_y + 3\hat{A}_z) \qquad (3.2.7)$$

which designate the internal points. The diamond lattice may not be reduced to a simple lattice. It should be noted that the diamond lattice is a relatively empty lattice, the interstitial volume being approximately 66% of the unit cell. This emptiness is characteristic of covalently bound crystals, since bonds may be formed between unlike atoms, and thus between different atomic electronic states. Bonds may also be formed between two, three or more electrons. The covalent bonds between these different electrons are highly directional, as is shown in Figure 3.15. We see that an *sp* bond is linear and an sp^2 bond is one in which the center atom is bonded to three atoms at the corners of an equilateral triangle the dsp^2 to corners of a square, the sp^3 (characteristic of Ge and Si) to corners of a regular tetrahedron, the d^2sp^2 and sp^3d^2 to corners of a regular octahedron, the d^4sp to corners of a trigonal prism.

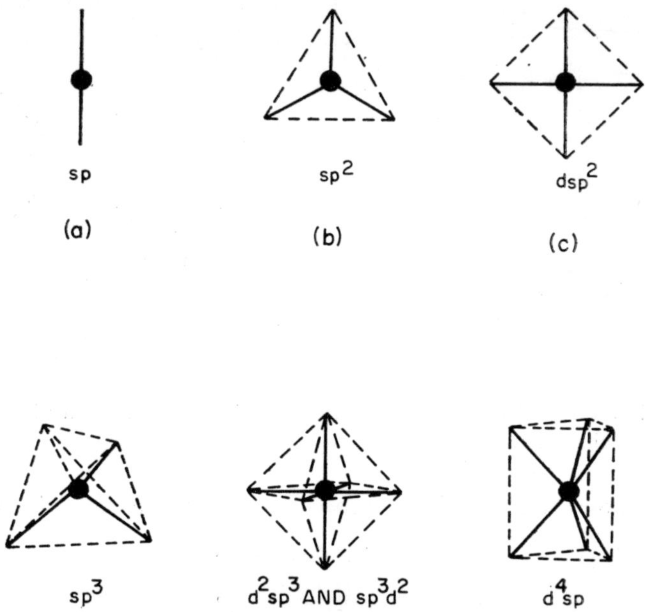

Figure 3.15 Bond configurations corresponding to some simple hybrid orbitals (a) *sp*-linear (b) sp^2-to corners of an equilateral triangle (c) dsp^2-to corners of a square (d) sp^3-to corners of a regular tetrahedron (e) d^2sp^3 and sp^3d^2-to corners of a regular tetrahedron (f) d^4sp-to corners of a trigonal prism (from *An Introduction to Cry tal Chemistry*, R. C. Evans, Cambridge University Press, 1966, p. 60.)

Table 3.6 lists some elements which show the diamond structure. Elements such as Ge and Si, from the fourth group of elements in the atomic chart, are typical monatomic semiconductors with the diamond structure.

Table 3.6 Some elements that crystallize with the diamond structure (from Wyckoff, p. 26)

Element	A (ångströms)
C	3.56679 (20°C)
Si	5.4370 (25°C)
	5.455 (1300°C)
Ge	5.65735 (20°C)
	5.65695 (18°C)
α-Sn	6.4912

The Zincblende Structure:

In the zincblende structure each Zn (or S) atom is surrounded by four S (or Zn) atoms at the corners of a regular tetrahedron. The unit cube is shown in Figure 3.16. Since the zincblende structure represents the tightest packing possible for the two different size ions in equal numbers, a few ionic compounds crystallize with this structure. This structure is characteristic of many covalently bound crystals, such as the III-V semiconducting compounds.

The zincblende structure is the particular case of the diamond structure in which one face-centered cubic lattice consists of atoms of one type, and the displaced face-centered cubic structure consists of atoms of another type.

Table 3.7 lists some of the more common compounds which have the zincblende structure.

Table 3.7 Compounds which crystallize with the zincblende structure

Crystal	A (ångström)
InSb	6.96
InAs	6.04
GaAs	5.63
GaP	5.35
CdS	5.82
ZnS	5.41

PROBLEM

Determine the coordinates of the Zn and of the S atoms in the unit cell of the zincblende structure.

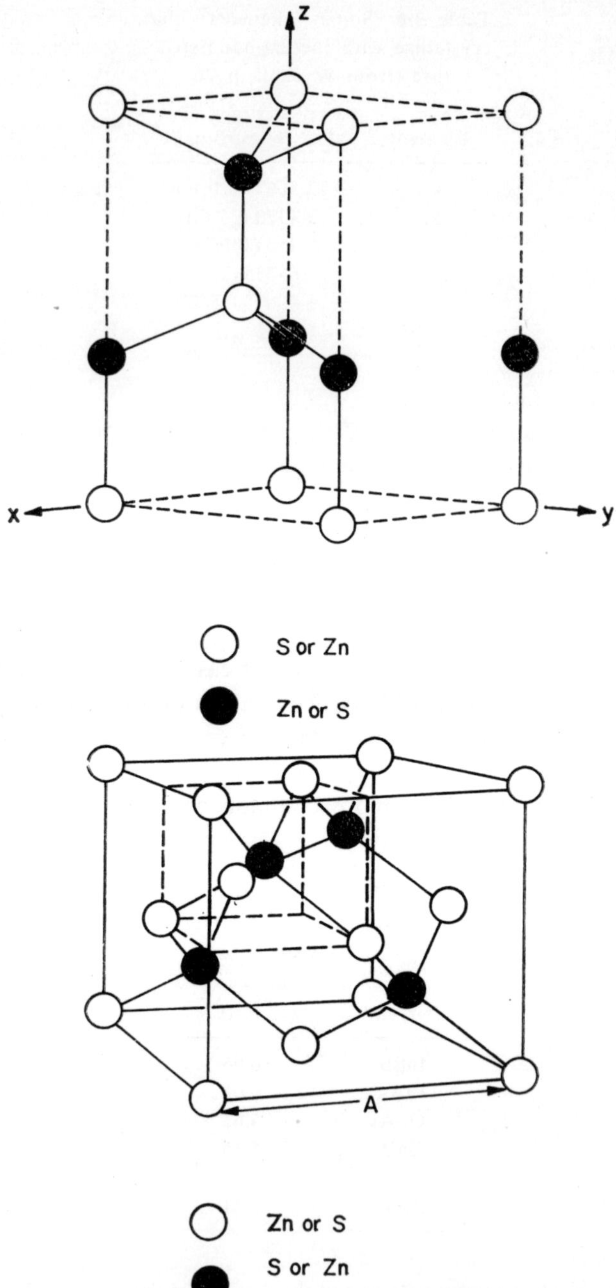

Figure 3.16 The (a) zincblende and (b) wurtzite structures (from Evans).

A structure closely related to the zincblende structure is the (hexagonal) wurtzite structure shown in Figure 3.16b. Although each atom is again tetragonally bonded to its four nearest neighbors, the difference between the structures occurs in the next nearest neighbor arrangements which, in the wurtzite structure, favors bonds with more ionicity. Some compounds which crystallize with the wurrzite structure are shown in Table 3.8. Note that because of the similarity between the wurtzite and zincblende structures, many compounds crystallize with both phases.

Table 3.8 Some wurztite structures (from Wyckoff, p. 112)

Crystal	*a* (ångströms)	*c* (ångströms)
CdS	4.134	6.749
CdSe	4.30	7.02
GaN	3.180	5.166
ZnS	3.811	6.234
ZnSe	3.98	6.53
ZnTe	4.27	6.99

Ionic Structures:

The Sodium Chloride Structure:

The cation-anion nature of ionic bonding produces compounds of the type *AB*. The simplest of the ionic structures is the Sodium Chloride structure. This is the structure of most of the alkali halides. Figure 3.17 shows the unit cell of the sodium chloride structure in which each chlorine atom is

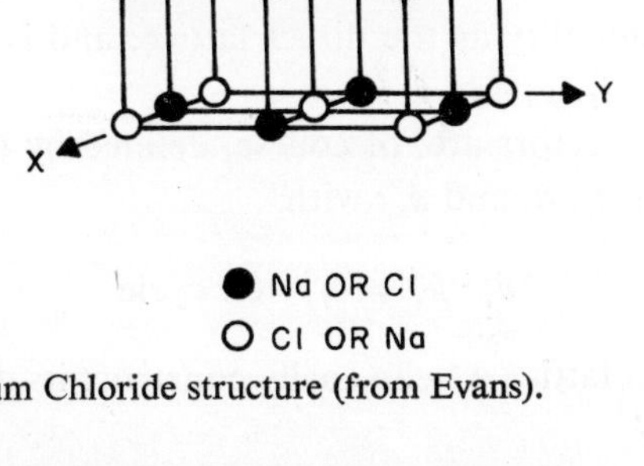

Figure 3.17 The sodium Chloride structure (from Evans).

surrounded by six sodium atoms. The NaCl structure may be thought of as two face-centered lattices displaced by one half of each of the basic lattice vectors, one lattice made up of Na^+ ions and the other made up of Cl^- ions.

The Cesium Chloride Structure

Many of the halides of cesium, such as CsBr, CsI and CsCl crystallize in the CsCl structure. This structure is a body-centered cubic structure with the central atom different from the atoms which make up the eight simple cubic lattice points. The unit cell of the CsCl structure is shown in Figure 3.18. Each Cs (or Cl) atom is surrounded by eight Cl (or Cs) atoms arranged at the corners of a cube.

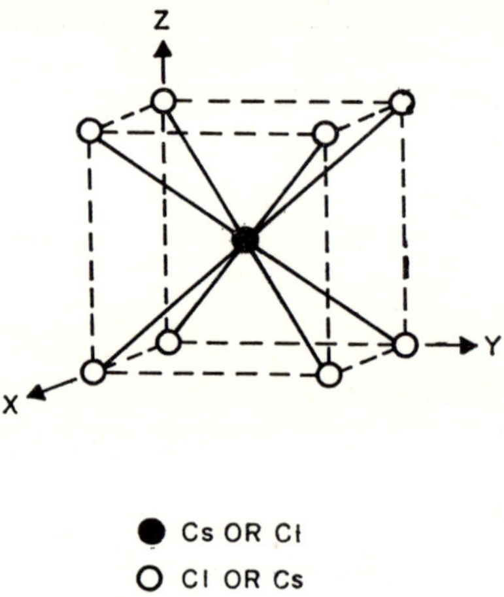

Figure 3.18 The caesium chloride structure (from Evans).

3.3 THE RECIPROCAL LATTICE

It is of great utility to define a dual space to our real three dimensional lattice space (our "direct" lattice). This dual lattice is chosen such that it is of the same dimensionality as the direct lattice, and is defined by the dual space lattice vectors $\hat{g}_i$, $\hat{g}_j$ and $\hat{g}_k$.

These dual lattice vectors are, of course, defined by their relations to the direct lattice vectors $\hat{a}_i$, $\hat{a}_j$ and $\hat{a}_k$, with

$$\hat{a}_i \cdot \hat{g}_j = \delta_{ij} \quad \text{et. cycle} \tag{3.3.1}$$

Thus for every direct lattice we can easily construct its dual (see 1.5).

One method of constructing a set of vectors which satisfy (3.3.1) is by the creation of the set of axial vectors

$$\hat{g}_i = \frac{\hat{a}_j \times \hat{a}_k}{\hat{a}_i \cdot \hat{a}_j \times \hat{a}_k}$$

$$\hat{g}_j = \frac{\hat{a}_k \times \hat{a}_i}{\hat{a}_i \cdot \hat{a}_j \times \hat{a}_k}$$

$$\hat{g}_k = \frac{\hat{a}_i \times \hat{a}_j}{\hat{a}_i \cdot \hat{a}_j \times \hat{a}_k} \qquad (3.3.2)$$

PROBLEM

Determine in general the volume of the unit cell of the dual lattice in terms of the volume of the direct lattice.

Since it can be shown that the volume of the unit cell of this dual lattice is in a reciprocal relation to the volume of the direct lattice, a dual lattice defined by equations (3.3.1) and (3.3.2) is called a *reciprocal* lattice. This reciprocal lattice contains all of the mathematical properties of the direct lattice. Thus it may be generated by successive unit translations of its unit cell. Any point of the reciprocal lattice may be obtained from any other point by means of a reciprocal lattice vector

$$\mathbf{G} = m_i\hat{g}_i + m_j\hat{g}_j + m_k\hat{g}_k \qquad (3.3.3)$$

where m_i, m_j and m_k are integers.

PROBLEM

Show that labelling a plane in the direct lattice by its Miller indices is identical with labelling it by the components of a reciprocal lattice vector and thus each point in the reciprocal lattice represents a set of lattice planes.

PROBLEM

Show that the distance between two consecutive parallel planes in a lattice is given by the inverse length of a basic reciprocal lattice vector.

PROBLEM

Show that two reciprocal lattice vectors $\mathbf{G}_1 = m_{1i}\hat{g}_i + m_{1j}\hat{g}_j + m_{1k}\hat{g}_k$ and $\mathbf{G}_2 = m_{2i}\hat{g}_i + m_{2j}\hat{g}_j + m_{2k}\hat{g}_k$ for which $m_{1i} : m_{1j} : m_{1k} = m_{2i} : m_{2j} : m_{2k}$ define the same set of crystal planes in the direct lattice.

From the definitions (3.3.1) and (3.3.2) we can determine that the body-centered cubic lattice has as its reciprocal lattice a face-centered cubic lattice whose basic vectors are

$$\hat{g}_i = \frac{1}{A}[0, 1, 1]$$

$$\hat{g}_j = \frac{1}{A}[1, -1, 0]$$

$$\hat{g}_k = \frac{1}{A}[1, 0, -1] \tag{3.3.4}$$

The unit cell of the reciprocal lattice has a volume $2/A^3$. It follows that the reciprocal of the face-centered cubic structure is the body-centered lattice.

It is easily shown that the reciprocal of a simple hexagonal lattice with lattice vectors $\hat{a}_i$, $\hat{a}_j$ and $\hat{a}_k$ as defined in Figure 3.4 is a simple hexagonal lattice with lattice constants $g_i = g_j = 2/\sqrt{3}a_i$, $g_k = 1/a_k$.

PROBLEM

Show that the cubic and simple hexagonal lattices are their own reciprocals.

Again, care must be taken when one wishes to take advantage of cubic symmetry and define a structure as a lattice with a basis. If this is not the most fundamental representation of the lattice, one may not generate a simple reciprocal lattice as we have done. It is beyond the scope of this book to determine the consequences of using nonfundamental representation. However, in the cases of the three dimensional diamond and the close-packed hexagonal lattices, in which their most basic representations are as lattices with a set of basis vectors, we need only state that the reciprocal lattice for the diamond structure is a body-centered cubic lattice which has a cube side equal to $2/A$. The reciprocal of the close-packed hexagonal structure is a simple hexagonal lattice structure.

3.4 BRAGG SCATTERING

In 1912, M. von Laue suggested that the reflections from a beam of X-rays incident on the surface of a single crystal would yield information about the symmetry of the reflecting lattice. Subsequent experiments by Friedrich and Knipping showed that the reflected beams produced a definite exposure pattern (such as that shown for a beam transmitted through Ge in Figure 3.19) when allowed to expose a photographic plate. W. Bragg[3] interpreted the experimental data by considering a beam of X-rays incident on a crystal at an angle θ to the surface as shown in Figure 3.20.

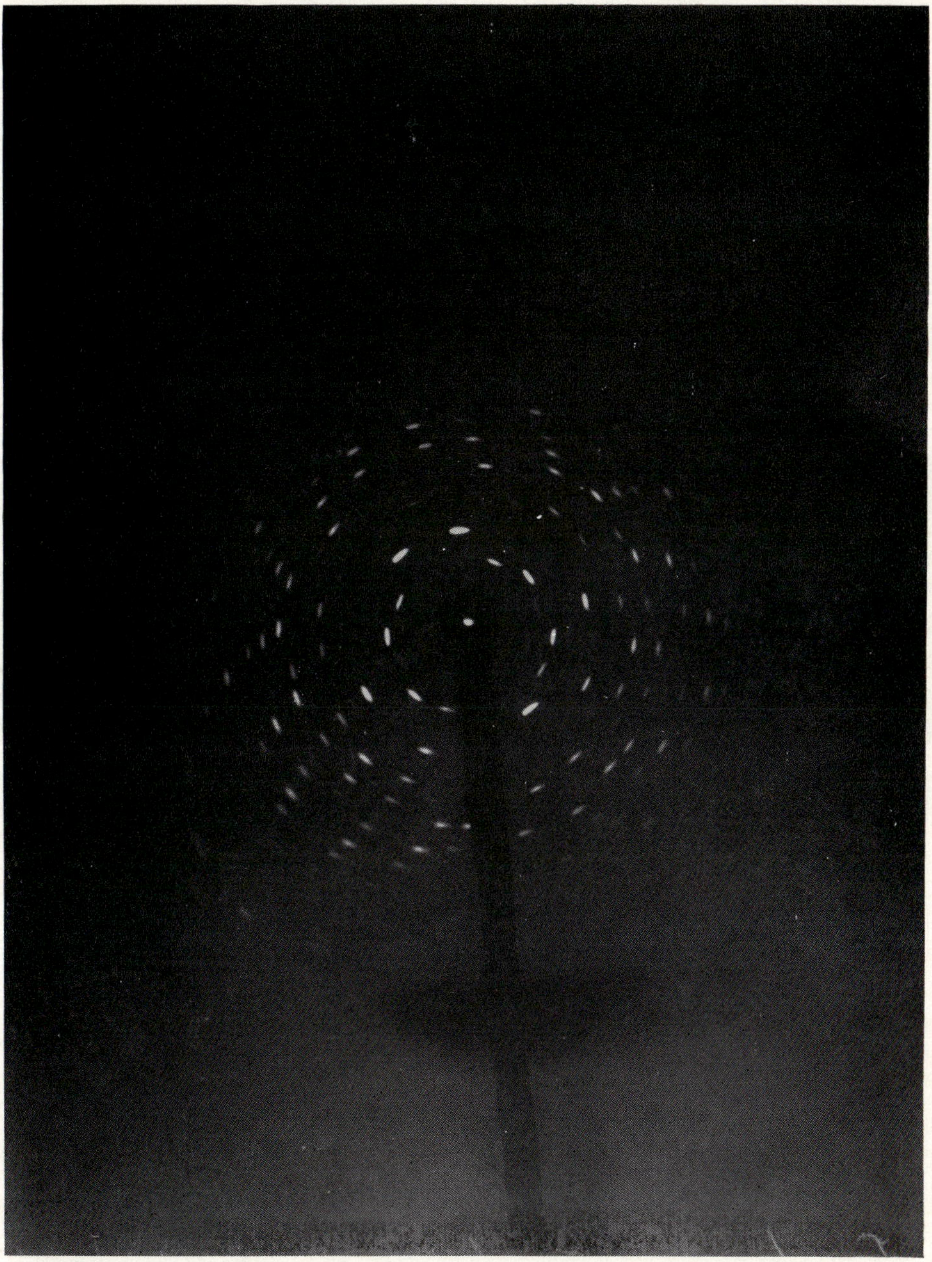

Figure 3.19 Spot pattern produced by 100 kV x-ray transmission parabel to the ⟨III⟩ axis of be (Courtesy of B. Paretzkin).

The condition for constructive interference of both the reflected and re-emitted waves from successive planes is

$$n\lambda = 2a \sin \theta \qquad n = 0, \pm 1, \pm 2 \ldots \tag{3.4.1}$$

where λ is the wavelength of the incident beam and a is the lattice spacing.

When incident normal to the crystal, a constructive interference is produced when

$$k = \frac{n\pi}{a} \qquad n = 0, \pm 1, \pm 2 \ldots \tag{3.4.2}$$

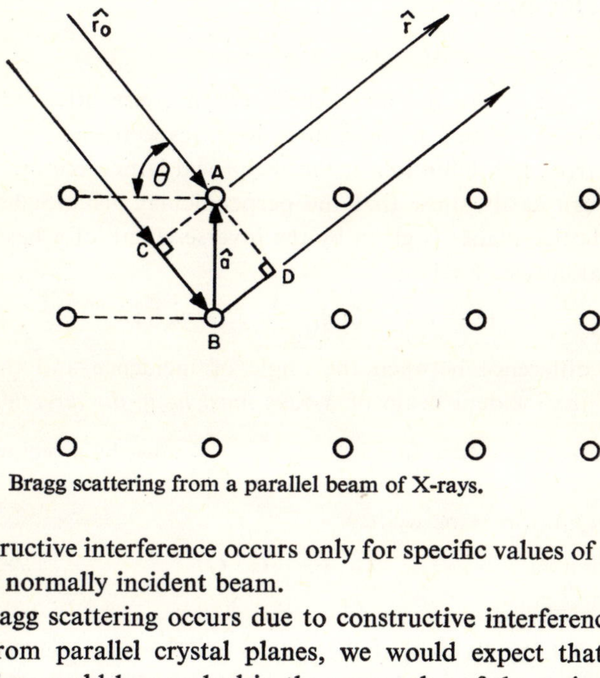

Figure 3.20 Bragg scattering from a parallel beam of X-rays.

Thus constructive interference occurs only for specific values of the momentum of the normally incident beam.

Since Bragg scattering occurs due to constructive interference of *X*-rays reflected from parallel crystal planes, we would expect that the Bragg scattering law could be couched in the vernacular of the reciprocal lattice, and that the directions corresponding to maximum reflection of incident *X*-rays are closely related to specific vector directions in the reciprocal lattice.

Let $\mathbf{r}_0$ and $\mathbf{r}$ be the directions of incidence and reflection, respectively, of a parallel beam of *X*-rays, as shown in Figure 3.20. If we define the lattice vector $\hat{a}$ as shown in Figure 3.20, we find

$$CB = (-\hat{a}) \cdot \hat{r}_0 = \hat{a}\hat{r}_0 \cos \theta$$

$$BD = \hat{a} \cdot \hat{r} = ar \cos \theta \tag{3.4.3}$$

For constructive interference to occur between the waves reflected from atoms A and B

$$\hat{a} \cdot (-\hat{r}_0 + \hat{r}) = n\lambda \tag{3.4.4}$$

Since (3.4.4 must hold for all lattice vectors, we may write specifically,

$$(\hat{a}_i)(\hat{r} - \hat{r}_0)_i = n_i\lambda$$

$$(\hat{a}_j)(\hat{r} - \hat{r}_0)_j = n_j\lambda$$

$$(\hat{a}_k)(\hat{r} - \hat{r}_0)_k = n_k\lambda \tag{3.4.5}$$

which yields, for example

$$(\hat{r} - \hat{r}_0)_k = \{n_k/a_k\}\,\lambda$$

where $\hat{a}_i$, $\hat{a}_j$ and $\hat{a}_k$ are the distances between consecutive lattice planes whose normals lie in the $\hat{\imath}$, $\hat{\jmath}$, and $\hat{k}$ directions, respectively.

However, from the definition of the reciprocal lattice vectors (3.3.1) and (3.3.2), one can easily show that the perpendicular distance between two consecutive lattice planes is given by the inverse length of a basic vector in the reciprocal lattice. We have

$$(\hat{r} - \hat{r}_0) = (n_i\hat{g}_i + n_j\hat{g}_j + n_k\hat{g}_k)\,\lambda \tag{3.4.6}$$

Thus the difference between the angle of incidence and the angle of reflection of the incident beam of X-rays *must lie in the direction of a reciprocal lattice vector*. More importantly, $\dfrac{\hat{r} - \hat{r}_0}{\lambda}$ must be a reciprocal lattice vector.

Equation (3.4.6) may be written

$$-\frac{\mathbf{G} \cdot \hat{r}}{\lambda} = \frac{\mathbf{G} \cdot \mathbf{G}}{2} \tag{3.4.7}$$

Equation (3.4.7) tells us that the values of $\hat{r}/\lambda$ which yield constructive interference of the reflected X-rays are such that, if they are plotted in the reciprocal space, $\mathbf{k}/(2\pi)$ must be of length $\mathbf{G}/2$. Plotting (3.4.7) in a reciprocal lattice space, we find that the values of $\hat{r}/\lambda$ which yield the so-called Bragg intensity maxima lie on a plane which is the perpendicular bisector of the reciprocal lattice vector **G**.

References

1. F. London and E. Eisenschitz, *Z. Physik*, **60**, 491 (1930).
2. A "close-packed" arrangement is a spatial arrangement of spheres which minimizes the unoccupied volume.
3. W. L. Bragg, *Proc. Cambridge Phil. Soc.*, **17**, 43, 1912.

CHAPTER 4

Energy Bands in Solids

4.1 HOMPOLAR COVALENT BONDING

We begin the study of the formation of energy bands in solids by performing an illustrative calculation which demonstrates the splitting of atomic energy levels and the subsequent formation of energy bands upon the formation of solids. We will utilize the variational method to determine the allowed states of a homopolar covalent bond. This treatment was originally formulated by Heitler and London[1] to explain valence bonding of the hydrogen molecule by the method of wave mechanics. It is an example of the Linear Combination of Atomic Orbitals (LCAO) method of describing molecular bonding.

We consider two similar atoms, A and B, each of which has an atomic configuration made up of a number of closed shells and one valence electron as shown in Figure 4.1. The total Hamiltonian for the system is

$$\mathscr{H} = \frac{p_1^2}{2m} + \frac{p_2^2}{2m} + \frac{e^2}{r_{A1}} + \frac{e^2}{r_{A2}} + \frac{e^2}{r_{B1}} + \frac{e^2}{r_{B2}} - \frac{e^2}{r_{12}} - \frac{e^2}{r_{AB}} \tag{4.1.1}$$

where m is the electronic mass.

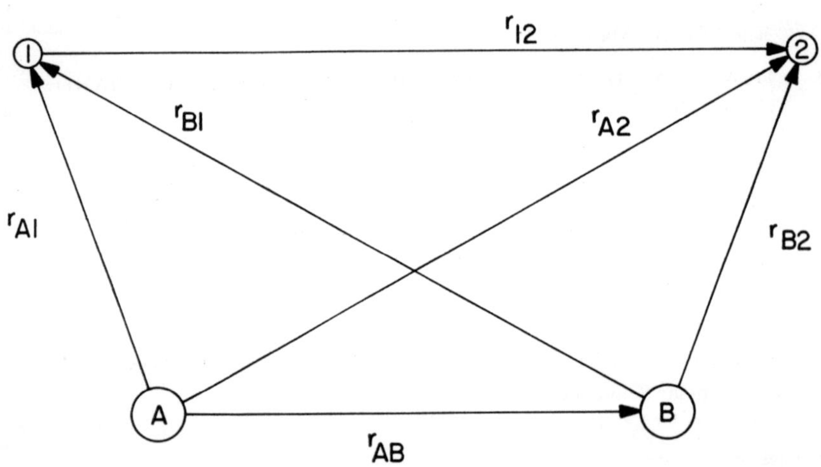

Figure 4.1 Configuration of two singly valent atoms used to illustrate homopolar covalent bonding.

Since both atoms A and B are assumed to be the same, their energies must be equal when the distance r_{AB} becomes very large. In this case the state of the system may be chosen as a product of the states of the individual atoms. If we choose $|\psi_{1A}\rangle$ as the state which represents electron 1 travelling about nucleus A and $|\psi_{2B}\rangle$ as the state in which electron 2 travels about nucleus B, then the product states $|\psi_T'\rangle = |\psi_{1A}\rangle\,|\psi_{2B}\rangle$ and $|\psi_T''\rangle = |\psi_{1B}\rangle\,|\psi_{2A}\rangle$ have the required two-fold degeneracy in energy. That is, at separations such as that $r_{AB} \approx r_{12} > r_{1A} \approx r_{2B}$ we get

$$\mathscr{H} = \left\{\frac{p_1^2}{2m} + \frac{e^2}{r_{A1}}\right\} + \left\{\frac{p_2^2}{2m} + \frac{e^2}{r_{2B}}\right\} = \mathscr{H}_1 + \mathscr{H}_2 \tag{4.1.2}$$

where $\mathscr{H}_1 = \mathscr{H}_2$.

$$\begin{aligned} \text{Thus } \mathscr{H}\,|\psi_T'\rangle &= (\mathscr{H}_1 + \mathscr{H}_2)\,|\psi_{1A}\rangle\,|\psi_{2B}\rangle \\ &= E_1'\,|\psi_{1A}\rangle\,|\psi_{2B}\rangle + E_2'\,|\psi_{1A}\rangle\,|\psi_{2B}\rangle \\ &= E_T'\,|\psi_{1A}\rangle\,|\psi_{2B}\rangle \end{aligned} \tag{4.1.3}$$

where $E_1' = E_2'$ and $E_1' + E_2' = E_T'$. A similar proof holds for $|\psi_T''\rangle$. Thus the states $|\psi_T'\rangle$ and $|\psi_T''\rangle$ are degenerate in their energy.

PROBLEM

Show that

$$A|\psi_T'\rangle + B|\psi_T''\rangle \tag{4.1.4}$$

is an eigenstate of $\mathscr{H}$.

We shall utilize the linear combination (4.1.4) and perform a variational calculation to determine the energy E of our system as a function of interatomic distance.

Utilizing condition (1.4.34), we find

$$0 = \begin{vmatrix} \langle\psi_T'|\,\mathscr{H}\,|\psi_T'\rangle - E & \langle\psi_T'|\,\mathscr{H}\,|\psi_T''\rangle - \langle\psi_T'|\psi_T''\rangle E \\ \langle\psi_T''|\,\mathscr{H}\,|\psi_T'\rangle - \langle\psi_T''|\psi_T'\rangle E & \langle\psi_T''|\,\mathscr{H}\,|\psi_T''\rangle - E \end{vmatrix} \tag{4.1.5}$$

We shall make use of the more concise notation

$$\begin{aligned} H_{12} &= \langle\psi_T'|\,\mathscr{H}\,|\psi_T''\rangle \\ H_{21} &= \langle\psi_T''|\,\mathscr{H}\,|\psi_T'\rangle \\ H_1 &= \langle\psi_T'|\,\mathscr{H}\,|\psi_T'\rangle \\ H_2 &= \langle\psi_T''|\,\mathscr{H}\,|\psi_T''\rangle \\ \Delta_{12} &= \langle\psi_T'|\psi_T''\rangle \\ \Delta_{21} &= \langle\psi_T''|\psi_T'\rangle \end{aligned} \tag{4.1.6}$$

Since the atoms are identical, $\Delta_{21} \equiv \Delta_{12} = \Delta$ and $H_{12} = H_{21} = H$. Thus

$$E^2(1 - \Delta^2) + E(-H_2 - H_1 + H\Delta + H\Delta) - H^2 + H_1H_2 = 0 \tag{4.1.7}$$

It is easily shown that

$$H_1 = 2E_1' + 2J - J' - \frac{e^2}{\delta r_{AB}} \tag{4.1.8}$$

$$H = 2\Delta E_1' + 2\sqrt{\Delta}\, K - K' + \frac{\Delta e^2}{\delta r_{AB}} \tag{4.1.9}$$

where

$$J = \left\langle \psi_T' \left| \frac{e^2}{\alpha' r_{A2}} \right| \psi_T' \right\rangle \tag{4.1.10}$$

represents the coulombic interaction of the valence electron of atom *B* with the nucleus and closed shells of atom *A*. Note also that

$$J' = \left\langle \psi_T' \left| \frac{e^2}{r_{12}} \right| \psi_T' \right\rangle \tag{4.1.11}$$

is the coulombic interaction between electrons 1 and 2, and

$$K = \left\langle \psi_T' \left| \frac{e^2}{r_{A1}} \right| \psi_T'' \right\rangle \tag{4.1.12}$$

is the resonance function (sometimes called an exchange function) which represents the interaction of electron one with both of the allowed states $|\psi_T'\rangle$ and $|\psi_T''\rangle$. We can define

$$K' = \left\langle \psi_T' \left| \frac{e^2}{r_{12}} \right| \psi_T'' \right\rangle \tag{4.1.13}$$

as the electron-electron interaction between the two states $|\psi_T'\rangle$ and $|\psi_T''\rangle$.

Equation (4.1.5) yields

$$E = \frac{-(-H_1 - H_2 + 2H\Delta)}{2(1 - \Delta^2)} \pm \frac{\sqrt{(-H_2 - H_1 + 2\Delta H)^2 - 4(1 - \Delta^2)(-H^2 + H_1 H_2)}}{2(1 - \Delta^2)} \tag{4.1.14}$$

which may be written

$$E = \frac{2H_1 - 2H\Delta \pm (2\Delta H_1 - 2H)}{2(1 - \Delta^2)} \tag{4.1.15}$$

Thus we see that upon bringing two similar atoms together we find a splitting of the doubly degenerate energy *E*. Let us first examine the energy E_A associated with the plus sign.

$$E_A = \frac{H_1 - H}{(1 - \Delta)} \tag{4.1.16}$$

Looking at the energy E_s associated with the negative sign, we find

$$E_S = \frac{H_1 + H}{(1 + \Delta)} \tag{4.1.17}$$

Thus the total system consisting of two atoms and two electrons has at small interatomic distances two energy allowed levels (exclusive of spin).

We must now find the two allowed state functions for the system. Using equation (1.4.31) we get

$$A(H_1 - E) + B(H - \Delta E) = 0 \tag{4.1.18}$$

Using

$$E = E_A = \frac{H_1 - H}{1 - \Delta} \tag{4.1.19}$$

we find $A = -B$. The allowed wave function corresponding to the energy E_A is

$$|\psi_A\rangle = A[|\psi_T'\rangle - |\psi_T''\rangle] \tag{4.1.20}$$

PROBLEM

Determine that the wave function corresponding to E_s is

$$|\psi_s\rangle = A\{|\psi_T'\rangle + |\psi_T''\rangle\}$$

We must now determine the values of the constants A and B. Normalizing $|\psi_A\rangle$ yields

$$A = \frac{1}{\sqrt{2 - 2\Delta}} \tag{4.2.21}$$

PROBLEM

Show by normalizing $|\psi_S\rangle$ that

$$A = \frac{1}{\sqrt{2 + 2\Delta}}$$

Thus the two allowed states for a system of two atoms bonded by the homopolar bond are

$$|\psi_A\rangle = \frac{1}{\sqrt{2 - 2\Delta}} \{|\psi_T'\rangle - |\psi_T''\rangle\} \tag{4.1.22}$$

$$|\psi_S\rangle = \frac{1}{\sqrt{2 + 2\Delta}} \{|\psi_T'\rangle + |\psi_T''\rangle\} \tag{4.1.23}$$

The labels S and A stand for symmetric and antisymmetric states and refer to the fact that, under exchange of electron coordinates

$$\begin{aligned} |\psi_A\rangle &\Rightarrow -|\psi_A\rangle \\ |\psi_S\rangle &\Rightarrow +|\psi_S\rangle \end{aligned} \tag{4.1.24}$$

A plot of the wave functions $|\psi_A\rangle$ and $|\psi_s\rangle$ in the Schrödinger representation is shown in Figure 4.2.

Figure 4.3a shows a splitting of the original degenerate energy level for the two distant molecules into two separate levels E_A and E_s as the inter-

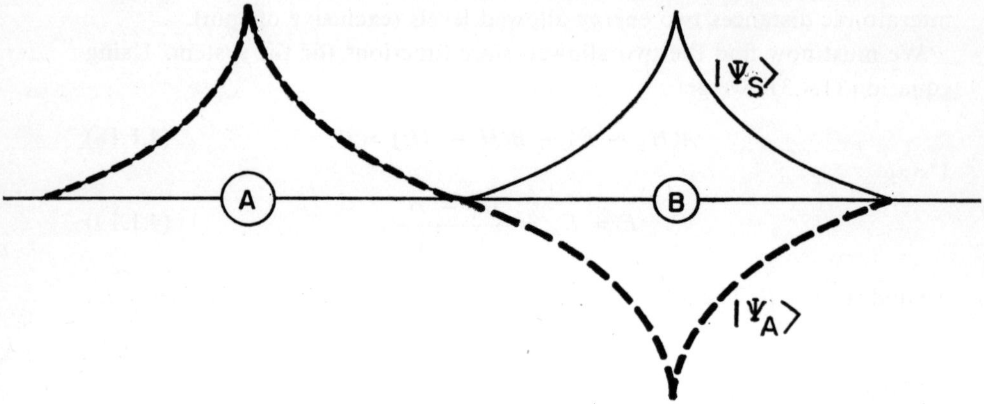

Figure 4.2 Graph of symmetric ($|\psi_s\rangle$) and antisymmetric ($|\psi\rangle_A$) states resulting from LCAO calculation for two singly hydrogen atoms.

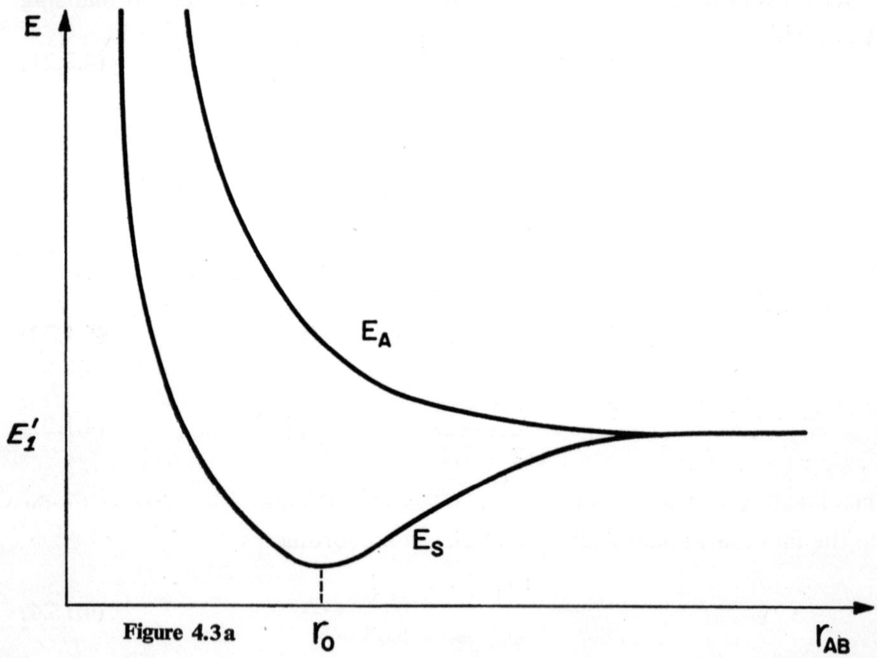

Figure 4.3 a

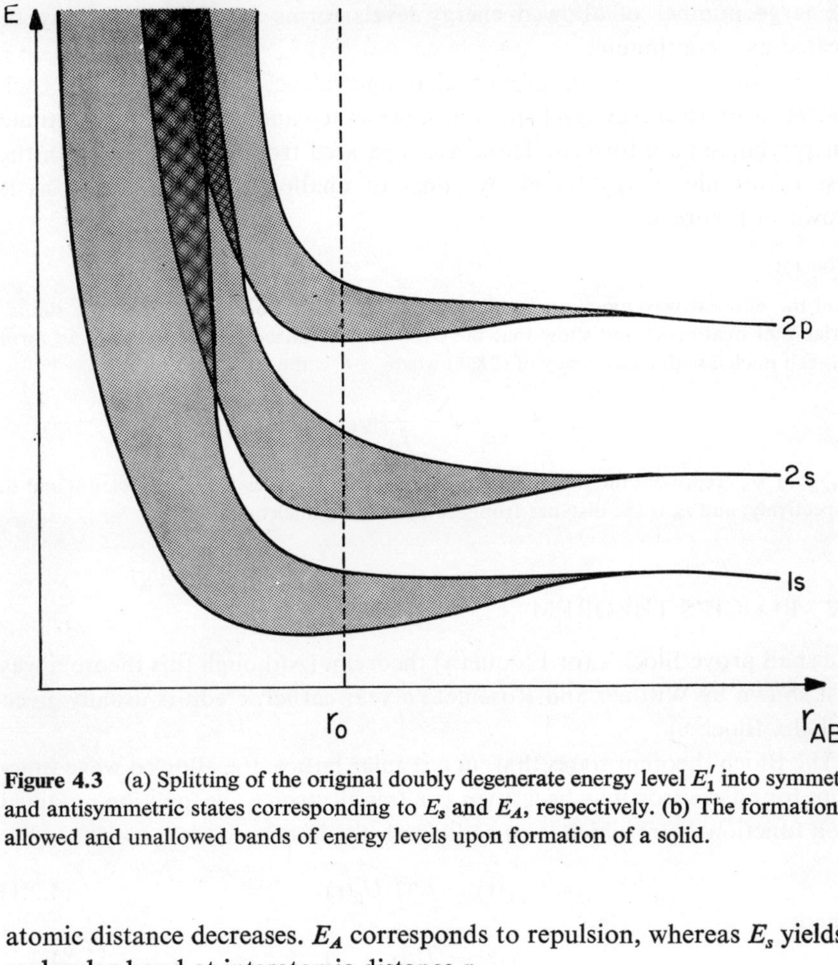

Figure 4.3 (a) Splitting of the original doubly degenerate energy level E_1' into symmetric and antisymmetric states corresponding to E_s and E_A, respectively. (b) The formation of allowed and unallowed bands of energy levels upon formation of a solid.

atomic distance decreases. E_A corresponds to repulsion, whereas E_s yields a molecular bond at interatomic distance r_0.

It should be noted that our calculation has been carried out exclusive of electron spin and that the homopolar bond resulting from the interaction between the orbital states of the valence electrons of the two atoms has yielded two distinct energy states. Had we chosen to include spin in our calculation, we would find that splitting of the energy degeneracy produces four allowed energy levels, only one of which corresponds to a stable bond.

The obvious extension of this two atom homopolar bond is that splitting of energy level degeneracy leads to a number of distinct states equal to the number of degenerate states being split. Thus, when a crystal (about 10^{23} atoms-cm^{-3}) is formed (e.g. from a collection of homopolar bonds) the large number of allowed energy levels forms a band which may be treated as a continuum.

When one takes account of more than one valence electron per atom each discrete atomic energy level splits its degeneracy and a number of continua (energy bands) are formed. These are separated from each other, as in the case of atomic energy levels, by zones of unallowed energy levels, as is shown in Figure 4.3 b.

PROBLEM

Find the allowed wavefunctions for the singly ionized hydrogen-molecule by a similar variational treatment[2] and show that the electron is localized and hops back and forth between nucleii with a frequency of $(2K/h)$ where $\pm K$ is the resonance energy

$$K = \langle \psi_A | - \frac{e^2}{r_B} | \psi_B \rangle,$$

$|\psi_A\rangle$ and $|\psi_B\rangle$ represent the states in which the electron is associated with nucleii A and B, respectively, and r_B is the distance from the electron to nucleus B.

4.2 BLOCH'S THEOREM

We shall prove Bloch's (or Floquet's) theorem. (Although this theorem was first shown by Wittmer and Rosenfeld a year earlier, credit is usually given to Felix Bloch[3].)

The Bloch theorem states that, in a regular lattice, the allowed wave functions for an electron may be written as a free electron wave function modified by a function $U_k(\mathbf{r})$ which is periodic with the lattice

$$\psi_{\mathbf{k}}(\mathbf{r}) = e^{i\mathbf{k}\cdot\mathbf{r}}\, U_{\mathbf{k}}(\mathbf{r}) \qquad (4.2.1)$$

where

$$U_{\mathbf{k}}(\mathbf{r}) = U_{\mathbf{k}}(\mathbf{r} + \hat{a}) \qquad (4.2.2)$$

To prove this we utilize the translation operator Γ which has the property that when it acts on a quantity in the space representation it evaluates the quantity at a point displaced from where it found it by one lattice site

$$\Gamma\psi(\mathbf{r}) = \psi(\mathbf{r} + \hat{a}) \qquad (4.2.3)$$

In order for (4.2.1) to represent an electron in the lattice, it must give the right energy eigenvalues. We look first at the commutation relation between Γ and the Hamiltonian operator $\mathscr{H}$ for an electron in a periodic lattice

$$(\mathscr{H}\Gamma - \Gamma\mathscr{H})\, \psi_{(x)} \qquad (4.2.4)$$

The first term yields

$$\mathscr{H}\Gamma\psi_{(x)} = \mathscr{H}\psi_{(x+a)} \tag{4.2.5}$$

Since all lattice sites are equivalent, $\psi(_x) = \psi_{(x+a)}$. We note that

$$-\frac{\hbar^2}{2m}\frac{\partial^2}{\partial x^2} = -\frac{\hbar^2}{2m}\frac{\partial^2}{\partial(x+a)^2} \tag{4.2.6}$$

and

$$V(\mathbf{r}) = V(\mathbf{r} + \hat{a})$$

Thus

$$\mathscr{H}\Gamma\psi_{(x)} = E_{(x)}\psi_{(x)} \tag{4.2.7}$$

and

$$\Gamma\mathscr{H}\psi_{(x)} = E_{(x+a)}\psi_{(x+a)} = E_{(x)}\psi_{(x)} \tag{4.2.8}$$

and $\mathscr{H}$ commutes with the translating operator. As we have stated, the chances are excellent that an eigenfunction of one will also be an eigenfunction of the other.

It is easily proven that a Bloch type wave function is an eigenfunction of the translation operator

$$\begin{aligned} \Gamma\Psi(\mathbf{r}) &= \Gamma e^{i\mathbf{k}\cdot\mathbf{r}}\, U_{\mathbf{k}}(\mathbf{r}) \\ &= e^{i\mathbf{k}.\hat{a}}\, e^{i\mathbf{k}\cdot\mathbf{r}}\, U_{\mathbf{k}}(\mathbf{r}) \\ &= e^{i\mathbf{k}\cdot\hat{a}}\Psi(\mathbf{r}) \end{aligned} \tag{4.2.9}$$

It is therefore an eigenfunction of the $\mathscr{H}$ operator. Thus (4.2.1) is an allowed wave function for an electron is a crystal. We should remember too that a sum of Bloch type wave functions is, therefore, also an eigenfunction of the Hamiltonian operator.

4.3 ELEMENTARY CALCULATIONS OF BAND STRUCTURE

The Small Perturbation Model of an Electron in a Solid

In a good conductor the electrons are very loosely bound. The Sommerfeld model assumed that the conducting electrons could be described as though they were a gas of free electrons travelling in the periodic potential set up by the positively charged lattice sites. This model is the simplest model of the electrons in a solid. Surprisingly enough, as we shall see, it accounts for many of the conduction properties of metals. For many purposes, however, it is naive to assume that the electrons in a crystal behave as though they are free. In a solid, electrons may be either weakly or strongly bound to their parent atoms, and this fact must be taken into account when attempting to describe the allowed motion of an electron in a crystalline lattice. We will first perform a calculation in which the allowed motion of an electron is assumed to be only slightly influenced by the periodic potential of the lattice and determine the effect of the ionic potential $V(\mathbf{r})$ on the allowed free electron energy spectrum (Figure 1.5).

Assume that we have a three dimensional crystal with sides $L_x = Ma_x$, $L_y = Na_y$ and $L_z = Ra_z$ where a_x, a_y and a_z are the lattice constants and M, N and R are integers. The state of an electron in the lattice is given by the ket vectors $|F_{+k}\rangle$ and $|F_{-k}\rangle$ where $|F_{+k}\rangle$ and $|F_{-k}\rangle$ represent positively and negatively travelling free electron states.

The change in energy of a system due to a small perturbation is given by the energy eigenvalues of equation (1.4.14). The secular equation is written

$$0 = \begin{vmatrix} \langle F_{+\mathbf{k}}|V(\mathbf{r})|F_{+\mathbf{k}}\rangle - E & \langle F_{+\mathbf{k}}|V(\mathbf{r})|F_{-\mathbf{k}}\rangle \\ \langle F_{-\mathbf{k}}|(V(\mathbf{r})|F_{+\mathbf{k}}\rangle & \langle F_{-\mathbf{k}}|V(\mathbf{r})|F_{-\mathbf{k}}\rangle - E \end{vmatrix} \tag{4.3.1}$$

We define

$$\langle F_{\pm\mathbf{k}}|V(\mathbf{r})|F_{\pm\mathbf{k}}\rangle = V_{\pm\mathbf{k},\,\pm\mathbf{k}} \tag{4.3.2}$$

which becomes in Schrödinger notation

$$V_{\pm\mathbf{k},\,\pm\mathbf{k}} = \int_0^{L_1}\int_0^{L_2}\int_0^{L_3} \psi_+^*(\mathbf{r})\, V(\mathbf{r})\, \psi_+(\mathbf{r})\, d\mathbf{r} \tag{4.3.3}$$

where

$$\psi_+(\mathbf{r}) = Ae^{i\mathbf{k}\cdot\mathbf{r}} \tag{4.3.4}$$

is the free wave state. We find

$$V_{+\mathbf{k},\,+\mathbf{k}} = A^*A \int_0^{L_1}\int_0^{L_2}\int_0^{L_3} V(\mathbf{r})\, d\mathbf{r} \tag{4.3.5}$$

Similarly

$$V_{-\mathbf{k},\,-\mathbf{k}} = B^*B \int_0^{L_1}\int_0^{L_1}\int_0^{L_3} V(\mathbf{r})\, d\mathbf{r} \tag{4.3.6}$$

where

$$\psi_-(\mathbf{r}) = Be^{-i\mathbf{k}\cdot\mathbf{r}} \tag{4.3.7}$$

Assuming that our wave functions are normalized within the crystal, there is no reason why the probability of finding a rightward travelling wave should be any different than that of a leftward travelling wave. Thus

$$\int_0^{L_1}\int_0^{L_2}\int_0^{L_3} \psi_+^*(\mathbf{r})\psi_+(\mathbf{r})\, d\mathbf{r} = \int_0^{L_1}\int_0^{L_2}\int_0^{L_3} \psi_-^*(r)\, \psi_-(\mathbf{r})\, d\mathbf{r} \tag{4.3.8}$$

which yields

$$|A|^2 = |B|^2 = \frac{1}{\text{Vol.}} \tag{4.3.9}$$

where Vol. is the volume of the crystal. From (4.3.5) and (4.3.6) we find

$$V_{+\mathbf{k},+\mathbf{k}} = V_{-\mathbf{k},-\mathbf{k}} = \frac{1}{\text{Vol.}} \int_0^{L_1}\int_0^{L_2}\int_0^{L_3} V(\mathbf{r})\,d\mathbf{r} \tag{4.3.10}$$

Equation (4.3.10) is the integral of the ionic potential over the entire crystal lattice, divided by the volume of the lattice. This defines an average potential due to the ionic cores. Since we can take our zero of potential anywhere, we will define it such that (4.3.10) is equal to zero.

It is easily shown that

$$V_{\mathbf{k},-\mathbf{k}} = \frac{1}{\text{Vol.}} \int_0^{L_1}\int_0^{L_2}\int_0^{L_3} e^{-2i\mathbf{k}\cdot\mathbf{r}}\,V(\mathbf{r})\,d\mathbf{r} \tag{4.3.11}$$

and

$$V_{-\mathbf{k},+\mathbf{k}} = \frac{1}{\text{Vol.}} \int_0^{L_1}\int_0^{L_2}\int_0^{L_3} e^{+2i\mathbf{k}\cdot\mathbf{r}}\,V(\mathbf{r})\,d\mathbf{r} \tag{4.3.12}$$

Provided that $V(\mathbf{r})$ is a real function of $\mathbf{r}$, we get

$$V_{+\mathbf{k},-\mathbf{k}} = V_{-\mathbf{k},+\mathbf{k}} \tag{4.3.13}$$

and the secular equation becomes

$$E^2 - |V_{+\mathbf{k},-\mathbf{k}}|^2 = 0 \tag{4.3.14}$$

It is easily shown that an integral of the form

$$\int_0^{L_1}\int_0^{L_2}\int_0^{L_3} e^{\pm i\mathbf{k}'\cdot\mathbf{r}}V(\mathbf{r})\,d\mathbf{r} \tag{4.3.15}$$

where $V(\mathbf{r})$ is a periodic function with period a (Figure 4.4) is equal to zero unless $k' = \dfrac{2n\pi}{a}$ where $n = \pm 1, \pm 2, \ldots$

Assuming that $V(\mathbf{r})$ may be expressed as the product of three independent periodic functions

$$V(\mathbf{r}) = V(x)V(y)V(z) \tag{4.3.16}$$

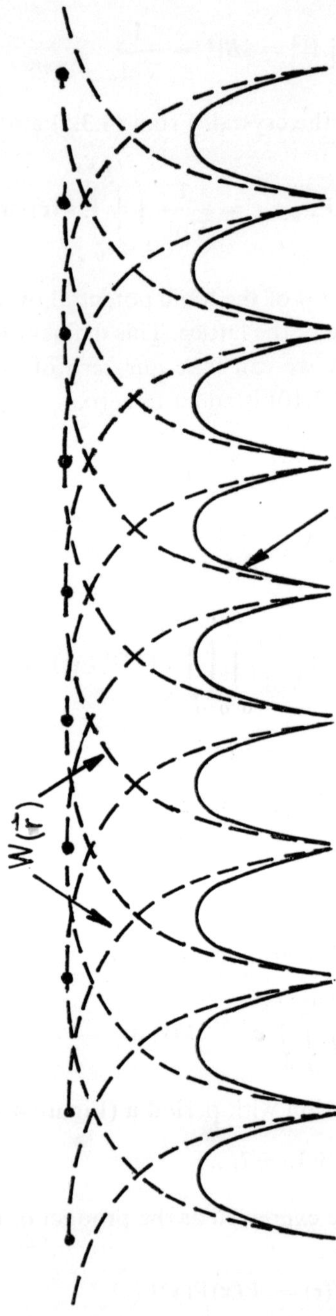

Figure 4.4 The addition of atomic potentials $W(\mathbf{r})$ to produce the periodic crystalline potential $V(\mathbf{r})$.

equation (4.3.15) may be written

$$\int_0^{L_x} V(x)\, e^{ik_x'x}\, dx \int_0^{L_y} V(y)\, e^{ik_y'y}\, dy \int_0^{L_z} V(z)\, e^{ik_z'z}\, dz \tag{4.3.17}$$

Let us first examine the leftmost term of equation (4.3.17). Since $V(x)$ is periodic, we can break the integral into a sum of many similar contributions. We find

$$\int_0^{L_x} e^{ik_x'x}\, V(x)\, dx = \sum_{J=0}^{J=m-1} \int_{Ja_x}^{(J+i)a_x} e^{ik_x'x} V(x)\, dx \tag{4.3.18}$$

where $L_x = ma_x$.

We now make the transformation $x = X + Ja_x$ and integrate between the first two lattice sites, which yields

$$\sum_{J=0}^{J=m-1} e^{ik'Ja_x} \int_0^{a_x} e^{ik'X}\, V(X + Ja_x)\, dX \tag{4.3.19}$$

The term

$$\sum_{J=0}^{J=m-1} e^{ik'Ja_x} = 1 + e^{ik'_xa_x} + e^{2ik'_xa_x} + \dots + e^{(m-1)ik'_xa_x} \tag{4.3.20}$$

is a geometrical series given by the quotient S_m where

$$S_m = \frac{1 - e^{ik'_xa_x(m)}}{1 - e^{ik'_xa_x}} \tag{4.3.21}$$

We shall assume that there are so many atoms in the chain along the x direction that $m - 1 \approx m$. Let us examine

$$\lim_{\substack{m\to\infty \\ k'_xa_x \neq 2n\pi}} \frac{1}{m}\left\{\frac{1 - e^{ik'_xa_xm}}{1 - e^{ik'_xa_x}}\right\} \tag{4.3.22}$$

$$\lim_{\substack{k'_xa_x \Rightarrow 2n\pi \\ m\Rightarrow\infty}} \frac{1}{m}\left\{\frac{1 - e^{ik'_xa_xm}}{1 - e^{ik'_xa_x}}\right\} \tag{4.3.23}$$

For all values of $k'_xa_x \neq 2n\pi$, $n = 0, \pm 1, \pm 2, \dots$, the numerator and the denominator of (4.3.22) are bounded functions and, thus, (4.3.22) is equal to zero. As k'_xa_x approaches $2n\pi$, (4.3.23) approaches 1 for all values of m and therefore the sum (4.3.20) is equal to m. Thus we may rewrite equation (4.3.11) as

$$V_{\mathbf{k},-\mathbf{k}} = \frac{1}{MNR} \int_0^{a_x}\int_0^{a_y}\int_0^{a_z} \frac{e^{\frac{in\pi x}{a_x}}\, e^{\frac{i\pi n_y}{a_y}}\, e^{\frac{i\pi n_z}{a_z}}\, V(x)V(y)V(z)\, dx\, dy\, dz}{a_xa_ya_z} \tag{4.3.24}$$

where Vol. $= MNRa_xa_ya_z$.

We see that the perturbation due to the small periodic crystalline potential has split the doubly degenerate free electron energy (corresponding to positively and negatively travelling waves) into two levels at the points $k_x = \dfrac{\pm\pi n_x}{a_x}$, $k_y = \dfrac{\pm\pi n_y}{a_y}$ and $k_z = \dfrac{\pm\pi n_z}{a_z}$. Thus at these points the energy of the electron is given by

$$\frac{\hbar^2\pi^2 n_x^2}{2ma_x^2} \pm \left| \frac{1}{a_x} \int_0^{a_x} e^{\frac{i\pi n_x}{a_x}} V(x)\, dx \right|_{\left\{k_x = \frac{n_x\pi}{a_x}\right\}} \tag{4.3.25}$$

$$\frac{\hbar^2\pi^2 n_y^2}{2ma_y^2} \pm \left| \frac{1}{a_y} \int_0^{a_y} e^{\frac{i\pi n_y}{a_y}} V(y)\, dy \right|_{\left\{k_y = \frac{n_y\pi}{a_y}\right\}} \tag{4.3.26}$$

$$\frac{\hbar^2\pi^2 n_z^2}{2ma_z^2} \pm \left| \frac{1}{a_z} \int_0^{a_z} e^{\frac{i\pi n_z}{a_z}} V(z)\, dz \right|_{\left\{k_z = \frac{n_z\pi}{a_z}\right\}} \tag{4.3.27}$$

Figure 4.5 shows a graph of the perturbed energy as a function of wave vector in the $\mathbf{k}_x$ direction. We see that, at the point $k_x = \pi n_x/a_x$ the free

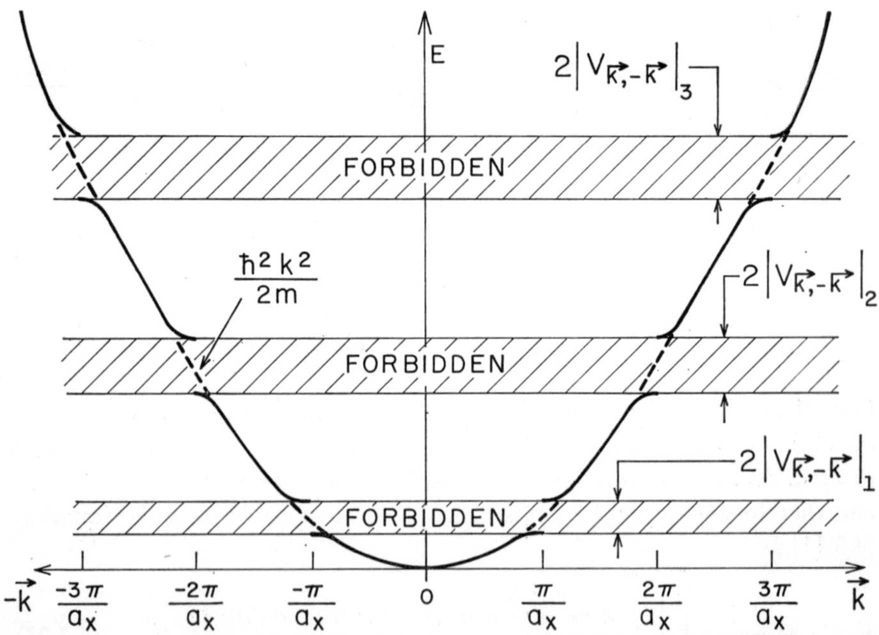

Figure 4.5 Perturbed energy *vs* **k** curve in one dimension for an electron loosely bound in a crystalline solid. Note alternating regions of allowed and unallowed energy (see Fig. 4.3 b).

electron energy parabola is split and there exists two allowed values of the energy a distance $|V_{k,-k}|_n$ above and below the free energy parabola.

In the energy intervals $|V_{k,-k}|_n$ there are no allowed energy states for the electron. These discontinuities in the allowed energy spectrum are called "forbidden energy gaps."

It is easily seen that the wave functions corresponding to electrons whose momenta correspond to $\pm\pi n_x/a_x$, etc., represent standing waves. Thus electrons with energies lying in these forbidden regions cannot propagate through the lattice and show up as Bragg reflection maxima on X-ray patterns.

PROBLEM

Show that, at the points $k_x = \dfrac{\pm\pi n_x}{a_x}$ wave functions of the type $\psi = Ae^{i\mathrm{k}_x\mathrm{x}} + Be^{-i\mathrm{k}_x\mathrm{x}}$ become standing waves, and that these waves carry no current.

The Tightly Bound Electron Model

In this section we will treat the electrons as if they experience a very strong coulombic attraction from each of the lattice sites such as found in ionic crystals.

We have examined the case of an electron in a lattice weakly interacting with the ionic potentials. In many cases, however, e.g., in the case of insulators, a weak lattice-electron potential is unrealistic. The case of strong atomic binding illustrates many examples of the simple one-electron methods and approximations used in the study of solids, and so, although our results will not be much different than for the weak binding case, we shall examine the tight binding model in some detail.

We imagine a three-dimensional rectangular lattice with lattice vectors $\hat{a}_x$, $\hat{a}_y$ and $\hat{a}_z$. The periodic crystalline potential $V(\mathbf{r})$ is taken explicitly as the sum of the atomic potentials $W(\mathbf{r})$ of the isolated atoms (at shown in Figure 4.4)

$$V(\mathbf{r}) = \sum_l w(\mathbf{r} - l\hat{a}_r), \qquad (4.3.28)$$

where the subscript l indicates that the sum is to be taken over all lattice sites.

We shall express our electronic wave function $\Psi_\mathbf{k}(\mathbf{r})$ in Bloch form

$$\Psi_\mathbf{k}(\mathbf{r}) = A \sum_j e^{i\mathbf{k}.j\hat{a}_r} U_\mathbf{k}(\mathbf{r} + j\hat{a}_r) \qquad (4.3.29)$$

where the Bloch function $U_{\mathbf{k}}(\mathbf{r} + j\,\hat{a}_r)$ is the atomic wave function for an isolated atom and thus has the required lattice periodicity

$$U_{\mathbf{k}}(\mathbf{r}) = U_{\mathbf{k}}(\mathbf{r} + m\hat{a}_r), \tag{4.3.30}$$

where $\hat{a}$ is a lattice translation in the $\mathbf{r}$ direction, and j and m are integers.

We express the Schrödinger equation for the Bloch-type function as

$$E\psi_{\mathbf{k}}(\mathbf{r}) = -\frac{\hbar^2}{2m}\nabla^2\psi_{\mathbf{k}}(\mathbf{r}) + V(\mathbf{r})\,\psi_{\mathbf{k}}(\mathbf{r}) \tag{4.3.31}$$

where m^* is the mass of the electron in the lattice.

The wave functions for the isolated atomic orbital must, of course, satisfy

$$E_0 U_{\mathbf{k}}(\mathbf{r}) = -\frac{\hbar^2}{2m}\nabla^2 U(\mathbf{r}) + W(\mathbf{r})U_{\mathbf{k}}(\mathbf{r}) \tag{4.3.32}$$

Equation (4.3.31) may be rewritten

$$A\sum_j e^{i\mathbf{k}.j\hat{a}_r}\left[EU_{\mathbf{k}}(\mathbf{r} + j\hat{a}_r) + \frac{\hbar^2}{2m}\nabla^2 U_{\mathbf{k}}(\mathbf{r} + ja_r) - V(\mathbf{r})U_{\mathbf{k}}(\mathbf{r} + j\hat{a}_r)\right] = 0 \tag{4.3.33}$$

Equation (4.3.33) becomes zero only if the quantity in brackets is zero. Thus each atomic orbital wave function is also an approximate solution to the crystal Hamiltonian. Subtracting (4.3.32) from (4.3.31) thus yields

$$E\psi_{\mathbf{k}}(\mathbf{r}) - E_0 U_{\mathbf{k}}(\mathbf{r}) = \frac{\hbar^2}{2m}[\nabla^2 U_{\mathbf{k}}(\mathbf{r}) - \nabla^2\psi_{\mathbf{k}}(\mathbf{r})] + V(\mathbf{r})\psi_{\mathbf{k}}(\mathbf{r}) - W(\mathbf{r})U_{\mathbf{k}}(\mathbf{r}) \tag{4.3.34}$$

$$A[E - E_0]\sum_j e^{i\mathbf{k}\cdot j\hat{a}_r}U_{\mathbf{k}}(\mathbf{r} + j\hat{a}_r) = A\sum_{j\neq 0} e^{i\mathbf{k}\cdot j\hat{a}_r}U_{\mathbf{k}}(\mathbf{r} + j\hat{a}_r)(V(\mathbf{r}) - W(\mathbf{r}+ja_r) \tag{4.3.35}$$

Multiplication of (4.3.35) from the left by

$$\psi^*_{\mathbf{k}}(\mathbf{r}) = A^*\sum_m e^{-ik\cdot m\hat{a}_r}U^*_{\mathbf{k}}(\mathbf{r} + m\hat{a}_r) \tag{4.3.36}$$

yields

$$A^*\sum_m e^{-i\mathbf{k}\cdot m\hat{a}_r}\,U^*_{\mathbf{k}}(\mathbf{r} + m\hat{a}_r)\,(E - E_0)\,A\sum_j e^{i\mathbf{k}\cdot j\hat{a}_r}\,U_{\mathbf{k}}(\mathbf{r} + j\hat{a}_r)$$

$$= A^*\sum_m e^{-i\mathbf{k}\cdot m\hat{a}_r}\,U^*_{\mathbf{k}}(\mathbf{r} + m\hat{a}_r)\,A\sum_{j'} V'(\mathbf{r})\,e^{i\mathbf{k}\cdot j'\hat{a}_r}\,U_{\mathbf{k}}(\mathbf{r} + j'\hat{a}_r) \tag{4.3.37}$$

where $V'_{(r)} = V_{(r)} - W_{(r+j\hat{a}_r)}$.

Integration of (4.3.37) over all of our crystal space which we may take, with little error, as extending from $+\infty$ to $-\infty$, we find

$$E - E_0 = \frac{|A|^2 \int_{-\infty}^{+\infty} d\mathbf{r} \sum_m \sum_{j'} e^{-i\mathbf{k}\cdot m\hat{a}_r} U_{\mathbf{k}}^*(\mathbf{r} + m\hat{a}_r)\, V'(\mathbf{r}) e^{i\mathbf{k}\cdot j'\hat{a}_r}\, U_{\mathbf{k}}(\mathbf{r} + j'\hat{a})}{\int_{-\infty}^{+\infty} \psi_{\mathbf{k}}^*(\mathbf{r})\, \psi_{\mathbf{k}}(\mathbf{r})\, d\mathbf{r}} \tag{4.3.38}$$

Making the substitution $-m + j' = p$ yields the allowed energy for an electron in the lattice

$$E - E_0 = \sum_p e^{i\mathbf{k}\cdot p\hat{a}_r} \int_{+\infty}^{-\infty} U_{\mathbf{k}}^*(\mathbf{r} + p\hat{a}_r)\, V'(\mathbf{r})\, U_{\mathbf{k}}(\mathbf{r})\, d\mathbf{r} \tag{4.3.39}$$

Again we have again assumed that the overlap of atomic orbital wave functions is negligible, i.e.,

$$\int_{-\infty}^{+\infty} \psi_{\mathbf{k}}^*(\mathbf{r})\, \psi_{\mathbf{k}}(\mathbf{r})\, d\mathbf{r} = |A|^2 N = 1 \tag{4.3.40}$$

where N is the number of atoms in our crystal.

Equation (4.3.39) affords an easy calculation of the allowed energy of an electron in a crystalline lattice. Let us investigate only the nearest neighbor interactions in a rectangular lattice, i.e., $p = 0, \pm 1$.

We shall define for the $\hat{x}$ direction

$$\alpha_x = \int_{-\infty}^{+\infty} U_{\mathbf{k}}^*(x + a_x)\, V'(x)\, U_{\mathbf{k}}(x)\, dx \tag{4.3.41}$$

$$\beta_x = \int_{-\infty}^{+\infty} U_{\mathbf{k}}^*(x)\, V'(x)\, U_{\mathbf{k}}(x)\, dx \tag{4.3.42}$$

$$\gamma_x = -\int_{-\infty}^{+\infty} U_{\mathbf{k}}(x + a_x)\, V'(x)\, U_{\mathbf{k}}(x)\, dx \tag{4.3.43}$$

Similar definitions hold for the $\hat{y}$ and $\hat{z}$ directions. Thus the energy may be written

$$E - E_0 = \alpha_x e^{-i\mathbf{k}_x\cdot\hat{a}_x} + \beta_x - \gamma_x e^{i\mathbf{k}_x\cdot\hat{a}_x} + \alpha_y e^{-i\mathbf{k}_y\cdot\hat{a}_y} + \beta y - \gamma_y e^{i\mathbf{k}_y\cdot\hat{a}_y} + \alpha_z e^{-i\mathbf{k}_z\cdot\hat{a}_z} + \beta_z - \gamma_z e^{i\mathbf{k}_z\cdot\hat{a}_z} \tag{4.3.44}$$

Under the assumption that the energy of an electron travelling in one direction is unaffected by the changes in energy occurring in orthogonal directions, we may write

$$E_x - E_{x_0} = \alpha_x e^{-i\mathbf{k}_x \cdot \hat{a}_x} + \beta_x - \gamma_x e^{i\mathbf{k}_x \cdot \hat{a}_x} \qquad \text{A}$$

$$E_y - E_{y_0} = \alpha_y e^{-i\mathbf{k}_y \cdot \hat{a}_y} + \beta_y - \gamma_y e^{i\mathbf{k}_y \cdot \hat{a}_y} \qquad \text{B} \qquad (4.3.45)$$

$$E_z - E_{z_0} = \alpha_z e^{-i\mathbf{k}_z \cdot \hat{a}_z} + \beta_z - \gamma_z e^{i\mathbf{k}_z \cdot \hat{a}_z} \qquad \text{C}$$

Let us consider the $\hat{x}$ direction only. For the case where $U_{\mathbf{k}}(x)$ is an asymmetric wave function $\alpha_x = -\gamma_x$. Neglecting E_0, the energy as a function of $\mathbf{k}_x$ becomes

$$E_{x_{\text{odd}}} = \beta_x - 2\gamma_x \cos \mathbf{k}_x \, \hat{a}_x \qquad (4.3.46)$$

and

$$E_{x_{\text{even}}} = \beta_x - 2\alpha_x \cos \mathbf{k}_x \hat{a}_x. \qquad (4.3.47)$$

Similarly

$$E_y|_{\text{odd}} = \beta_y - 2\gamma_y \cos \mathbf{k}_y \hat{a}_y$$

$$E_y|_{\text{even}} = \beta_y + 2\alpha_y \cos \mathbf{k}_y \hat{a}_y$$

$$E_z|_{\text{odd}} = \beta_z - 2\gamma_z \cos \mathbf{k}_z \hat{a}_z$$

$$E_z|_{\text{even}} = \beta_z + 2\alpha_z \cos \mathbf{k}_z \hat{a}_z \qquad (4.3.48)$$

The forms of the energy bands in the $\hat{x}$, $\hat{y}$ and $\hat{z}$ directions are shown for odd wave functions in Figure 4.6. Thus the assumption that electrons are tightly bound produces bands of allowed energies similar to those derived by assuming that the crystalline potential is very weak. Performing the above calculations for a number of valence electrons yields a number of allowed and unallowed energy bands.

We note that in the preceeding calculation we did not specify the form of the atomic potential except to demand that it be well localized about a particular lattice site. The width of an energy band, for example in the $\hat{x}$ direction, is $4\gamma_x$ (odd wave functions) where γ_x is given by (4.3.43). Since overlap is very small, γ_x will be small and the energy bands in our tight binding model will be narrow. The difference in energy between $E_{x_{\text{odd}}}$ and $E_{x_{\text{even}}}$ at $k = n\pi/a_x$ is the energy gap which can be shown to be approximately of the same magnitude as the band width itself. Thus our tight binding model is an approximation valid only for the case where the size of the energy gap is equivalent to the band width.

In the limiting case where the crystalline potential consists of a number of equally spaced delta functions, the limiting case of the Kronig-Penny Model,[4] equivalent results are obtained.

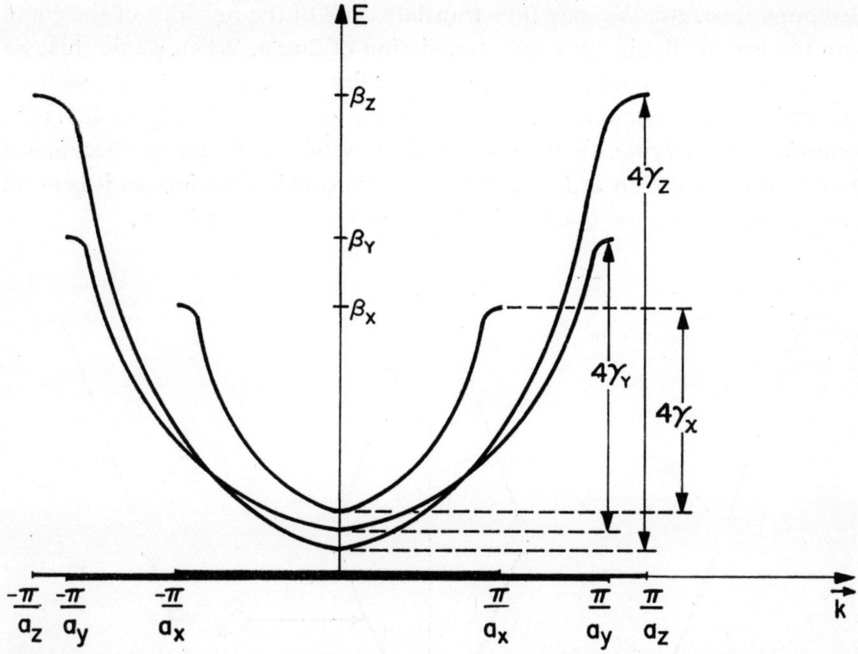

Figure 4.6 Allowed energy bands produced by the tight binding calculation for an odd wave function.

4.4 BRILLOUIN ZONES

It is useful to note that changing the value of k by $\pm 2n\pi/a_r$ will not change the lattice periodicity of the Bloch function

$$\psi_k(\mathbf{r}) = e^{i\mathbf{k}\cdot l\hat{a}_r}\, U_k(\mathbf{r}) \qquad \text{a}$$

$$\psi_{(k\pm 2n\pi/a_r)}(\mathbf{r}) = e^{i\mathbf{k}\cdot l\hat{a}_r}\, [U_{(k\pm 2n\pi/ar)}(\mathbf{r})] \qquad \text{b} \tag{4.4.1}$$

where $U_{(k\pm 2n\pi/a_r)}(\mathbf{r})$ retains the lattice periodicity of $U_k(\mathbf{r})$. We have seen in both the loosely bound (4.3.25) and the tightly bound (4.3.46) electron models that for any $U_k(\mathbf{r})$ periodic in a_r, $E(k)$ is periodic with a period $2n\pi/a_r$. Thus we may map our E vs. **k** curves into any interval of width $2n\pi/a_r$ without the loss of any information.

Referring to the E vs **k** curve for the weak binding model (Figure 4.5), it is generally agreed that the first set of allowed energy levels ranging from $k = \pm\pi/a$ is called the first Brillouin zone. The energy levels lying between the intervals $k = -2\pi/a, -\pi/a$ and $k = +2\pi/a, +\pi/a$ make up the second

Brillouin zone, etc. We may thus translate each of the portions of the curve into the first Brillouin zone by a translation of $2n\pi/a$. When we do this, we get the curve shown in Figure 4.7. The solid lines within $k = \pm\pi/a$ construct the Reduced Representation. The representation shown in Figure 4.5 is the *extended zone scheme*. In the reduced zone scheme, the energy becomes a many valued function of **k** rather than a monotonic function, as it is at all points except the zone boundaries in the extended zone scheme.

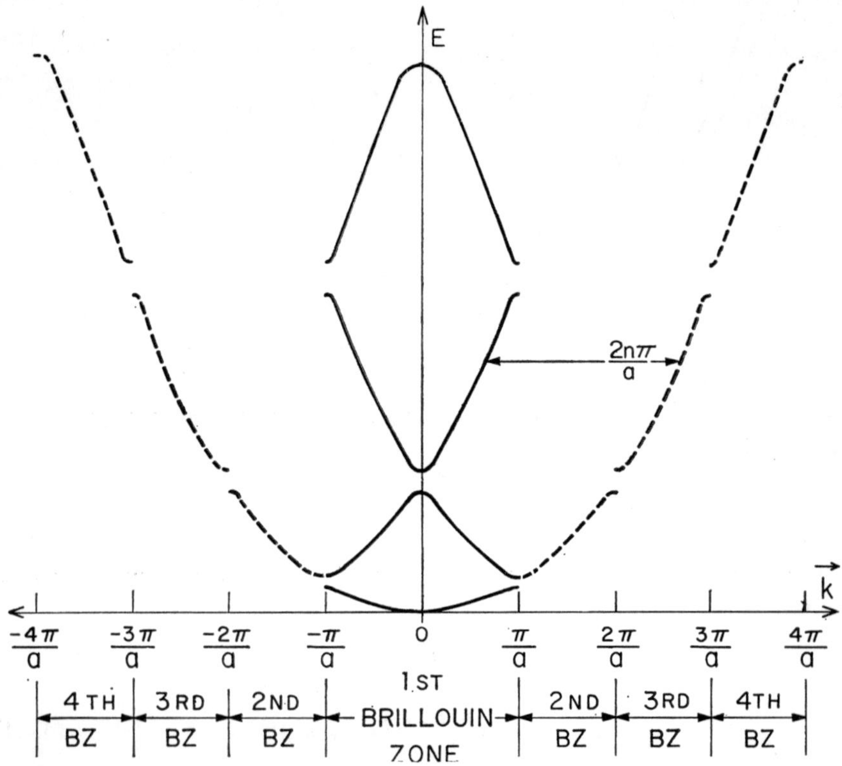

Figure 4.7 The reduced representation.

We found that the edges of the Brillouin zones could be correlated with the occurrence of constructive interference of reflected *X*-rays. We would expect therefore that there is some correlation between the volumes enclosed by the planes in reciprocal lattice space from which Bragg reflection occurs

$$\frac{\mathbf{G}\cdot\mathbf{r}}{\lambda} = \frac{|G|^2}{2} = \frac{\mathbf{G}\cdot\mathbf{k}}{2\pi} \tag{4.4.2}$$

and our Brillouin zones. Equation (4.4.2) tells us that scattering occurs from planes which are the perpendicular bisectors of our reciprocal lattice vectors.

The construction of the Brillouin zones in reciprocal space is attributed to Leon Brillouin.[5] The first Brillouin zone in **k** (or reciprocal lattice space) space is found in the following manner: In two dimensions, the first Brillouin zone is the area enclosed by the lines which are the bisectors of the smallest **k** vectors of the lattice. The second zone is the area enclosed by these lines and the lines which are the perpendicular bisectors of the next largest allowed **k** vectors. In Figure 4.8 the construction of the first five Brillouin zones is shown for two dimensional hexagonal lattice. The first 10 zones are drawn for the two dimensional square lattice in Figure 4.9.

It is possible, however, for the perpendicular bisector of the reciprocal lattice vector which determines the $(n + 1)^{\text{th}}$ Brillouin zone to intersect the area th enclosed by the reciprocal lattice vector which determines the nth Brillouin zone. In this case the nth zone encompasses the similar area.

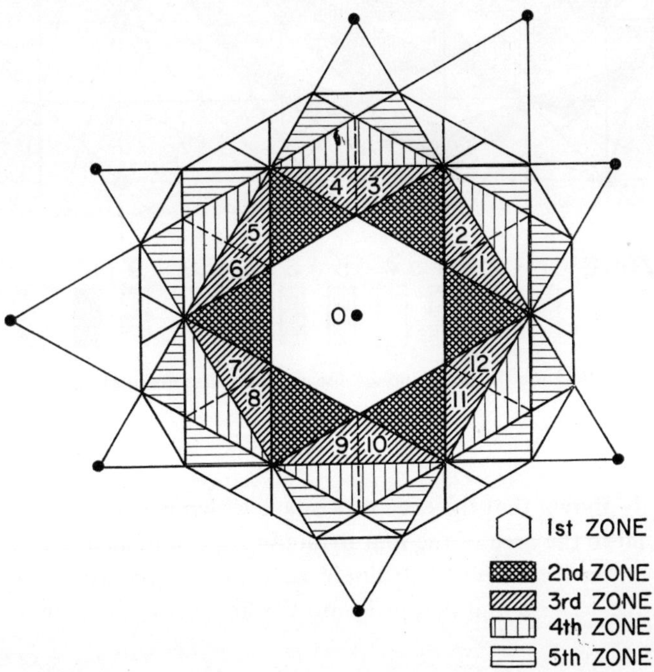

Figure 4.8 The first five Brillouin zones for the two dimensional hexagonal lattice (from L. Brillouin, *Wave Propogation in Periodic Structures*, used with permission of McGraw Hill, New York, 1946, p. 124).

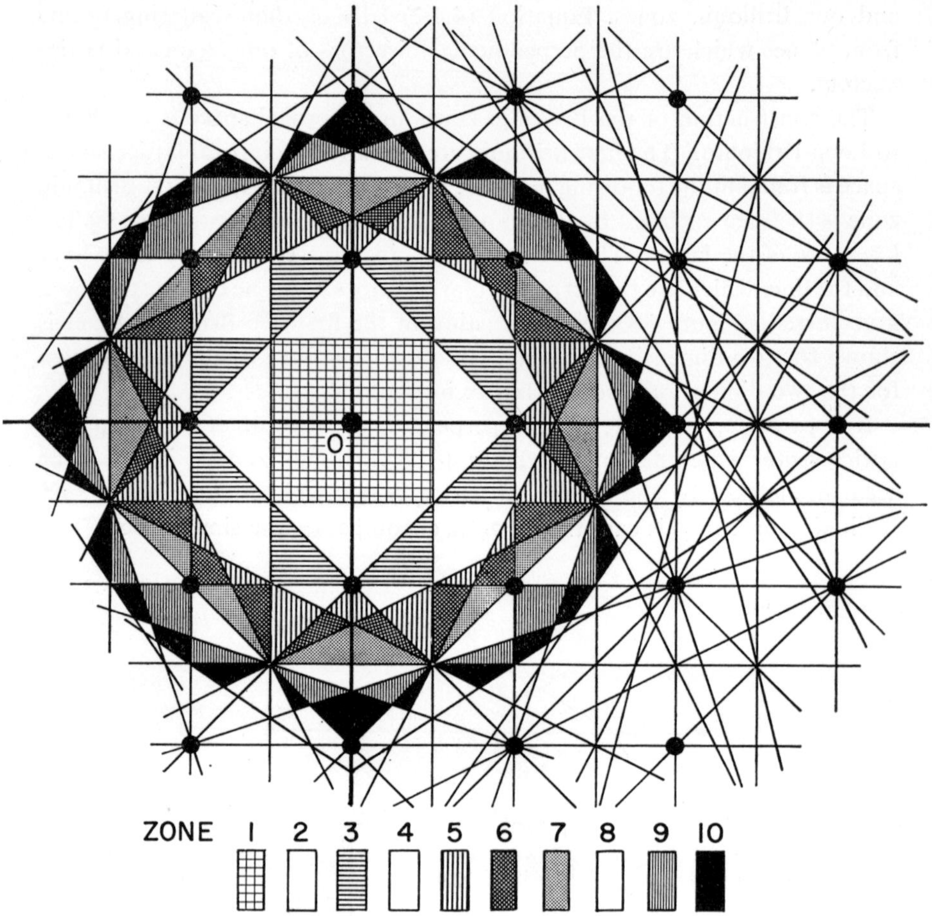

Figure 4.9 The first ten Brillouin zones for the two dimensional square lattice (after Brillouin).

It is easily shown that the area of **k** space which is encompassed by each zone is equal to the area of the first Brillouin zone and, as one would expect from our earlier treatment of Brillouin zones, each successive zone may be translated by $\pm 2n\pi/a$ and mapped onto the first Brillouin zone.

PROBLEM

Show how the second, third, fourth Brillouin zones for the two dimensional hexagonal lattice may be reduced to the first zone.

In three dimensions, the first Brillouin zone is the volume enclosed by the planes that are the perpendicular bisectors of the basic vectors forming our three dimensional **k** space. The second Brillouin zone is the volume enclosed between the planes which define the first Brillouin zone and the volume enclosed by the planes which are the perpendicular bisectors of the next largest **k** vectors. Similar constructions define the higher order Brillouin zones.

PROBLEM

It may be shown in general that the volume of a Brillouin zone is equal to the volume of the unit reciprocal lattice cell. Prove this for the first Brillouin zone of the simple hexagonal lattice.

The first four Brillouin zones and their reductions into the first Brillouin zone are shown in Figure 4.10 for the simple cubic lattice. The first Brillouin zone for the face-centered cubic lattice is shown in Figure 4.11. The symbols indicate particular types of symmetry properties at different points in the zone. The volume of this zone is $4/A^3$ where A is the edge of the unit cell of the direct lattice. Note that this Brillouin zone is similar to the fourth Brillouin zone of the simple cubic lattice.

The first Brillouin zone for the body-centered cubic lattice is shown in Figure 4.12. The second zone is similar to the fourth zone for the simple cubic lattice. The volume of each zone for the body-centered cubic lattice is $2/A^3$.

The first and second Brillouin zones for the simple hexagonal lattice are shown in Figure 4.13. The volume is $2/\sqrt{3}a^2C$.

The more complicated diamond and zincblende structures have Brillouin zones which are identical with those of the face-centered cubic lattice, and those of the close-packed hexagonal lattice are identical with the simple hexagonal lattice. In general, the Brillouin zones of a lattice with basis vectors are the same as those for the simple lattice without basis vectors.

PROBLEM

For a cubic crystal whose sides are of length La_x, Ma_y and Na_z show, using the Born-Von Karman periodic boundary conditions

$$e^{i\mathbf{k}\cdot\mathbf{r}} = e^{i\mathbf{k}\cdot[L\hat{a}_x + M\hat{a}_y + N\hat{a}_z]}$$

that the density $\delta(\mathbf{k})$ of allowed values of **k** per unit volume is a constant and is equal to $1/8\pi^3$. Argue that this result leads to the conclusion that there are as many allowed **k** values in a Brillouin zone (or in a unit reciprocal cell) as there are unit cells in the crystal volume.

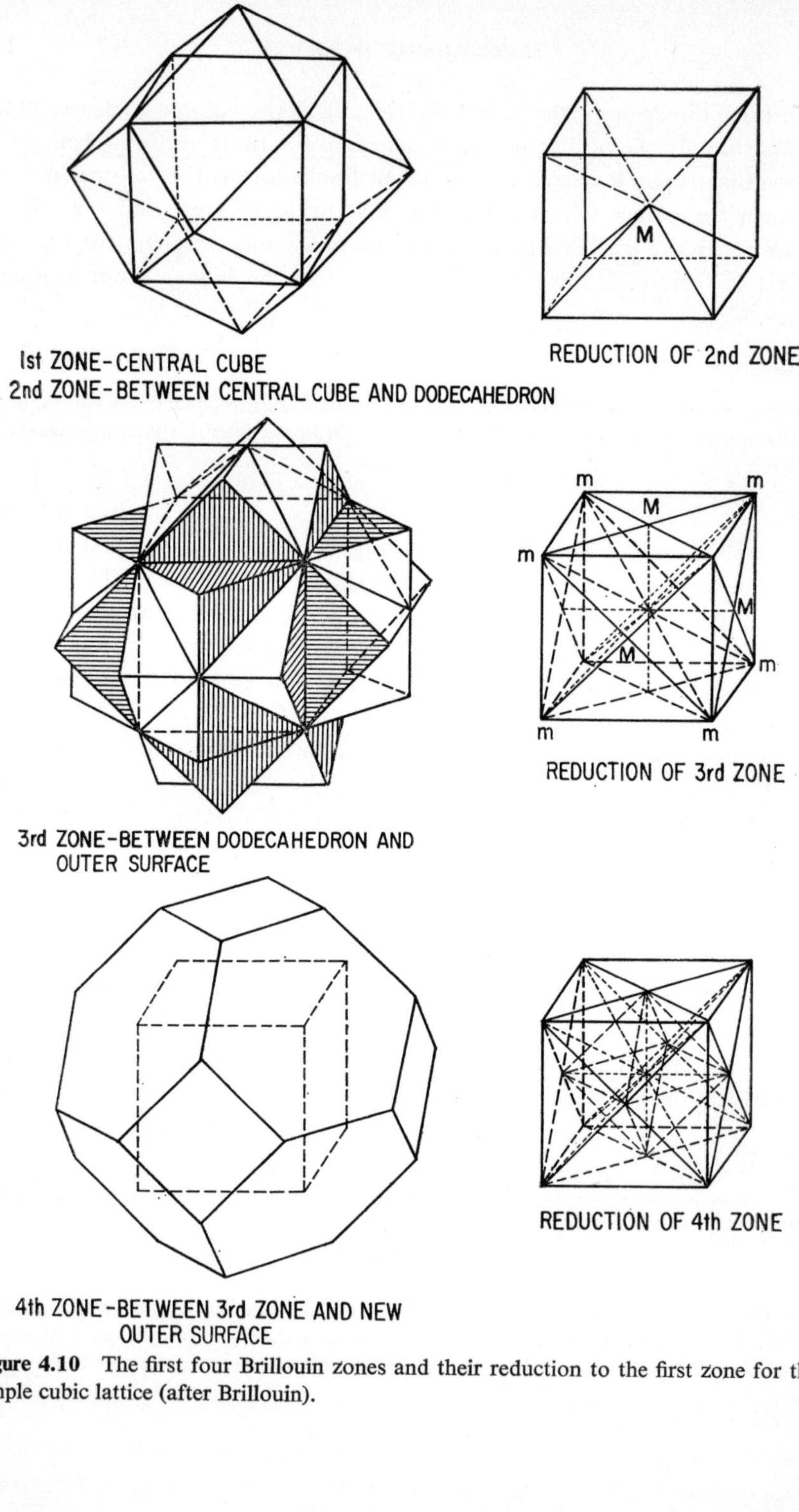

Figure 4.10 The first four Brillouin zones and their reduction to the first zone for the simple cubic lattice (after Brillouin).

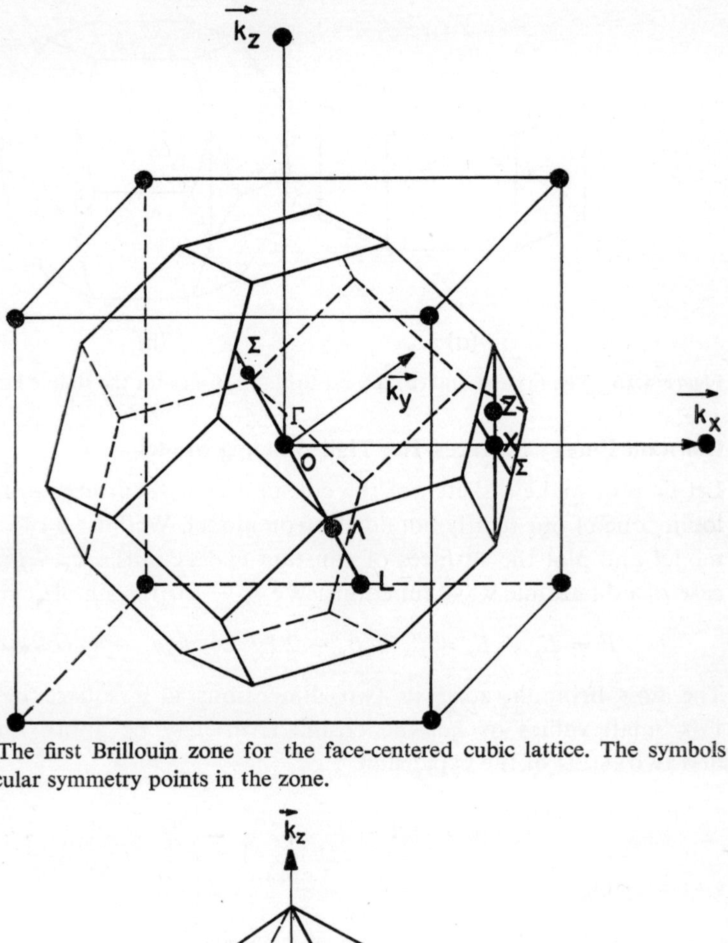

Figure 4.11 The first Brillouin zone for the face-centered cubic lattice. The symbols indicate particular symmetry points in the zone.

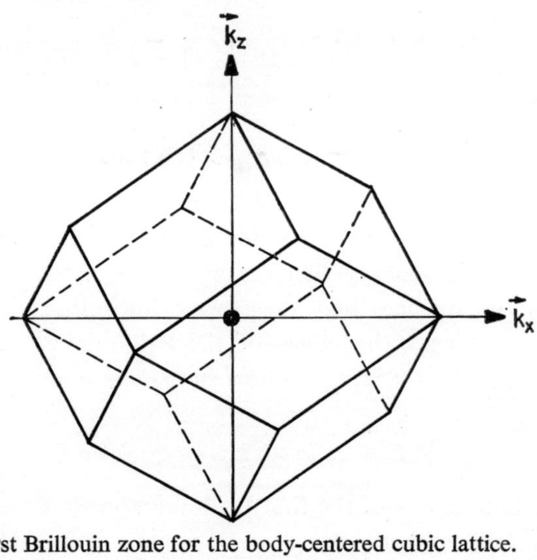

Figure 4.12 The first Brillouin zone for the body-centered cubic lattice.

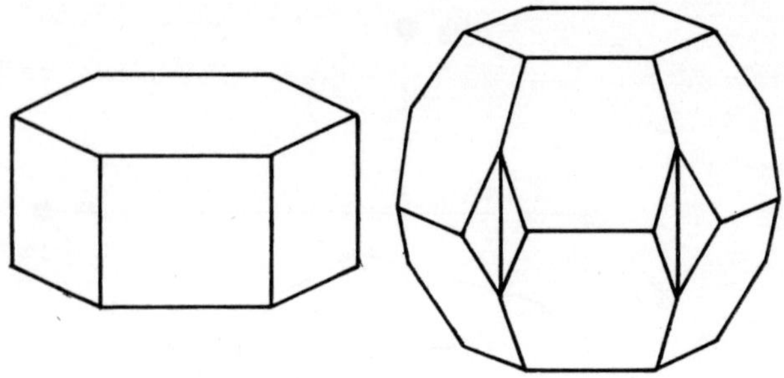

Figure 4.13 The (a) first and (b) second Brillouin zones for the simple hexagonal lattice

Constant Energy Surfaces-The Tight-Binding Model

Let us now make a sketch of the constant energy surfaces in the first Brillouin zone of our tightly bound electron model. Well use a two dimensional model and plot the surfaces of constant energy in the $\hat{k}_x$, $\hat{k}_y$ plane for the case of odd atomic wave functions. We have for the simple cubic lattice

$$E = E_x \neq E_y = \beta_x + \beta_y - 2\,\{\gamma_x \cos k_x a_x + \gamma_y \cos k_y a_y\} \qquad (4.4.3)$$

The first Brillouin zone in two dimensions is a square of sides $2\pi/a$. For small values of ka, the cosine term may be approximated by the first two terms of the expansion

$$\cos x = 1 - \frac{x^2}{2!} + \frac{x^4}{4!} - \frac{x^6}{6!} + \dots . \qquad (4.4.4)$$

which yields

$$E = (\beta_x + \beta_y - 2\gamma_x - 2\gamma_y) + \gamma_x k_x^2 a^2 + \gamma_y k_y^2 a^2 \qquad (4.4.5)$$

We shall take the first term in (4.4.5) equal to our zero of energy since it may be assumed to have no dependence on $\hat{k}_x$ or $\hat{k}_y$. Thus at very small **k** values, the constant energy surfaces are ellipses. Note that in particular cases the energy surfaces may degenerate into circles. We look again at (4.4.3), this time near the edges of the Brillouin zone. For values near $k_x = \pi/a$ and $k_y = \pi/a$, the gradient in **k** space of our energy surface becomes

$$\nabla_k E = 2\gamma_x a \sin k_x a + 2\gamma_y a \sin k_y a \qquad (4.4.6)$$

Thus on the boundaries of the first Brillouin zone $\mathbf{V}_k\mathbf{E}_x$ and $\mathbf{V}_k\mathbf{E}_y$ are equal to zero and the constant energy surface intersects the boundary of the

Brillouin zone at right angles. In three dimensions the surfaces of constant energy at small **k** values are ellipsoids of revolution. In extremely simple cases these may degenerate into spherical energy surfaces.

Thus, the constant energy surfaces become distorted for larger **k** values as shown in Figures 4.14a and 4.14b.

PROBLEM

For our tight binding model in 3-dimensions show that in the first Brillouin zone

$$\frac{d}{dk_i} E(\mathbf{k}) = -\frac{d}{dk_i} E(-\mathbf{k})$$

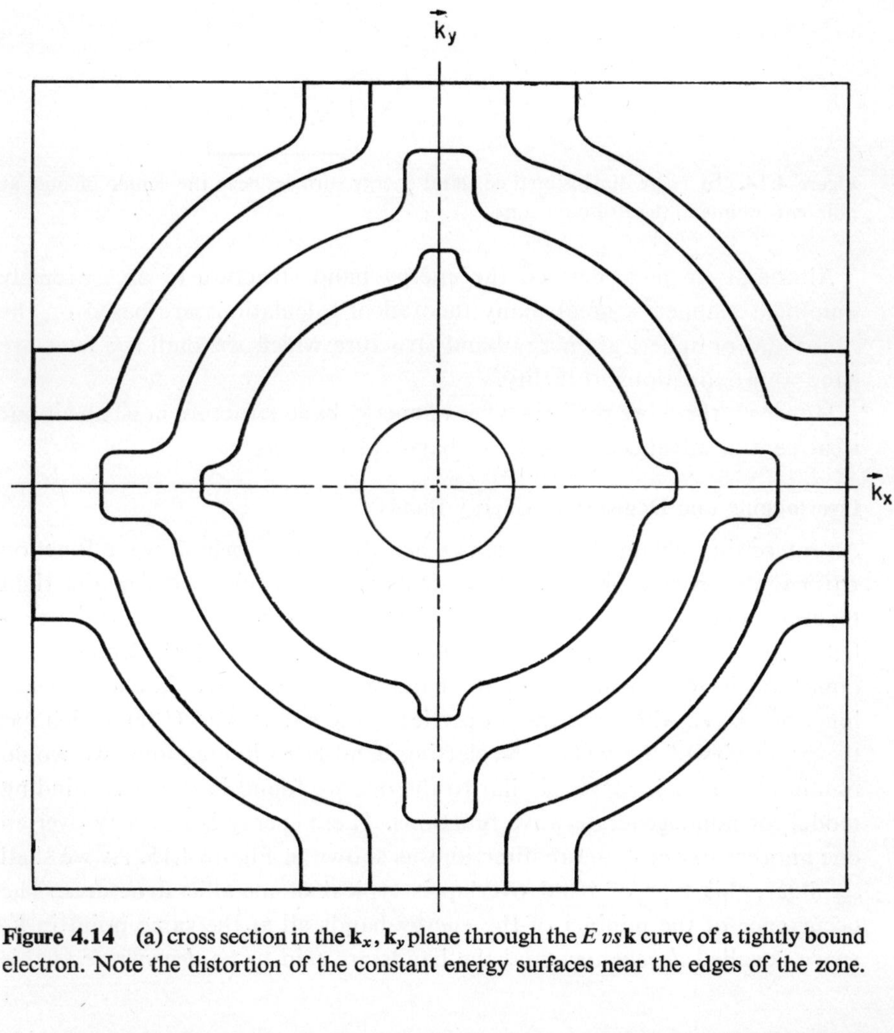

Figure 4.14 (a) cross section in the $\mathbf{k}_x$, $\mathbf{k}_y$ plane through the E *vs* $\mathbf{k}$ curve of a tightly bound electron. Note the distortion of the constant energy surfaces near the edges of the zone.

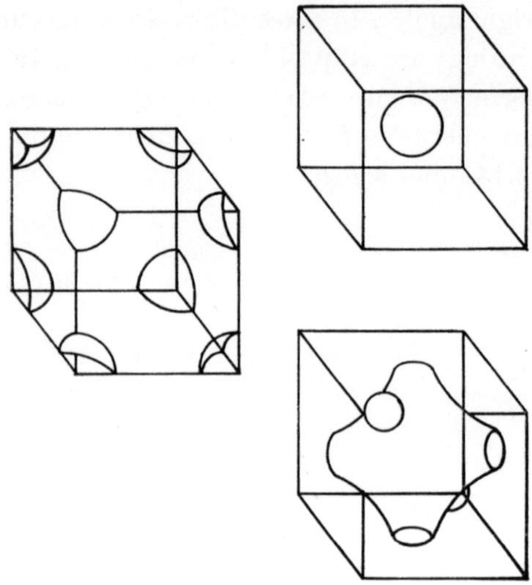

Figure 4.14. (b) Three dimensional constant energy surfaces near the center of and at different points in the Brillouin zone.

Although we have derived the energy band structure in an extremely simplified manner, a great many theoretical calculations are based on the ellipsoidal or spherical energy band structure which we shall see later are good approximations to reality.

However, the more realistic types of energy band structure must await our treatment of advanced methods of band calculations.

Overlapping and Degenerate Energy Bands

An interesting phenomenon occurs when the atomic orbital wave function $U_{\mathbf{k}}(\mathbf{r})$ is degenerate in energy. For instance, we could perform the tight binding calculation for a hypothetical element, which has an atomic configuration $1s^2 2s^2 2p^2$. Assuming that only the $2s^2$ and $2p^2$ electrons are important in the calculation of the band structure, we would choose wave functions $U_s(\mathbf{r})$, which has the properties of the s state, and $U_p(\mathbf{r})$, which has the properties of the p state. Neglecting band-band interactions, we would obtain four E vs $\mathbf{k}$ curves similar to the one we found in our tight binding model for nondegenerate wave functions. These energy bands may overlap one another in one or more directions as shown in Figure 4.15. As we shall see later, this type of band overlap is typical of metallic structures. The occurrence of the minima of the energy bands all at the same point in $\mathbf{k}_x$ space is called *degeneracy*, and the bands are said to be *degenerate* at this point. In addition, it is only due to the nature of our simple calculation that the energy bands show minima and maxima at the origin of k space. As we shall see later, actual calculations for many solids yield minima at other points in the Brillouin zone.

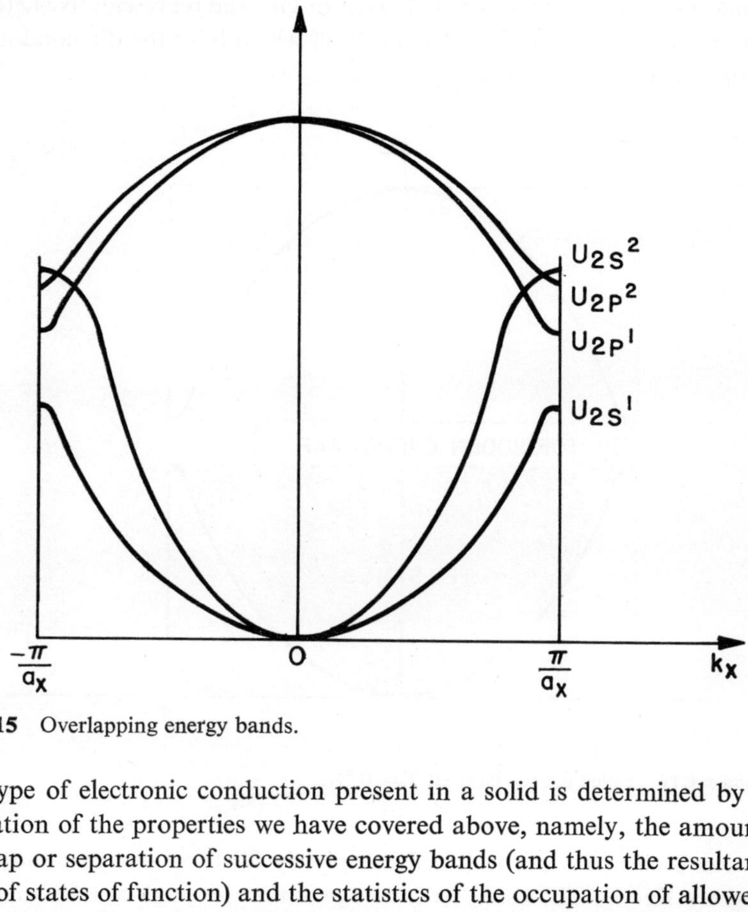

Figure 4.15 Overlapping energy bands.

The type of electronic conduction present in a solid is determined by a combination of the properties we have covered above, namely, the amount of overlap or separation of successive energy bands (and thus the resultant density of states of function) and the statistics of the occupation of allowed energy levels. Before we proceed to discuss the statistics which govern the method of filling the allowed energy bands, we can formulate some quite general deductions from our previous discussions.

A simple Bravais lattice contains one atom per unit cell and a Brillouin zone contains 2 allowed states per unit cell of the crystal (considering orbital states only). If the atoms making up the solid contain an even number of electrons, and no energy band overlap exists, invocation of the Pauli exclusion principle at zero degrees Kelvin yields the fact that all of the zones occupied by electrons are completely filled. For example, in the case of an atom which has two valence electrons and which crystalizes in a simple Bravais lattice, we find that the first Brillouin zone is completely filled. And the second completely empty, as shown in Figure 4.16. This is the case for homopolar semiconductors such as Si or Ge, and representative heteropolar semiconductors such as GaAs and InSb which have the diamond and zinc-blende structures, respectively, and thus two atoms per unit cell.

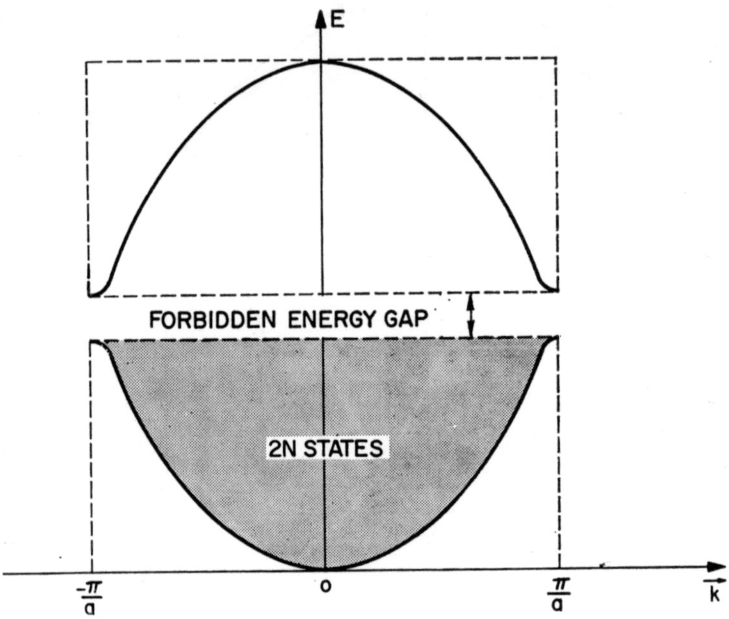

Figure 4.16 Filled energy band at $T = 0\,°K$.

Our 8 electrons per unit cell completely fill these energy bands at zero degrees Kelvin. The usual terminology is that the highest fully or partially occupied bands of crystal are called the *valence* bands. The first band unoccupied at 0°K is the *conduction* band. The first forbidden energy gap between the valence and conduction bands is the gap usually referred to as the *energy gap*. This is because it requires a great deal of energy to empty the first valence band and thus the effects of the second forbidden energy gap go unseen at ordinary temperatures. The type of conduction in this situation is dependent on the amount of energy required to ionize carriers from the filled valence band to the unfilled conduction band, where there are so many

Table 4.1 Band gaps of some typical semiconductors (from E. L. Kern and E. Earlywine, Semiconductor Products and Solid State Technology, October, 1965)

Group	Material	Energy Gap (eV)	Direct or Indirect
		α 2.0	D
III	B	β 1.53	D
IV	Ge	0.66	I
	Si	1.107	I
	SiC	2.8	I
V	P	1.45	D
VI	S	2.65	D
	Se	1.8	D
	Te	0.32	D
	BN	α 4.6	I
	BP	6.0	I
	BAs	5	D
	AlN	3.8	D
	AlP	3.0	I
	AlAs	2.16	I
	AlSb	1.6	I
	GaN	3.3	D
	GaP	2.25	I
III-V	GaAs	1.47	D
	GaSb	0.68	D
	InN	2.5	D
	InP	1.27	D
	InAs	0.36	D
	InSb	0.17	D
	ZnS	3.58	D
	ZnSe	2.67	D
	ZnTe	2.26	D
	CdS	2.59	D
II-VI	CdSe	1.7	D
	CdTe	1.50	D

Group	Material	Energy Gap (eV)	Direct or Indirect
	HgS	2.0	D
	HgSe	0.6	D
	HgTe	0.025	
Lead	PbS	0.37	I
Salts	PbSe	0.26	I
	PbTe	0.29	I
	Mg_2Si	0.77	
	Mg_2Ge	0.74	
	Mg_2Sn	0.36	
II-IV	Ca_2Si	1.9	
	Ca_2Sn	0.9	
	Ca_2Pb	0.46	
	ZnSb	0.53	
	ZnAs	0.90	
	Zn_3As_2	0.93	
II-V	CdSb	0.46	
	$CdAs_2$	1.0	
	Cd_3As_2	0.13	
	Bi_2S_3	1.3	
	Bi_2Se_3	0.28	
	Bi_2Te_3	0.20	
	Sb_2S_3	1.7	
V-VI	Sb_2Se_3	1.2	
	Sb_2Te_3	0.30	
	As_2Se_3	1.6	
	As_2Te_3	1.0	
VIII-V	$PtSb_2$	0.07	
III-VI	In_2S_3	2.28	
	In_2Se_3	1.25	
	In_2Te_3	1.0	

unfilled allowed energy levels that the electron may be treated as if it were almost free.

Thus, at zero degrees Kelvin, pure semiconductors are insulators. As the temperature is raised, electrons become capable of crossing the energy gap and the material becomes less resistive (until lattice scattering begins to increase the resistivity as in metals.)

We are now able to understand, at least qualitatively, one of the properties typical of semiconductors and insulators, i.e., the decrease in electrical resistance with temperature until scattering processes become dominant. Table 4.1 gives the energy gaps of some typical representative semiconductors.

For the case in which each atom has an odd number of electrons, and no energy band overlap occurs, the highest occupied band must only be partially filled. Thus the application of only a small electric field will give carriers enough energy to move about in the unfilled energy levels of the partially occupied band. Thus we would expect atoms with an odd number of valence electrons and *no* band overlap to be dominated by metallic conduction. This is the case, e.g., for the metallic atoms of group I and III of the periodic table. For the case of the group II metals, which usually crystallize in the close-packed hexagonal structure, metallic conduction is due to the overlap of Brillouin zones which completely obliterates the energy gap.

The Fermi Surface

At absolute zero in temperature, the electrons in a solid will occupy the lowest energy levels available in accordance with the Pauli exclusion principle. Thus, all of the energy levels at absolute zero will be filled up to an energy E_f (the Fermi Energy at zero degrees Kelvin). Above this level no allowed energy levels will be filled. This is true for all of the cases we have just discussed.

The surface in **k** space which corresponds to the condition at $T = 0°K$ that all of the allowed energy levels within its volume are filled and no energy levels outside of the volume are occupied, is called the *Fermi surface* of the material.

In the case of some insulators and semiconductors, in which no overlap of energy bands (and thus no overlap of Brillouin zones occurs) the Fermi surface has the shape of the largest completely occupied Brillouin zone. In some metals the Fermi surface may be quite a complicated structure such as is shown in Figure 4.17 for aluminium.

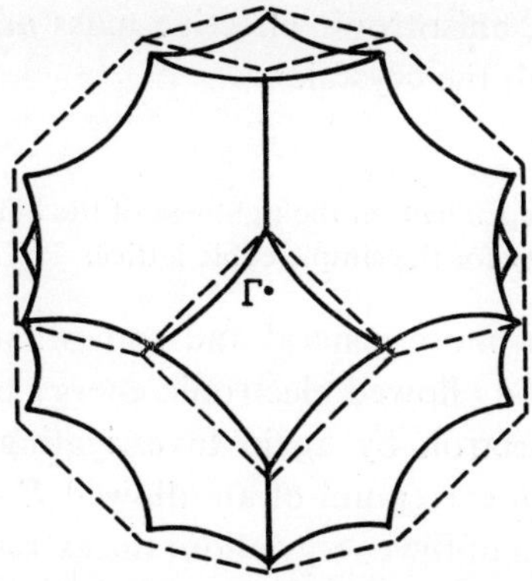

Figure 4.17 A Fermi surface in the second Brillouin zone of Aluminium (W. A. Harrison, *Phys. Rev.*, **118**, 1190, 1960).

The Effective Mass

Now that we've investigated, on a simple model, the shape of the energy bands encountered in simple solids and shown how these energy bands arise, it is advisable to stop for a moment and collect our thoughts on the matter. We have found, for both the loose and tight binding models, that for small values of **k**, the electron energy varies as k^2, i.e., in a manner similar to the allowed energy for a free electron,

$$E \propto \frac{\hbar^2 k^2}{2m} \tag{4.4.7}$$

Of course, for a free electron, the mass m is an isotropic quantity.

We may analogize between the motion of a free electron and an electron in a strong crystalline potential by comparing the energy versus **k** curves for the tight binding model near $\mathbf{k} = 0$.

$$\begin{aligned} E_{T.D.} = {} & \beta_x + \beta_y + \beta_z + \gamma_x[-2 + k_x^2 a_x^2] + \gamma_y[-2 + k_y^2 a_y^2] \\ & + \gamma_z[-2 + k_z^2 a_z^2] \end{aligned} \tag{4.4.8}$$

with those of the free electron

$$E_F = \frac{\hbar^2}{2m}(k_x^2 + k_y^2 + k_z^2) \tag{4.4.9}$$

When we compare components in **k** space, we find that

$$\frac{2\gamma_x a_x^2}{\hbar^2} = \frac{1}{m_x^*}; \quad \frac{2\gamma_y a_y^2}{\hbar^2} = \frac{1}{m_y^*}; \quad \frac{2\gamma_z a_z^2}{\hbar^2} = \frac{1}{m_z^2} \tag{4.4.10}$$

Thus, we can define an anisotropic effective mass m_i^* for an electron near $\mathbf{k} = 0$ travelling through the crystal.

PROBLEM

How is the effective mass dependent on the tightness of the binding? Show that the effective mass is a scalar quantity for the simple cubic lattice.

Let us define, now, a more general and somewhat more useful analogy between the shape of the allowed electronic energy bands and the intrinsic properties of a free electron by again investigating the behaviour of an electron located near an extremum of an allowed E vs $\mathbf{k}$ curve. Let us perform a Taylor expansion of the energy about the extremum E_0 of a particular energy band in $\mathbf{k}$ space. We may write

$$E| = E_0 + \sum_{i=x,y,z} \frac{dE(k)}{dk_i}(k_i - k_0) + \frac{1}{2} \sum_{i=x,y,z} \sum_{j=x,y,z} \frac{d^2E(k)}{dk_i dk_j}[k_i - k_j] + \ldots \tag{4.4.11}$$

Since we chose to expand about an extremum, the term including $dE(k)/dk_i$ is equal to zero. Thus we get for the allowed energy near a band extremum

$$E = E_0 + \frac{1}{2} \sum_{i=x,y,z} \sum_{j=x,z,z} \frac{d^2E(k)}{dk_i dk_j}[k_i - k_j]^2 + \ldots \tag{4.4.12}$$

If we compare (4.4.12) with the expression for the energy of a free particle, we find that an effective mass tensor may be defined in terms of the curvature of the E vs $\mathbf{k}$ surfaces

$$\frac{1}{m_{ij}^*} = \frac{1}{\hbar^2} \sum_{i=x,y,z} \sum_{j=x,y,z} \frac{d^2E(k)}{dk_i dk_j} \tag{4.4.13}$$

Of course, by a suitable transformation to the principal axes of the crystal, the tensor may be diagonalized. In the particular case where the effective mass tensor has been diagonalized

$$\frac{1}{m_i^*} = \frac{1}{\hbar^2} \frac{d^2E(k)}{dk_i^2} \tag{4.4.14}$$

Near an extremum the energy surface is usually very nearly ellipsoidal. For an ellipsoidal constant energy surface, with major axis $\mathbf{k}_x$ and minor axes $\mathbf{k}_y$ and $\mathbf{k}_z$ we call m_x^* the longitudinal effective mass and m_y^* and m_z^* the transverse effective masses. Note that in the case of spherical energy surfaces the effective mass reduces to a scalar quantity.

It is interesting to note the peculiarity inherent in our definition (4.4.14). First, the effective mass may become negative. Thus, if a force **F** is applied to an electron located near an energy band extremum which is a maximum, the electron will act as if it has a negative mass and thus will be accelerated in a direction opposite to the direction of the applied force. This leads to the concept of a mobile positive particle or *hole*. It is, of course, true for both the minima and the maxima type of band extrema, that the resultant acceleration of an electron in a crystal lattice may be in a direction other than that of the applied force.

PROBLEM

Show that for a force **F** applied to an electron near an extremum of energy, the acceleration $\mathbf{A}_i$ that the electron undergoes is

$$\mathbf{A}_i = \frac{1}{\hbar^2} \sum_{j=x,y,z,} \frac{d^2E(k)}{dk_i\, dk_j} \mathbf{F}_j \tag{4.4.15}$$

Note that for the case of the Lorentz force $q/C\,(\mathbf{V} \times \mathbf{B})$ no work is done ($\mathbf{FV} = 0$). Does the above still hold? **V** is the velocity of the electron and **B** is the magnetic induction field.

Many experiments do not yield specific information concerning each of the effective masses, but rather determine a so-called average effective mass for the energy surface. For the case of ellipsoidal energy surfaces where energy of an electron at any time is representable by a single point in **k** space, it is quite simple to calculate the so-called *density of states effective mass* m_d^*. Near $\mathbf{k} = 0$ we find from (4.4.9)

$$dE(k) = \hbar^2 \left[\frac{k_x\, dk_x}{m_x^*} + \frac{k_y\, dk_y}{m_y^*} + \frac{k_z\, dk_z}{m_z^*} \right] \tag{4.4.16}$$

The total number of distinct energy states per Brillouin zone is clearly the number of allowed values of **k**

$$E = \hbar^2 \left[\sum_{i=x,y,z} \int_{-\frac{\pi n_i}{a_i}}^{+\frac{\pi n_i}{a_i}} \frac{ki\, dki}{m_i^*} \right] \tag{4.4.17}$$

Thus the number of electronic states within a Brillouin zone is

$$E = \frac{\hbar^2}{2} \sum_{i=x,y,z} \left(\frac{n_i \pi}{a_i} \right)^2 \frac{1}{m_i^*} \tag{4.4.18}$$

where we have divided by two because counting both + and − values of k_x, k_y and k_z yields a degeneracy of 2 in the energy.

Thus we have transformed our E, which was a function of **k**, into a function of n_x/a_x, n_y/a_y and n_z/a_z and the effective masses m_x^*, m_y^* and m_z^*. If we construct a vector N such that

$$\mathrm{N} = \hat{\imath}\,\frac{n_x(m_x^* m_y^* m_z^*)^{1/6}}{a_x m_x^{*1/2}} + \hat{\jmath}\,\frac{n_y(m_x^* m_y^* m_z^*)^{1/6}}{a_y m_y^{*1/2}} + \hat{k}\,\frac{n_z(m_x^* m_y^* m_z^*)^{1/6}}{a_z m_z^{*1/2}} \tag{4.4.19}$$

we find

$$E(k) = \frac{\hbar^2 \pi^2 N^2}{2m_d^*} \tag{4.4.20}$$

where m_d^* is defined as $(m_x^* m_y^* m_z^*)^{1/3}$ and is called the density of states effective mass. In this space too, each value of E is represented by a single point, and so the number of allowed energy states with energy less than a certain energy E is just the number of points in the positive octant of a sphere of radius N.

Thus we find

$$\frac{\pi N^3}{6} = \int_0^E g(E)\,\mathrm{d}E \tag{4.4.21}$$

Substituting (4.4.20) into (4.4.21) and integrating, we find again the parabolic density of states

$$g(E)\,\mathrm{d}E = \left(\frac{2m_d^*}{\hbar^2}\right)^{3/2} \frac{1}{4\pi^2} E^{1/2}\,\mathrm{d}E \tag{4.4.22}$$

PROBLEM

Show, for the tight binding model that the density of states function, near the corners of the Brillouin zone is also parabolic. (NOTE: this corresponds to the top of an allowed energy band.) What is the density of states effective mass in this case?

It can be shown by an extension of a theorem by Van Hove,[6] that, in general, the density of states function $g(E)$ must be of the form as shown by the dashed curve in Figure 4.18a (These arguments are beyond the scope of our work). We have plotted the parabolic density of states function with a dashed line. We note that, as expected, the parabolic approximation is valid near $E = 0$ but quickly deviates from this behaviour.

In general, a density of states function which represents a non-degenerate band must contain the following singularities: (1) a minimum approached as $E^{1/2}$, (2) two saddle points (B and C) and (3) a maximum approached as $(E_D - E)^{1/2}$.

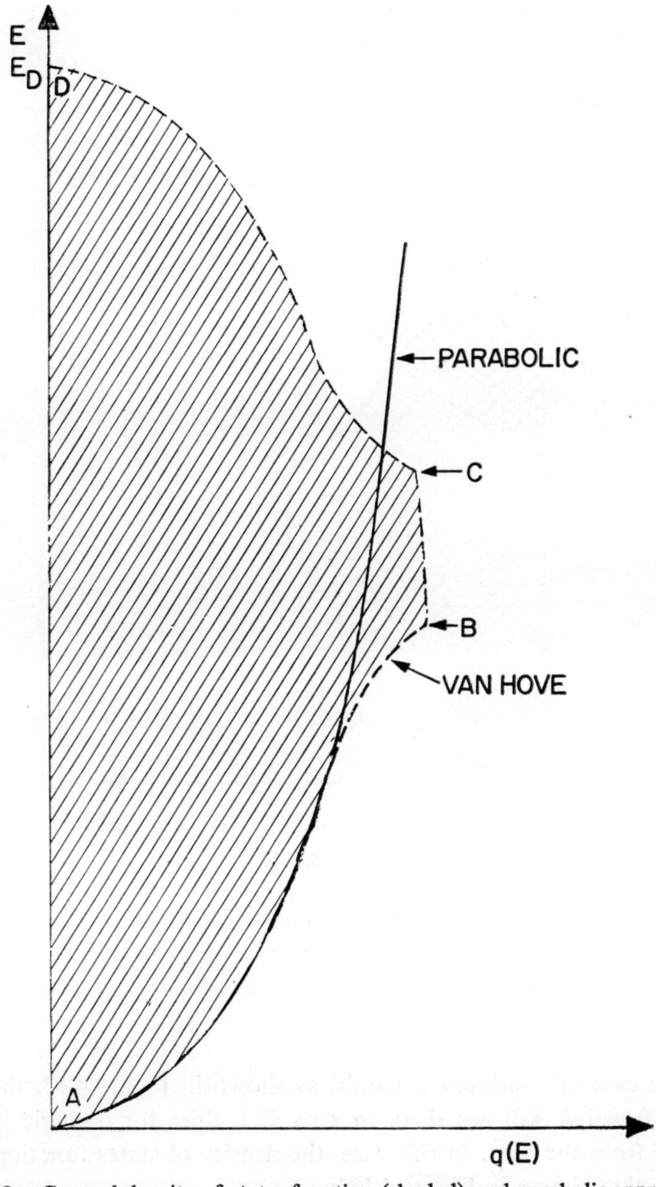

Figure 4.18a General density of states function (shaded) and parabolic approximation.

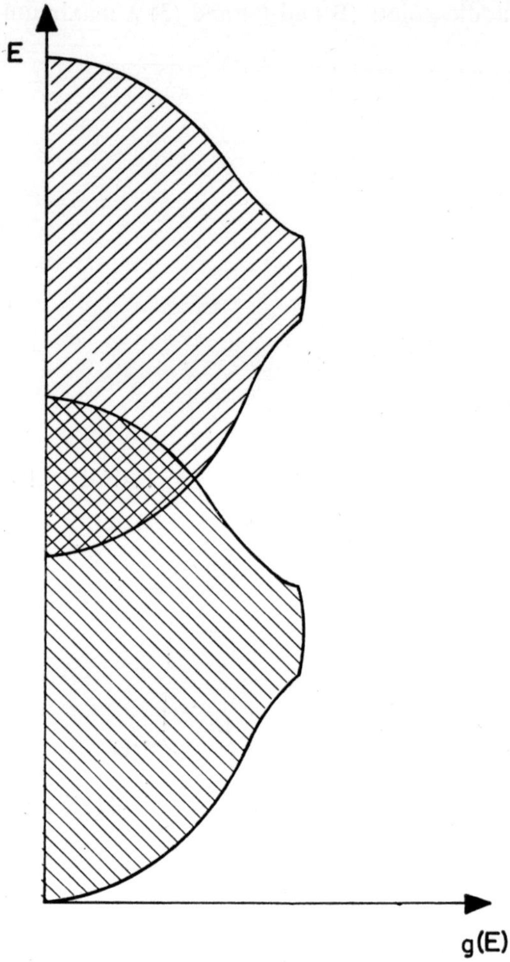

Figure 4.18b

Figure 4.18b Density of states resulting from overlapping bands.

For the case of overlapping bands, as shown in Figure 4.15, the density of states function will not drop to zero as it does for a single band well separated from the next. In this case, the density of states function may be similar to that shown in Figure 4.18b.

4.5 ADVANCED METHODS FOR THE CALCULATION OF BAND STRUCTURE

We have examined some of the more elementary methods by which we could learn the fundamentals of the allowed energy levels in solids. We know, however, that the cases of extremely small lattice potential and of extremely tight binding are only approximations which serve to demonstrate that the general behaviour of the allowed motions of electrons in solids is relatively independent of the particular lattice potential chosen. In reality, we know that an electron travelling in the periodic potential of a crystalline lattice will experience a lattice potential whose strength varies from material to material. The tight binding calculation, although it depends on the atomic state of the outer valence electrons and the crystalline structure, does not take into account solids whose atomic valence electron wave functions do not characterize the conducting electron. We must then find some middle ground, upon which we can begin a study of the motion of electrons in solids with a suitable description of the wave function of a travelling electron.

We will now investigate a few of the more advanced methods of band structure calculations and the results of their applications to typical solids.

The Methods of Hartree and Fock

We have discussed the theory of energy bands in crystals from a one-electron point of view. In the weak binding model we derived the general shape of the energy bands and the relation of the energy gap to the strength of the ionic potential. The tight binding model further revealed the modifications in the allowed energy states of a single tightly bound electron arising from the imposition of the Bloch condition. These simplified calculations, although they serve to acquaint us with the first principles of the theory of solids in the extreme cases of tight and loose binding, are however not useful for actual calculations, since there are many forces, such as electron-electron interactions, which we have not yet taken into account. This section is intended to show how some of these interactions are taken into account to form a background for our study of the methods actually used to perform band structure calculations.

The Hartree Method

In 1928, Hartree[7] proposed that, when one is solving for the energy of a system which includes many electrons, each electron should satisfy an equation of the form (1.5.1) in which the Hamiltonian $\mathscr{H}_i$ contains a term which represents the interaction of the ith electron with all of the other electrons. This interaction term may be represented as the average interaction $V_{i,j}$ of this electron with the Coulomb fields of the other electrons

$$V_{i,j} = \sum_{i \neq j} A_i^2 \langle \xi_j | V_{i,j} | \xi_j \rangle \tag{4.5.1}$$

where $V_{i,j}$ is the Coulomb operator for the ith and j^{th} electrons and A_j is a normalizing constant. Thus the ith electron Hamiltonian may be written

$$\mathscr{H}_i = T_i + V_i(\mathbf{r}) + e^2 \sum_j A_j^2 \left\langle \xi_j \left| \frac{1}{r_{i,j}} \right| \xi_J \right\rangle \tag{4.5.2}$$

$V_i(r)$ is the ionic potential (which is the only term we have dealt with in our preceeding models).

Thus one must solve a set of N integro-differential equations of the from (1.5.1), one for each of the N electrons in the system. Since these equations are coupled by the dependence of the ith energy eigenvalue of all of the other single particle states, one must use an iterative procedure to solve for the best possible solutions for both the energy eigenvalues and the single particle states. Thus this method is usually called the Hartree Self-Consistent Field method. One chooses a set of particular normalized states, say $|\xi_{0j}\rangle$, for use in (4.5.1). Equation (1.5.1) is then solved for an approximate E_i. The procedure is then repeated with successive $|\xi_j\rangle$s repeated in V_{ij} as many times as it takes to produce a final solution which differs negligibly from the proceeding one.

PROBLEM

Show that for a simple product state of the form $|\xi_T\rangle = \prod_i^N |\xi_i\rangle$, the energy of Coulomb-interaction is

$$-\frac{e^2}{2} \sum_{i,j} \langle \xi_i | \langle \xi_j | V_{i,j} | \xi_i \rangle \, | \xi_j \rangle \, A_i^2 A_j^2 \tag{4.5.3}$$

where A_i, A_j are the normalization constants for the states $|\xi_i\rangle$ and $|\xi_j\rangle$, respectively.

The Hartree self-consistent field method has been applied quite successfully in the calculations of atomic wave functions and in the theory of metals.

The Hartree-Fock Method

It was suggested by both Slater[8] and Fock[9] that the Hartree method outlined above be modified to include the more realistic case on an antisymmetric determinantal wave function of the form (1.5.11).

By using the method of Lagrange Multipliers to include the orthogonality constraint, one may determine, for the case of orthogonal antisymmetric states, that the Hartree Hamiltonian must be modified to yield

$$\left[\mathscr{H}_i - \sum_{\substack{j \\ \text{parallel} \\ \text{spins only}}} \langle \xi_j(\mathbf{r}_2)|\, V_{i,j}\, |\xi_i(\mathbf{r}_2)\rangle\, |\xi_j(\mathbf{r}_1)\rangle\right] |\xi_i(\mathbf{r}_1)\rangle = E\, |\xi_i(\mathbf{r}_1)\rangle \tag{4.5.4}$$

where $\mathscr{H}_i$ is the Hartree Hamiltonian. The above is the Hartree-Fock equation for the ith electron state in the field of N-1 mutually orthogonal electrons.

The term which must be added to the Hartree Hamiltonian

$$\sum_{\substack{j \\ \text{parallel} \\ \text{spins only}}} \langle \xi_j(\mathbf{r}_2)|\, V_{i,j}\, |\xi_i(\mathbf{r}_2)\rangle\, |\xi_j(\mathbf{r}_1)\rangle \tag{4.5.5}$$

is an exchange function of the type we have dealt with before. It is a measure of the correlation between the ith and j^{th} single particle states, i.e., if both the ith and j^{th} electrons have the same spin coordinates, the exchange term is a measure of their potential energy of interaction.

The Fock equations too must by solved by an iterative technique. They are, of course, too complicated to be useful for the calculation of crystal wave functions, but have been used successfully in the calculation of atomic wave functions.

Problem

Show that the total energy of a system representable by the Fock equation is not merely the sum of the energies of the individual electrons, but is, in Schrödinger notation

$$E = \sum_i E_i - \frac{1}{2}\sum_{i,j} \iint \frac{e^2}{r_{12}} |\psi_j(\mathbf{r}_2)|^2\, |\psi_i(\mathbf{r}_1)|^2\, d\mathbf{r}_1\, d\mathbf{r}_2$$

$$+ \frac{e^2}{2} \sum_{\substack{i,j \\ \text{all parallel} \\ \text{spins}}} \iint \frac{\psi_i^*(\mathbf{r}_1)\, \psi_j^*(\mathbf{r}_2)\, \psi_i(\mathbf{r}_2)\, \psi_j(\mathbf{r}_1)\, d\mathbf{r}_1\, d\mathbf{r}_2}{r_{12}} + \frac{1}{2}\sum_{a \neq b} \frac{e^2}{R_{a,b}} \tag{4.5.6}$$

where $e^2/R_{a,b}$ is the ionic potential. What is the significance of each of the above terms?

It has been shown by Koopmans[10] that the negative of the ith energy eigenvalue is just the energy required to remove the ith electron from a crystalline material in which the spin-orbital wave function has a spatial portion which is of the Bloch type. Similarly, the energy required to shift an electron from the state $|\xi_j\rangle$ to the state $|\xi_i\rangle$ is $(E_i - E_j)$.

The Wigner-Seitz Cell

We have seen in our treatment of the tight binding model that, due to the periodicity of the Bloch wave function we need determine the wave function only inside a unit cell of the crystal lattice.

It is ofttimes more convenient to define a type of unit cell in the direct lattice which has a lattice point as the center of symmetry. Unfortunately, the unit cells of most lattice structures are not of this form.

Wigner and Seitz[11] have developed a method by which one can construct a cell within the direct lattice which is symmetric about a lattice point and from which the entire crystalline volume may be built through successive

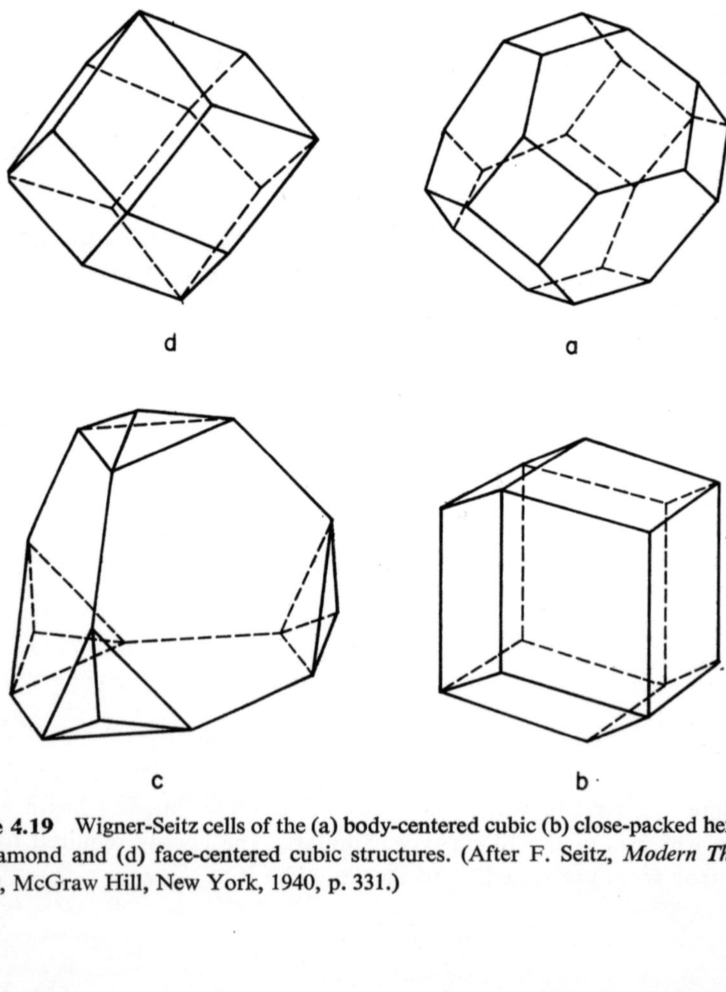

Figure 4.19 Wigner-Seitz cells of the (a) body-centered cubic (b) close-packed hexagonal (c) diamond and (d) face-centered cubic structures. (After F. Seitz, *Modern Theory of Solids*, McGraw Hill, New York, 1940, p. 331.)

translations. This ***Wigner-Seitz Cell*** is defined as the smallest volume enclosed by the planes which perpendicularly bisect the vectors joining a chosen atom to its nearest neighbors in the direct lattice.

It is the direct lattice analogue of the construction of the first Brillouin zone in **k** space for simple Bravais lattices. In fact, for a simple Bravais lattice, the Wigner-Seitz cell for the direct lattice is identical to the first Brillouin zone in a **k** space lattice whose structure is the same as that of the direct lattice. For example the Wigner-Seitz cell for the face-centered cubic lattice is the same as the first Brillouin zone for the body-centered cubic lattice, and vice versa.

The Wigner-Seitz cell is the basis for a convenient method by which one can perform simple band calculations which are not very sensitive to the actual crystal structure. This method is called the Cellular Method, and was used by Wigner and Seitz to determine the band structure of metallic sodium. We shall perform an illustrative calculation making use of this method when we deal with the phenomenon of impurity band conduction.

It can be shown for a simple Bravais lattice that the Wigner-Seitz cell has all of the symmetry properties of the direct lattice, and that every point within the Wigner-Seitz cell of an atom lies closer to that atom than to any other in the crystal. The Wigner-Seitz cells for the body-centered cubic, the close-packed hexagonal, diamond and the face-centered cubic structures are shown in Figure 4.19.

A fundamental difference occurs, however, between the Wigner-Seitz cell and the first Brillouin zone for the case of a lattice with a basis, such as the diamond or the close-packed hexagonal structures. The generation of the entire lattice from its Wigner-Seitz cell may require not only simple translation, but rotation as well.

At times, it is convenient to approximate the Wigner-Seitz Cell by a sphere. In this approximation the radius r_s of the Wigner-Seitz sphere is

$$r_s = \left(\frac{3}{4\pi N}\right)^{1/3} \tag{4.5.7}$$

for a crystal of N atoms per cubic cm.

PROBLEM

What is the ratio of the volume of the actual Wigner-Seitz cell to that of the Wigner-Seitz sphere for the face-centered cubic structure?

Boundary Conditions for the Bloch Type Electronic Wave Functions at Faces of the Wigner-Seitz Cell

If the vectors $\mathbf{r}$ and $\mathbf{r} + \hat{a}$, where $\hat{a}$ is a lattice translation, refer to opposite faces of a Wigner-Seitz cell, as seen in Figure 4.20 we can easily show that the Bloch functions $\psi_{\mathbf{k}}(\mathbf{r})$ and $\psi_{\mathbf{k}}(\mathbf{r} + \hat{a})$ must satisfy the condition

$$\psi_{\mathbf{k}}(\mathbf{r} + \hat{a}) = e^{i\mathbf{k}\cdot\hat{a}}\,\psi_{\mathbf{k}}(\mathbf{r}) \tag{4.5.8}$$

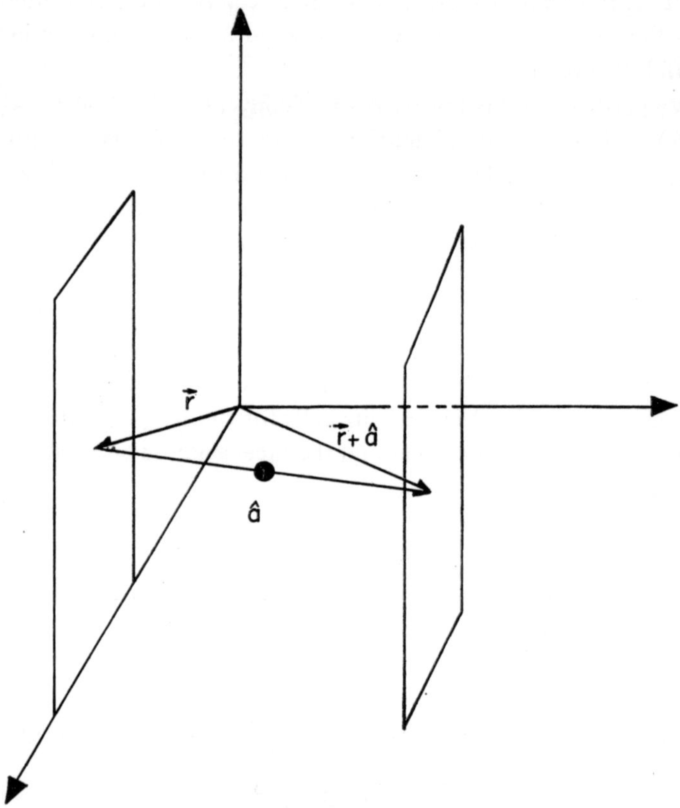

Figure 4.20 Opposite faces of a Wigner-Seitz Cell.

PROBLEM

Prove (4.5.8).

If we perform a Taylor expansion of both sides of (4.5.8) about the point $(\mathbf{r} + \hat{a})$, we find

$$\Delta\hat{a}\cdot[\nabla\psi_{\mathbf{k}}(\mathbf{r} + \hat{a}) - e^{i\mathbf{k}\cdot\hat{a}}\,\nabla\psi_{\mathbf{k}}(\mathbf{r})] = 0 \tag{4.5.9}$$

Thus we also obtain the condition

$$\nabla\psi_{\mathbf{k}}(\mathbf{r} + \hat{a}) = e^{i\mathbf{k}\cdot\hat{a}}\,\nabla\psi_{\mathbf{k}}(\mathbf{r}) \tag{4.5.10}$$

It is, in practice, quite a difficult task to make general use of the boundary conditions (4.5.8) and (4.5.9) for all values of **k** at the Wigner-Seitz surface. Instead, several approximations have arisen, the most useful of which is to satisfy these conditions only at certain points at the Wigner-Seitz sphere for particular wave functions. We shall examine the particular cases of the boundary conditions for both symmetric and antisymmetric atomic wave-functions for the points $\mathbf{k} = 0$ and $\mathbf{k} = \dfrac{n\pi}{\hat{a}}$.

We take first the case of a wave function $\psi_{\mathbf{k}}^{s}(\mathbf{r})$ which is symmetric in its space coordinates about the nucleus (Figure 4.21). In order for the wave functions for each of the atomic spheres to be identical at $\mathbf{k} = 0$, (4.5.10) becomes

$$\nabla\psi^{s}(\mathbf{r} + \hat{a}) = \nabla\psi^{s}(\mathbf{r}) \tag{4.5.11}$$

However, the gradients of $\psi^{s}(\mathbf{r} + \hat{a})$ and $\psi^{s}(\mathbf{r})$ must be directed oppositely. Thus, in order to have continuity of the electronic wave function, it must also be true that

$$\nabla\psi^{s}(\mathbf{r} + \hat{a}) = -\nabla\psi^{s}(\mathbf{r}). \tag{4.5.12}$$

To simultaneously satisfy (4.5.11) and (4.5.12), we find the necessary condition

$$\nabla\psi^{s}(\mathbf{r}) = \nabla\psi^{s}(\mathbf{r} + \hat{a}) = 0. \tag{4.5.13}$$

In a similar manner, (4.5.8) yields, at the point $k = \pi/a$

$$\psi^{s}_{\frac{\pi}{a}}(\mathbf{r} + \hat{a}) = -\psi^{s}_{\frac{\pi}{a}}(r) \tag{4.5.14}$$

from which we obtain the condition

$$\psi^{s}_{\frac{\pi}{a}}(\mathbf{r}) = 0. \tag{4.5.15}$$

We now consider the case of a wave function $\psi_{k}^{a}(\mathbf{r})$ antisymmetric about the nucleus (Figure 4.20). The requirement for continuity of the Bloch wave functions yields the condition

$$\psi_{0}^{a}(\mathbf{r}) = 0 \tag{4.5.16}$$

For the edges of the Wigner-Seitz cell we find

$$\nabla\psi^{a}_{\frac{\pi}{a}}(\mathbf{r} + \hat{a}) = 0 \tag{4.5.17}$$

we shall use the above conditions in section 5.1.

Figure 4.21 Wave functions symmetric ($\psi_k^s(\mathbf{r})$) and antisymmetric ($\psi_k^a(\mathbf{r})$) about lattice sites in a crystal.

Plane Wave Expansion

An important consequence of the method by which we defined our reciprocal lattice is that a travelling wave may be now written in terms of its Fourier component reciprocal lattice vectors. This will prove to be a valuable technique for making detailed calculations.

From Fourier analysis, we know that a function $f(\mathbf{r})$, periodic, e.g., with lattice periodicity such that

$$f(\mathbf{r}) = f(\mathbf{r} + \hat{a}), \tag{4.5.18}$$

where $\hat{a}$ is a unit translation in the $\mathbf{r}$ direction, may be expressed as a Fourier series of harmonic components

$$f(\mathbf{r}) = \sum_n A_n e^{2\pi i \mathbf{k}\cdot\mathbf{r}}, \tag{4.5.19}$$

where the nth harmonic component represents a plane wave.

We may write

$$f(\mathbf{r}) = \sum_n A_n e^{2\pi i [k_x x + k_y y + k_z z]}, \tag{4.5.20}$$

To satisfy the conditions

$$f(x + a_x) = f(x),$$
$$f(y + ay) = f(y),$$
$$f(z + a_z) = f(z), \tag{4.5.21}$$

we set

$$k_x = \frac{n_x \pi}{a_x}$$
$$k_y = \frac{n_y \pi}{a_y}$$
$$k_z = \frac{n_z \pi}{a_z} \tag{4.5.22}$$

where n_x, n_y and n_z are integers.

Thus the quantities k_x, k_y and k_z vary inversely as the direct lattice constants and are, respectively, n_x, n_y and n_z unit reciprocal lattice translations, i.e.,

$$k_x = n_x g_x$$
$$k_y = n_y g_y$$
$$k_z = n_z g_z \tag{4.5.23}$$

Thus we may write (4.5.19) as

$$f(\mathbf{r}) = \sum_n A_n e^{2\pi i[\mathbf{G}\cdot\mathbf{r}]} \tag{4.5.24}$$

where **G** is a vector of the reciprocal lattice. The coefficients A_n are of course given by

$$A_n = \frac{1}{\hat{a}_x \cdot \hat{a}_y \cdot \hat{a}_z} \int_{\text{unit cell}} f(\mathbf{r}) e^{-2\pi i(\mathbf{G}\cdot\mathbf{r})}\, d\mathbf{r} \tag{4.5.25}$$

Thus, *the dual space which we have defined as the reciprocal lattice space is identical with the space we have been calling "k-space" or momentum space.*

The simplest approximation for the wave function of an electron travelling in a lattice is a linear combination of plane waves, each of which satisfies the Bloch condition, e.g.

$$e^{i[\mathbf{k}+\mathbf{G}]\cdot\mathbf{r}} \tag{4.5.26}$$

where **G** is a reciprocal lattice vector. Thus we could choose the linear combination wave function

$$\Phi_{\mathbf{k}}(\mathbf{r}) = \sum_{\mathbf{G}} A(\mathbf{G}) e^{i(\mathbf{k}+\mathbf{G})\cdot\mathbf{r}} \tag{4.5.27}$$

as an approximate solution to our problem of describing the state of an electron travelling in a crystal.

Because of the large strength of the crystalline potential close to the lattice atoms, however, the character of the wave function changes drastically in this region. In the region close to the lattice sites the wave function takes on an atomic character and oscillates rapidly. At large distances from the lattice atoms it takes on the aspect of a slowly modulated free electron (see Figure 4.22). Thus to characterize the true behaviour of the electron,

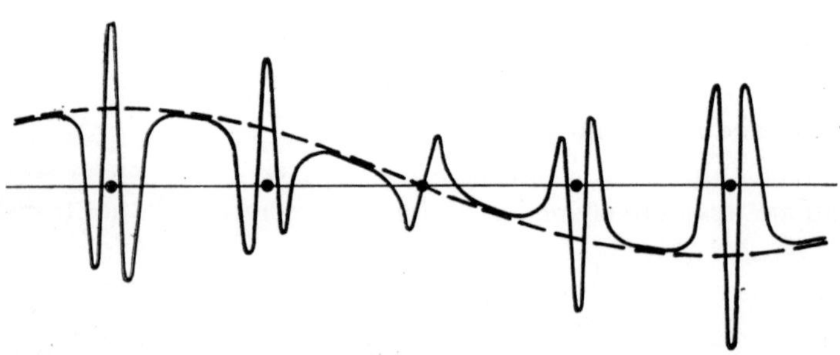

Figure 4.22 Wave function of an electron in a crystal showing rapid oscillations near the attice sites and plane wave behaviour between.

one must use many terms in the series (4.5.27). In practice the convergence is found to be poor and intractable for many calculations.

Perhaps a more important reason for the nonutility of the plane wave expansion is the fact that the expansion will in general converge to the lowest energy state for a particular wave vector **k**. Thus, although one hopes to solve for the valence and conduction bands of a crystal, the expansion will converge to one of the bands produced by the inner core electrons. The fact that a plane wave is a good first approximation in many metals is due to the large spatial extent of the slowly varying portion of the electronic wave function as opposed to the very small regions of oscillatory behaviour. Thus it is necessary to find a different method by which we can easily approximate the character of the electron throughout the entire crystal space.

The Orthogonal Plane Wave Method (OPW)

In 1940, C. Herring described a method by which one could properly characterize the state of an electron both near and far from the atomic lattice sites by the addition of an atomic-like character to the plane wave function close to sites.[12]

The technique begins somewhat similarly to the one described in the preceding section. The valence electron state is approximated by a combination of plane waves. These plane wave states, in the present case, however, are chosen such that they are orthogonal to the eigenstates of the atomic core electrons. (These core eigenstates may be approximately determined by a Hartree, or Hartree–Fock, calculation.) In this case, of course, the lowest root of a variational calculation will tend toward the correct energy eigenvalue, i.e., the lowest valence state. In addition, the intentional application of an atomic-like character near the nucleus must produce more rapid convergence than the expansion of plane waves to an accurate valence state.

The method suggested by Herring is based on the following assumptions:

(1) In the region near the nucleus, the atomic potential is so strong that the states near the nucleus will not differ appreciably in energy from one another.

(2) In general, the nth valence electron state $|\phi_{m\mathbf{k}}(\mathbf{r})\rangle$ is too rapidly varying to be quickly approximated by a few terms of an expansion. Our calculations, however, will be approximately valid if we construct states which represent our valence electron bands orthogonal not to the true $|\phi_{m\mathbf{k}}(\mathbf{r})\rangle$ but to states $|\phi'_{m\mathbf{k}}(\mathbf{r})\rangle$ which closely approximate the $|\phi_{m\mathbf{k}}(\mathbf{r})\rangle$.

These assumptions allow for the substitution of an approximate valence state which is somewhat more slowly varying that the true $|\phi_{m\mathbf{k}}(\mathbf{r})\rangle$ and thus will converge quickly with the use of a small number of $\mathbf{k}$ values. Thus we could modify (for example) the sharp peak of an s state by adding to it a linear combination of multiples of $1s$ and $2s$ states.

We construct our approximate valence electron states $|\phi'_{m\mathbf{k}}(\mathbf{r})\rangle$ in the following manner

$$|\phi'_{m\mathbf{k}}(\mathbf{r})\rangle = |\phi_{m\mathbf{k}}(\mathbf{r})\rangle - \sum_j a_j |\Omega_j\rangle \tag{4.5.28}$$

where the state $|\Omega_j\rangle$ is the jth core atomic orbital state. Thus our rapidly varying valence band state is modified by the deletion of a few rapidly varying terms. We may now represent the approximate state $|\phi'_{m\mathbf{k}}(\mathbf{r})\rangle$ by a few terms of the Fourier series of plane wave states

$$|\phi'_{m\mathbf{k}}(\mathbf{r})\rangle = \sum_{\mathbf{G}} A_{\mathbf{G}} e^{i(\mathbf{k}+\mathbf{G})\cdot\mathbf{r}} \tag{4.5.29}$$

Substitution of equation (4.5.29) into equation (4.5.28) yields

$$|\phi_{m\mathbf{k}}(\mathbf{r})\rangle = \sum_j a_j |\Omega_j\rangle + \sum_{\mathbf{G}} A_{\mathbf{G}} e^{i(\mathbf{k}+\mathbf{G})\cdot\mathbf{r}} \tag{4.5.30}$$

where our valence electron eigenstate $|\phi_{mk}\rangle$ is assumed to be orthogonal to the $|\Omega_j\rangle$. It follows that

$$|\phi_{m\mathbf{k}}(\mathbf{r})\rangle \approx \sum_{\mathbf{G}} A_{\mathbf{G}}\, e^{i(\mathbf{k}+\mathbf{G})\cdot\mathbf{r}} \tag{4.5.31}$$

where now the plane wave states must also be orthogonal to each of the core states $|\Omega_j\rangle$. State (4.5.31) is then utilized in the calculation of the band structure.

Although the above OPW method has proven quite valuable for the calculation of energy band parameters in semiconductors such as Ge and Si, it fails badly in the calculation of lattice bonding energies as we shall see later.

The Augmented Plane Wave Method (APW)

Another successful technique has been devised in order to properly describe both the atomic and plane wave nature of the conducting electron near and far from a lattice sites, respectively. Originally proposed by Slater[13], it has been extended by the work of Saffren and Slater.[14] The method has become quite tractable and has been successful in predicting the energy band structure of metals.[15]

In this scheme, a Wigner-Seitz type sphere is constructed about each lattice site. In order to characterize the wave function far from the nucleus, the radius of this sphere is assumed to be somewhat smaller that the radius of the actual Wigner-Seitz sphere and, as a consequence, the interstitial volume is slightly larger. In the regions interstitial to the Wigner-Seitz type spheres, the potential is assumed to be constant and the electronic state is approximated by a plane wave state. Inside of the spheres, in the region near to the nucleus, the electronic state is again assumed to be of an atomic nature. These wave functions are then joined smoothly at the surfaces of the spheres. The resulting Bloch type wave function is then used in a variational type calculation to determine the allowed energy as a function of **k**.

PROBLEM

For a crystal lattice, show that if an atomic orbital state (valid inside of a Wigner-Seitz type sphere) is joined continuously with a plane wave state (valid outside of the Wigner-Seitz type sphere) at the surface of the sphere, the first normal derivative of the resulting wave function must be discontinuous at the surface.
Hint! The solution $\varphi(\mathbf{r})$ of the Schrödinger equation inside the sphere is

$$\phi(\mathbf{r}) = \sum_{l,m} C_{l,m} R_{(r,l)} Y_{l,m}(\theta, \varphi) \tag{4.5.32}$$

When this solution is matched to the plane wave solution in terms of spherical harmonics

$$e^{i\mathbf{k}\cdot\mathbf{r}} = \sum_{l=0}^{\infty} (2l+1)\, i^l \frac{R_l(\mathbf{r})}{kr} P_l(\cos\theta) \tag{4.5.33}$$

one finds

$$C_{l,m} = (2l+1)\, i^l \left[\frac{j_l(k, R_s)}{R_l(R_s E)}\right] Y^*_{l,m}(\theta', \varphi') \tag{4.5.34}$$

The k · p Method

The **k** · **p** method is an extremely convenient perturbation method by which one can determine very accurately the shape of the E vs **k** curves near an energy extremum.

The Schrödinger equation of the Bloch function $U_{\mathbf{k}}(\mathbf{r})$ is

$$\left(\frac{p^2}{2m} + V(\mathbf{r})\right) U_{\mathbf{k}}(\mathbf{r}) + \frac{\hbar}{m} \mathbf{k} \cdot \mathbf{p} U_{\mathbf{k}}(\mathbf{p}) = E U_{\mathbf{k}}(\mathbf{r}) \tag{4.5.35}$$

where $p = -i\hbar\nabla_{\mathbf{r}}$ is the crystal momentum and

$$E = E_k - \frac{\hbar^2 k^2}{2m^*} \tag{4.5.36}$$

PROBLEM

Prove equation (4.5.35) Equation 4.5.35 may be written

$$\mathscr{H}_0 U_{\mathbf{k}}(\mathbf{r}) + \mathscr{H}_1 U_{\mathbf{k}}(\mathbf{r}) = E U_{\mathbf{k}}(\mathbf{r}) \tag{4.5.37}$$

One can solve (4.5.37), with the perturbing Hamiltonian $\mathscr{H}_i$ deleted, quite easily near $\mathbf{k} = 0$. The effect of the $\mathbf{k} \cdot \mathbf{p}$ term may then be treated by the usual perturbation technique. In a crystal in which there is inversion symmetry through the center of the Brillouin zone (as in the diamond lattice), the point $\mathbf{k} = 0$ must correspond to an extremum of the energy. Because of this symmetry, the first order term in the energy is equal to zero. The second order corrections to the energy, then are given by the matrix elements

$$\langle U_{j,i}| \frac{\hbar}{m^*} \mathbf{k} \cdot \mathbf{p} | U_{j,k}\rangle = \sum_{\substack{m \\ l \neq j}} \frac{\langle U_{ji}| \frac{\hbar}{m^*} \mathbf{k} \cdot \mathbf{p} | U_{l,m}\rangle \langle U_{l,m}| \frac{\hbar \mathbf{k} \cdot \mathbf{p}}{m^*} | U_{j,s}\rangle}{E_j - E_l} \tag{4.5.38}$$

where the state U_{ji} is the jth energy state, and m refers to the number of degeneracies.

The above neglects the spin-orbit interaction which introduces a term into $\mathscr{H}$ of the form

$$\frac{\hbar^2}{4m^2c^2} \sigma \cdot [\nabla \mathbf{V} \times \mathbf{k}] \tag{4.5.39}$$

where σ is the spin operator.

Kane has determined the band structures of Ge and Si[16] and of Indium Antimonide[17] by the above method, and found large deviations from the parabolic behaviour predicted by the tight binding model.

The Pseudopotential Method[18]

Another perturbation method useful for the determination of energy bands in solids is the pseudopotential method. As before, we expand the true electron state $|\psi\rangle$ in both plane wave states $|\mathbf{k}\rangle$ and core states $|\phi_i\rangle$.

$$|\psi\rangle = \sum_k a_k |\mathbf{k}\rangle + \sum_i a_i |\phi_i\rangle \tag{4.5.40}$$

The Schrödinger equation for an electron travelling in the lattice becomes

$$-\frac{\hbar^2}{2m} \nabla^2 |x\rangle + V|x\rangle + \sum_i (E_i - E) a_i |\phi_i\rangle = E|x\rangle \tag{4.5.41}$$

where E_i is the energy of the ith core state and we have written the plane wave portion of (4.5.40) as $|x\rangle$. Multiplication of (4.5.41) from the left by $\langle\phi_j|$ yields

$$a_j = -\langle\phi_j|x\rangle \tag{4.5.42}$$

Substituting (4.5.42) into (4.5.41) yields the pseudopotential equation

$$-\frac{\hbar^2}{2m}\nabla^2 |x\rangle + W|x\rangle = E|x\rangle \tag{4.5.43}$$

where

$$W = V + \sum_i (E - E_i)\,|\phi_i\rangle\,\langle\phi_i| \tag{4.5.44}$$

is the ***pseudopotential*** which has the advantage that it is a small term. The energy E will always be greater than the core state energy E_i, thus $(E - E_i)$ tends to cancel the negative ion core attraction $V(\mathbf{r})$. $|x\rangle$ is called the pseudo wavefunction. In the region between the ion cores, it is an exact solution to (4.5.43) with W removed. The effect of the pseudopotential may then be taken into account via perturbation techniques.

Although (4.5.43) is an exact equation, it is easily shown that replacement of the pseudopotential by any other will still yield an exact equation. If we let a new W' be of the form

$$W' = V + \sum_i G_i\,|\phi_i\rangle\,\langle\phi_i| \tag{4.5.45}$$

where G is an arbitrary function, we find substitution into (4.5.43) and multiplication from the left by $\langle\psi|$ yields

$$\langle\psi|\,\mathscr{H}\,|x\rangle + \sum_i G_i\,\{\langle\psi|\phi_i\rangle\,\langle\phi_i|x\rangle\} = E'\,\langle\psi|x\rangle \tag{4.5.46}$$

The true state $\langle\psi|$ is orthogonal to all $|\phi_i\rangle$, thus the second term drops out no matter what the form of G_i. This yields again an exact equation

$$E\langle\psi|x\rangle = E'\,\langle\psi|x\rangle \tag{4.5.47}$$

Thus there is no absolute pseudopotential, but a class of pseudopotentials which yields useful results. This is, of course, to be expected is we recall that the pseudopotential W must necessarily be only a small correction to a free state. The success of the pseudopotential method gives renewed faith in the loose binding model of conducting electrons.

4.6 BAND STRUCTURE OF SOME TYPICAL SEMICONDUCTORS

The detailed information concerning the band structure of semiconductors is not the result of a single calculation based upon one or another of the foregoing methods. Instead, the knowledge we have concerning the structure of the energy bands is a result both of independent calculations based on more than one of the models described above, and of very many types of experiments which are yet to be discussed.

For this reason, one must make a choice as to whether to present the detailed band pictures of some typical semiconductors in the present section, or to wait until the relevant experiments have been discussed and present the pictures as a whole.

I have chosen to do the former. Those aspects of the band structures which are derived from experiments not yet discussed will be discussed fully in later sections.

Homopolar Semiconductors

1. Germanium and Silicon

Theoretical calculations based on the OPW and $\mathbf{k} \cdot \mathbf{p}$ methods have been carried out for Ge by Herman[19] and for Ge and Si by Cardona and Pollack[20] respectively.

Figures 4.23 a and 4.23 b show the E vs $\mathbf{k}$ curves for Ge and Si, respectively, both of which have the diamond structure. These drawings are schematic and represent the compilation of both experimental and theoretical investigations. They include the effect of spin-orbit interaction.

In both Ge and Si the maxima of the valence band falls at $\mathbf{k} = 0$. The valence bands of these crystals are actually composed of three separate bands, two of which are degenerate at $\mathbf{k} = 0$. The third is separated at $\mathbf{k} = 0$ from these by being lowered 0.28 eV and 0.044 eV in the case of Ge and Si, respectively, by the effect of the spin-orbit interaction. In the absence of the spin-orbit interaction, the two upper valence bands shown in the figure are degenerate at all points in k space with one another for both Ge and Si.

If we use the form of the valence bands near the extrema of the two degenerate valence bands as given by Dresselhause, *et al.*[21] and Elliot[22], the energy is given by

$$E(k) = -\frac{\hbar^2}{2m^*}\left[Ak^2 \pm \sqrt{B^2k^4 + C^2(k_x^2k_y^2 + k_y^2k_z^2 + k_z^2k_x^2)}\right] \tag{4.6.1}$$

where A, B and C are constants and $|\mathbf{k}| = \sqrt{k_x^2 + k_y^2 + k_z^2}$. We can get a feeling for the shapes of the degenerate valence bands near $\mathbf{k} = 0$ by the information in Table 4.2.[23] The negative sign refers to the highest valence band and the positive sign to the next highest for both Si and Ge. Note that the highest valence band is extremely anisotropic, whereas the next highest is nearly spherical.

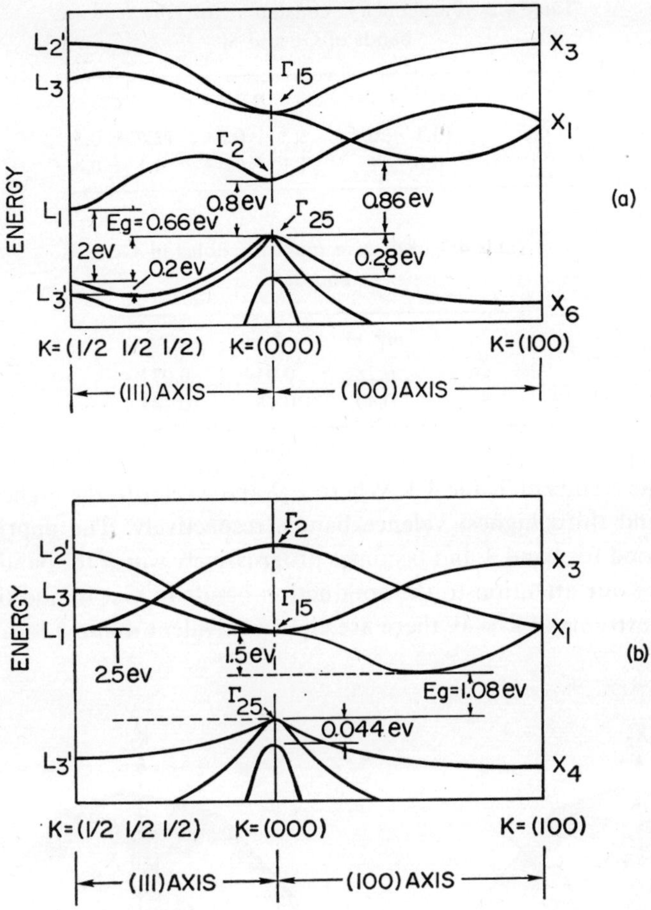

Figure 4.23 The energy band structure of (a) Ge and (b) Si.

From our previous results dealing with the effective mass of charge carriers in crystals, we would expect the warping of the degenerate valence band energy surfaces near $\mathbf{k} = 0$ to manifest itself in different effective masses in these bands. By using the assumption of spherical isotropy for each of the degenerate bands, i.e.,

$$E_k = -\frac{\hbar^2}{2m^*}\left[A \pm \sqrt{B^2 + C^2/6}\right]^{\frac{1}{2}} k^2 \qquad (4.6.2)$$

Table 4.2 Parabolicity constants for the valence bands of Ge and Si

	A	B	C
Ge	13.1 ± 0.4	8.3 ± 0.6	12.5 ± 0.5
Si	4.0 ± 1	1.1 ± 0.4	4.1 ± 0.4

Table 4.3 Effective mass for holes in Ge and Si

	m_1^*/m	m_2^*/m	m_3^*/m
Ge	0.28	0.049	0.077
Si	0.49	0.16	0.245

we find the results of Table 4.3. Where 1, 2 and 3 refer to the highest, second highest and third highest valence bands, respectively. The approximation is quite good for band 3, but becomes progressively worse for bands 2 and 1.

Turning our attention to the conduction bands in Ge, in addition to the required extreme at $\mathbf{k} = 0$, there are eight equivalent minima located in the

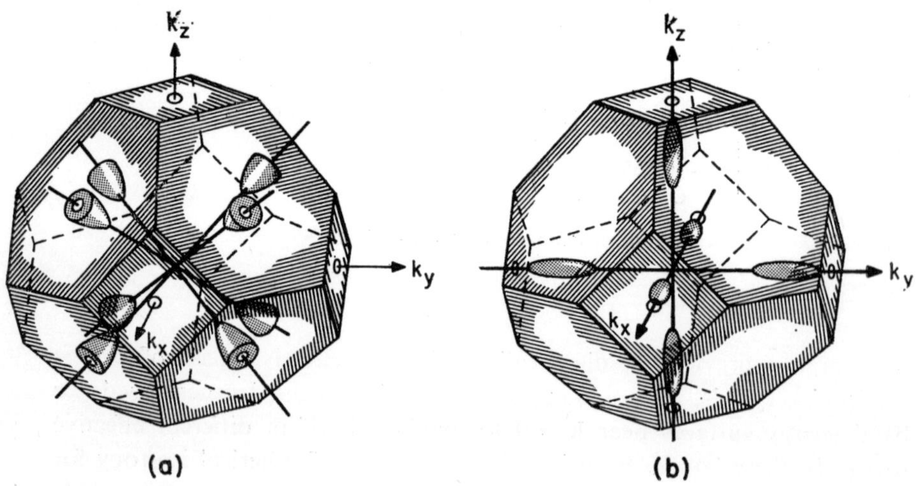

Figure 4.24 (a) The eight constant energy conduction band ellipsoids along the $\{111\}$ axes in Ge. (b) The six constant energy conduction band ellipsoids along the $\{100\}$ axis in Si. (From J. P. McKelvey, *Solid State and Semiconductor Physics*, Harper and Row, New York, 1966, p. 295.)

(111) directions at the edge of the Brillouin zone (Figure 4.24a). These occur at the center of the hexagonal face of the Brillouin zone, and are the lowest extrema of the conduction band (0.66 above the valence band at room temperature). They define the so-called thermal energy gap. These extrema are, to a good approximation, ellipsoids of revolution. There are also 6 slightly higher (100) minima.

For Si there are six (equivalent) conduction band minima about 8 of the way along the Brillouin zone. These define the thermal energy gap of 1.08 eV at room temperature (Figure 4.24b).

The transverse and longitudinal effective mass ratios m_t^* and m_l^*, respectively, are as shown in Table 4.4 for electrons at the extrema[24] where m_d^* is the density of states effective mass $(m_l^* m_t^{*2})^{1/3}$.

Table 4.4 Longitudinal (m_l^*) and transverse (m_t^*) electron effective masses in Ge and Si. m_d^* is the density of states effective mass

	m_l^*/m	m_t^*/m	m_d^*/m
Ge	1.64 ± 0.003	0.0819 ± 0.003	0.22
Si	0.98 ± 0.004	0.19 ± 0.01	1.33

An interesting phenomena occurs when mixed crystals of Ge and Si are grown. As the percentage of Si in Ge is increased, as shown in Figure 4.25, the (111) minima rise rapidly. At about 15 atomic percent Si, the rapid rise of the (111) minima has brought it to the same energy as the more slowly moving (100) minima. Above 15 atomic percent Si, we see the slower rise of the (100) minima which remains the lowest-lying band for all higher concentrations of Si.

Heteropolar Semiconductors

Energy Band Structure of GaAs and GaP

GaAs is a typical heteropolar semiconductor which crystallizes in the zincblende structure. An early determination of the energy band structure of GaAs was made by Calloway[25], who determined the general structure of the bands by use of a perturbation technique applied to Herman's OPW results on Ge.[19] An exact determination of the band structure has been performed by Pollack, *et al.*[26]

The thermal energy gap lies at the center of the Brillouin zone 1.53 eV above the top of the valence bands (Figure 4.26a). A subsidiary conduction band minimum, 0.36 eV above the $k = 0$ point, is located at the (100) edge of the Brillouin zone. Symmetry considerations require that there be three equivalent minima of this type. The effective masses of the holes in the spin-orbit-split valence bands V_l and V_k have been measured to be 0.12 m and 0.68 m respectively.

In GaP, however, the smallest energy gap lies 0.35 eV below the direct band edge and is located on the (100) direction (in which case symmetry considerations require six equivalent minima within the zone or at the band edge as Figure 4.26b).

Mixed crystals of $GaAs_xP_{1-x}$ show the transition from direct to indirect band gap at approximately 45% GaP[27] as shown in Figure 4.27.

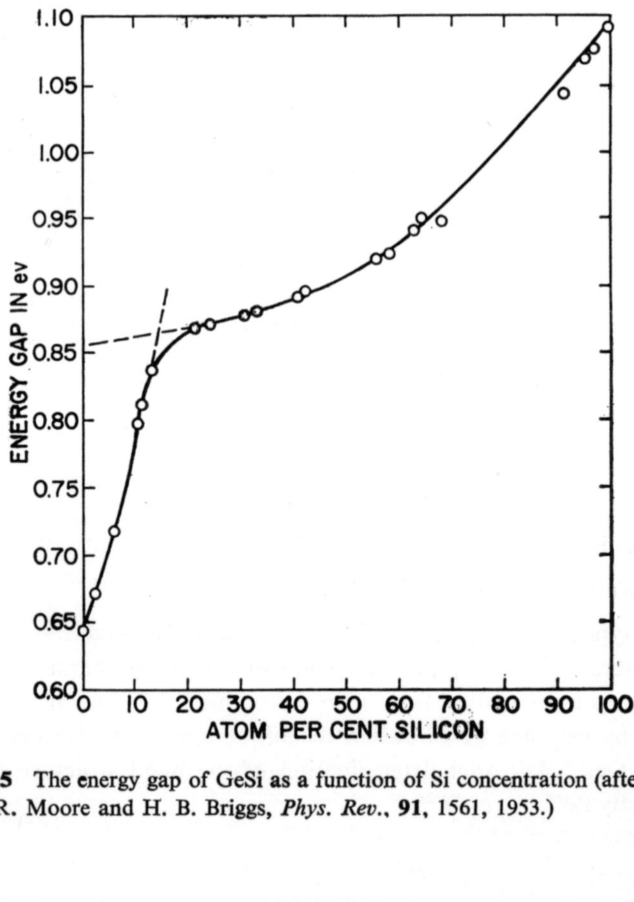

Figure 4.25 The energy gap of GeSi as a function of Si concentration (after R. Braunstein, A. R. Moore and H. B. Briggs, *Phys. Rev.*, **91**, 1561, 1953.)

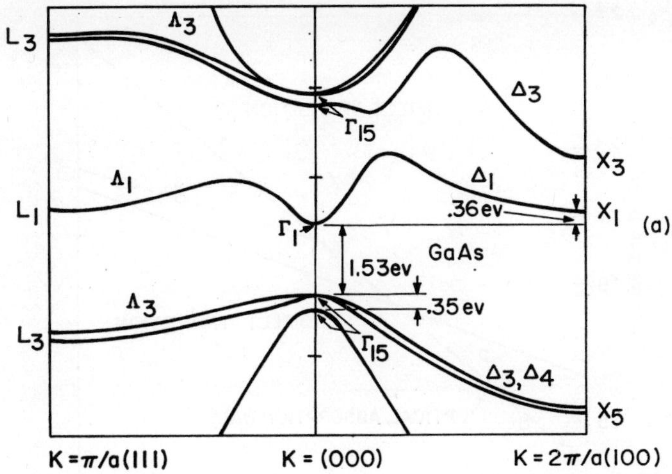

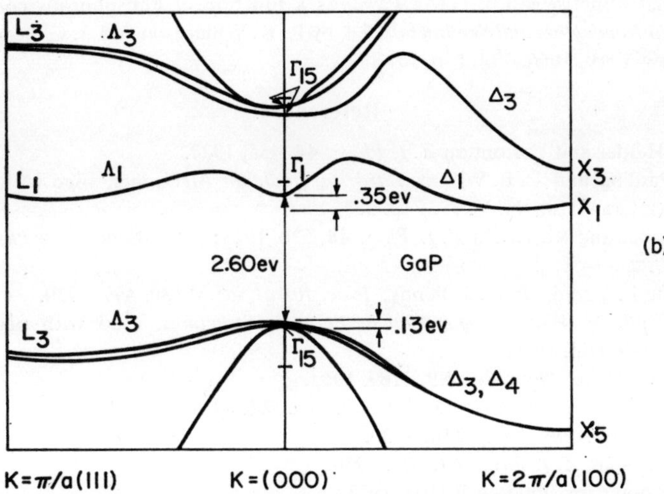

Figure 4.26 The energy band structure of (a) GaAs and (b) GaP. (after Pollack, et. al.)

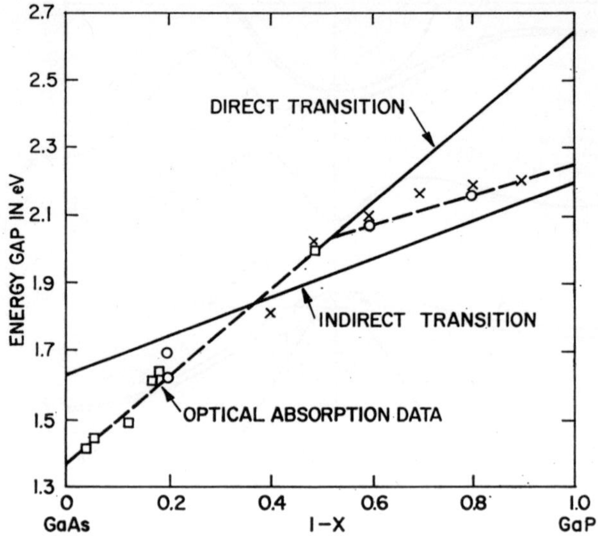

Figure 4.27 Energy gap of $GaAs_xP_{(1-x)}$ as a function of Phosphorous concentration. (From *Semiconductors and Semimetals*, Ed. by R. K. Willardson and A. C. Beer, Academic Press, New York, 1966, Vol. I, p. 151.)

References

1. W. Heitler and F. London *Z. f. Phys.*, **44**, 455, 1927.
2. L. Pauling and E. B. Wilson, *Intro. to Quantum Mechanics*, used with permission of McGraw-Hill, N. Y. 1935 (p. 327).
3. Wittmer and Rosenfeld *Z. f. Phys.* **48**, 530, 1928; Felix Bloch, *Z. f. Phys.*, **52**, 555, 1929.
4. R. de L. Kronig & W. J. Penny, *Proc. Royal Soc.* A**130**, 499, 1930.
5. L. Brillouin, *Wave Propogation in Periodic Structures*, used with permission of McGraw-Hill, 1946.
6. L. Van Hove, *Phys. Rev.* **89**, 1189, 1953.
7. D. R. Hartree, *Proc. Cambridge Phil. Soc.*, **24**, 89, 111, 1928.
8. J. C. Slater, *Phys. Rev.* **210**, 1930.
9. E. V. Fock, *Z. f. Phys.*, **61**, 126, 1930.
10. T. Koopmans, *Physica*, **1**, 104, 1933.
11. E. Wigner and F. Seitz, *Phys. Rev.*, **43**, 804, 1933; **46**, 509, 1934.
12. C. Herring, *Phys. Rev.*, **57**, 1169, 1940.
13. J. C. Slater, *Phys. Rev.*, **51**, 846, 1937.
14. M. M. Saffren and J. C. Slater, *Phys. Rev.*, **92**, 1126, 1953.
15. See for instance H. Schlosser, Ph. D. Thesis, Carnegie Inst. of Technology (unpublished) for the application of the APW to Lithium and Sodium.
16. E. O. Kane, *J. Phys.* Chem. Solids, **1**, 83, 1956.
17. E. O. Kane, *J. Phys.* Chem. Solids, **1**, 249, 1956.
18. This section follows the treatment given by W. A. Harrison, *Phys. Today*, **22**, 23, 1969.

19. F. Herman, *Phys. Rev.*, **93**, 1214, 1954.
20. M. Cardona and F. H. Pollack, *Phys. Rev.*, **142**, 530, 1966.
21. G. Dresselhaus, A. F. Kip and C. Kittel, *Phys. Rev.*, **95**, 568, 1954; **98**, 368, 1955.
22. R. J. Elliot, *Phys. Rev.*, **96**, 268, 1954; **96**, 280, 1954.
23. R. N. Dexter, H. J. Zeiger and B. Lox, *Phys. Rev.*, **104**, 637, 1956.
24. R. N. Dexter, H. J. Zeiger and B. Lox, *Phys. Rev.*, **104**, 637, 1956. See also Dresselhaus, Kip and C. Kittel, *Phys. Rev.*, **98**, 368, 1955.
25. J. Calloway, *Jnl. Electronics*, **2**, 330, 1957.
26. F. H. Pollack, C. W. Higgenbotham and M. Cardona, Proc. International Conf. on the Physics of Semiconductors, Kyoto, 1966. (*Jnl. Jap. Phys. Soc.*, Vol. 21, Supplement, 1966).
27. See for instance *Semiconductors and Semimetals*, Vol. I, p. 143, Academic Press, N. Y. 1966, Ed. by R. K. Willardson and A. C. Beer.

CHAPTER 5

Impurities and Disorder in Crystals

5.1 SUBSTITUTIONAL IMPURITIES

UNTIL NOW, we have been discussing the allowed energy states of perfect crystals. These are theoretically pure materials in which no unwanted impurity atoms or defects exist within the crystalline lattice. The present state of the art of fabricating single crystal materials, however, is not yet advanced enough to produce absolutely pure and impurity free materials. As a matter of fact, many of the desirable conducting properties of semiconducting materials arise from the intentional "doping" of crystals with impurities. For this reason, investigations into the galvanometric or optical properties of materials must invariably take into account the presence of unwanted impurities or defects. These impurities may either take the place of host atoms in the lattice or rest between lattice sites, in which cases they are called *substitutional* or *interstitial*, respectively. Both perturb the long range lattice order, and thus introduce perturbing potentials at the impurity location as shown in Figure 5.1. These disruptions of the ordered crystalline potential are localized regions in the crystal at which our requirement for Bloch-type periodicity is no longer valid. Thus to describe the states of an electron at these centers of perturbation we must turn to means other than those already discussed. Let us treat the simpler case of substitutional impurities.

For the case of homopolar semiconductors such as Ge or Si, each lattice site is, as we have found, bound to each of its four nearest neighbors by a "saturated" covalent bond. In the case where a substitutional impurity atom of valence $M > 4$ takes the place of one of the host atoms, the number of valence electrons in excess of the number required for bonding to each of the four nearest neighbors $(M - 4)$ will not be covalently bound. This situation for the case $M = 5$ (e.g., the $4s^2$, $4p^3$ electrons As in Si) is shown in Figure 5.2.

In treating this extra electron we can, as a first approximation, assume that it is bound to the impurity atom by its coulombic attraction to the nucleus and take into account the effect of the lattice on this binding force only by inclusion of the dielectric constant of the host lattice and an effective mass m^* for the bound electron.

We shall assume that the potential $V_{(r)}$ which attracts the extra electron to the nucleus is spherically symmetric and is of the form $V = -e/\varkappa r$, where $\varkappa$ is the dielectric constant of the host lattice.

The radial solution to (1.3.56) becomes (see reference 5, Chapter I)

$$R_{n,l}(r) = e^{\left[-\frac{rm^*e^2}{n\varkappa\hbar^2}\right]} \left[\frac{2rm^*e^2}{n\varkappa\hbar^2}\right]^l \mathscr{F}_{\left(l+1-n,\ 2l+2,\ \frac{2rm^*e^2}{n\varkappa\hbar^2}\right)} \tag{5.1.1}$$

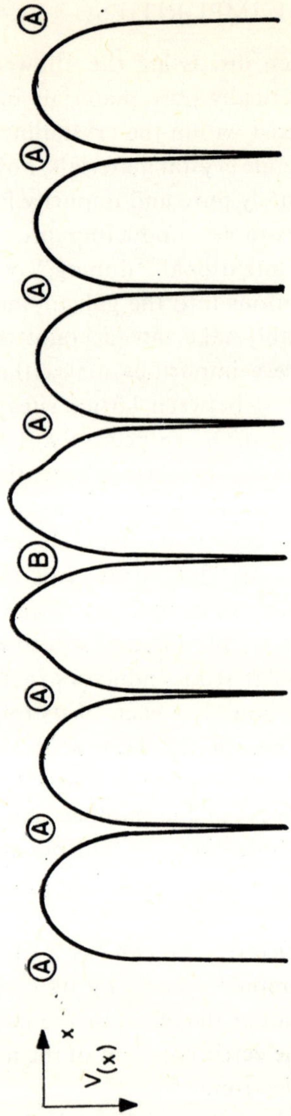

Figure 5.1 Crystalline potential perturbed by an impurity.

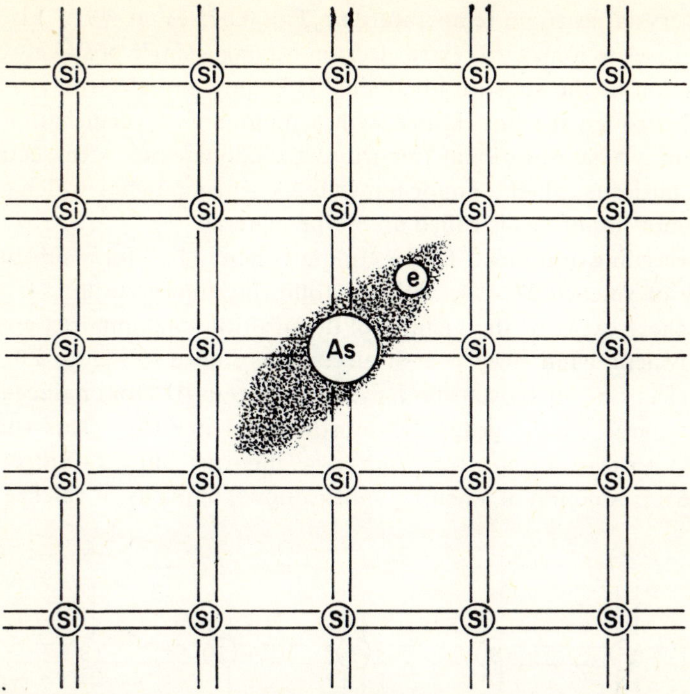

Figure 5.2 Unionized substitutional arsenic impurity (donor) ($m = 5$) in Si.

The binding energy E_B is given by

$$E_B = -\frac{e^4 m}{2\hbar^2 \varkappa^2 n^2} \tag{5.1.2}$$

where we have assumed perfect shielding due to the impurity core electrons.

This binding energy is of course similar to that obtained by the simple Bohr theory. However, in the present case, the binding energy is reduced by the square of the dielectric constant $\varkappa$ of the host lattice. The first ionization potential for phosphorous is 10.9 eV in vacuum. In Ge, ($\varkappa = 16$) the ionization potential calculated from (5.1.2) is 0.008 eV which is of the order of magnitude of observed values. It should be pointed out that the hydrogenic approximation for the impurity atom yields a binding energy which is relatively independent of the particular impurity chosen. It depends only on the dielectric constant of the host material.

Thus the effect of the dielectric constant is to lower the free state binding energy substantially, so that a small amount of thermal energy introduced into the crystal at room temperature ($k_0T = 0.025$ eV at 300°K) is enough in many cases to ionize this extra electron leaving a single positively charged impurity ion at the substitutional site. This unbound electron is now free to travel through the host lattice with a minimum of perturbation by the host atoms. A substitutional impurity which donates an extra electron into the host lattice is called a donor impurity. A semiconductor which contains many donor impurities is called an n-type material.

The preceding arguments follow "mutatis mutandis" for a substitutional impurity of valence $M < 4$; with the following interpretation: If we take $M = 3$, there exists in the vicinity of the substitutional impurity an unsaturated covalent band due to a Si electron as shown in Figure 5.3. This Si valence electron which orbits both the Si and group III atom induces a small positive charge on the group III substitutional impurity. This charge, in turn, acts to trap an electron from a neighboring Silicon–Silicon bond. Thus the introduction of a neutral substitutional impurity of valence $M < 4$

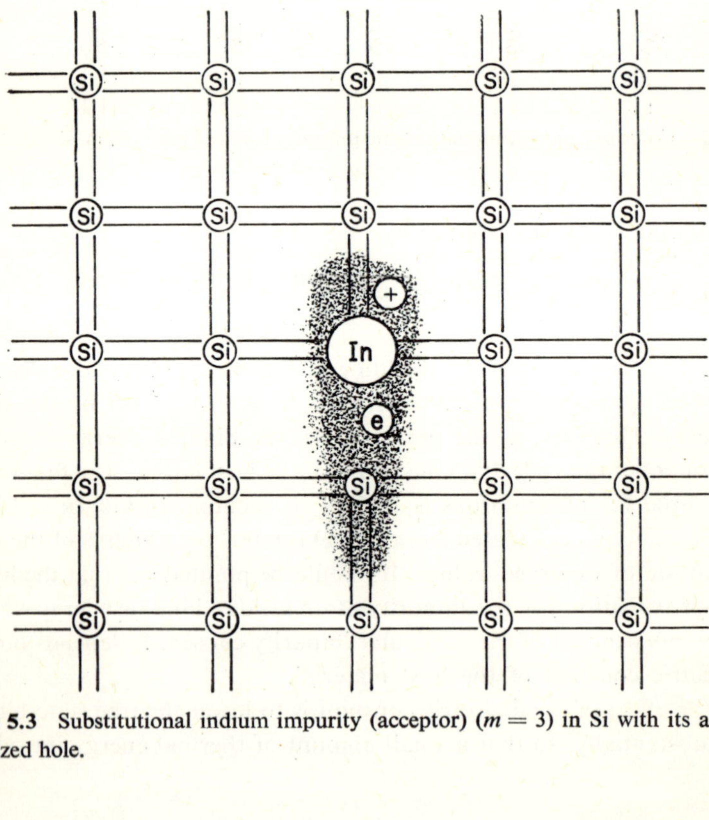

Figure 5.3 Substitutional indium impurity (acceptor) ($m = 3$) in Si with its associated unionized hole.

is equivalent to the introduction of an electronic potential well into which a neighboring electron from Si – Si bond can migrate. An impurity which tends to trap electrons in this manner is called an *acceptor* impurity. A semiconductor which contains many acceptor impurity atoms is called a *p-type* material. The acceptor impurity and extra electron combination is called an *ionized acceptor* and is negatively charged. The ionization energy of this acceptor impurity is most conveniently discussed in terms of the binding energy of the induced positive charge (a *hole*) which, in the process of ionization from the acceptor impurity, has migrated to the Si – Si bond from which the ionizing electron came. If we assume that the positive hole is bound in the coulombic field, of a negatively charged acceptor nucleus, we can solve for the ionization energy of the induced positive charge at the acceptor site again using a potential of the form $V = +e^2/\varkappa r$. Of course this binding energy will be of the form of (5.1.2). It should be noted that once the hole has made the transition from the acceptor impurity atom to a neighboring Si – Si bond, further migration of the hole from one Si – Si bond to another may be made without the expenditure of any energy because the initial state is indistinguishable from the final state. Thus, after ionization, the electron or hole is free to move through the crystal.

The zero in energy for impurity electrons is taken at the minimum of the conduction band at $\mathbf{k} = 0$, for above this level the electron will be free. We must remember, however, that to find the ionization energy E_a of the bound acceptor hole, we inverted our energy scale, thus the zero in hole energy is taken as the maximum of the valence band. This is quite plausible, since the energy necessary to ionize the acceptor hole must be equal to the energy of the ionizing electron. This ionizing electron migrates from one of the saturated covalent bonds which make up the valence band of the crystal.

A class of impurity atoms known as amphoteric impurities may act either as donors or acceptors. This behaviour may come about in three ways: (1) an impurity may act as a donor or acceptor depending on whether it is substitutional or interstitial. (2) In a compound semiconductor such as GaP, substitution of an impurity atom with $M = 4$ for either P ($M = 5$) or Ga ($M = 3$) will yield an acceptor or donor impurity site, respectively. (3) An impurity capable of multiple ionization may act as donor or acceptor depending on the degree of ionization. An example is Au which shows three acceptor levels and one donor level in Ge.

One may show that the screening of the coulomb potential of an impurity atom by the free electrons in the crystal reduces the effective range of the impurity potential to the form

$$V_{(r)} = \frac{e^2}{\varkappa r} \exp\left[-\frac{r}{\lambda}\right] \qquad (5.1.3)$$

where λ is the 'screening length' or the effective range of the perturbing potential and is given by

$$\lambda = 4\pi e^2 g_{(\mu)} \qquad (5.1.4)$$

where $g(\mu)$ is the density of states at the Fermi Energy.

The uncertainties in position and momentum of the electron associated with a substitutional impurity must, of course, obey the Heisenberg Uncertainty Principle i.e.,

$$\Delta \times \Delta p \geqq \hbar,$$

$$\Delta \times \geqq \frac{1}{\Delta k} \qquad (5.1.5)$$

Since a substitutional impurity is localized in space, the **k** values for its bound electron must have a large associated uncertainty Thus the introduc-

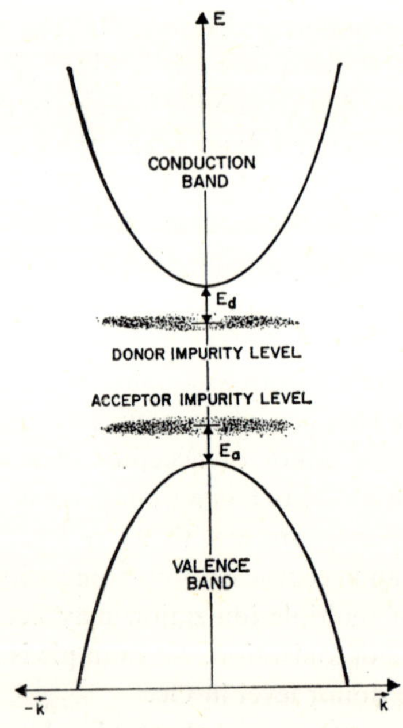

Figure 5.4 Donor and acceptor impurity energy levels with associated uncertainty in **k** space.

tion of substitutional impurities into a crystal lattice produces allowed discrete bound energy states within the forbidden energy gap. These levels are "smeared out" in **k** space as is shown in Figure 5.4.

PROBLEM:

What is she uncertainty Δ **k** about $\mathbf{k} = 0$ of the **k** value for the 3p electron of a phosphorous atom in Ge?

There is another quite useful, but less accurate, method by which the allowed energy levels of a crystal may be shown. The magnitude of the energy gap in a particular direction is plotted vertically vs distance in the direct lattice. Since the abscissa does not change direction, the energy gap remains constant in this scheme. A semiconductor containing both donor and acceptor substitutional impurities and which has an energy gap E_g is shown in Figure 5.5. This type of diagram is also called an energy band diagram. The observed binding energies of typical impurities in Ge, Si and GaAs are shown in Figure 5.6 using this scheme.

We should note, however, that the hydrogenic approximation which led us to equation (5.1.3) for donor and acceptor impurities, also allows for the existence of discrete *excited* impurity energy states. These discrete excited states of donor impurities may lie within the conduction band, just as the excited hole states of the acceptor impurities may lie within the valence band.

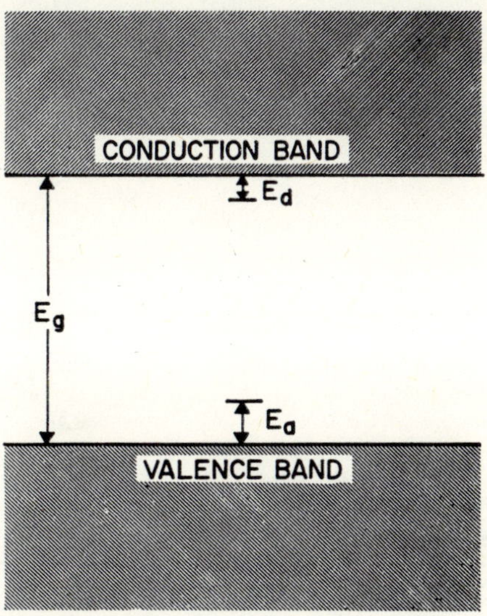

Figure 5.5 Energy *vs* distance type band diagram for a semiconductor with energy gap E_g which contains donor and acceptor impurities at E_d and E_a.

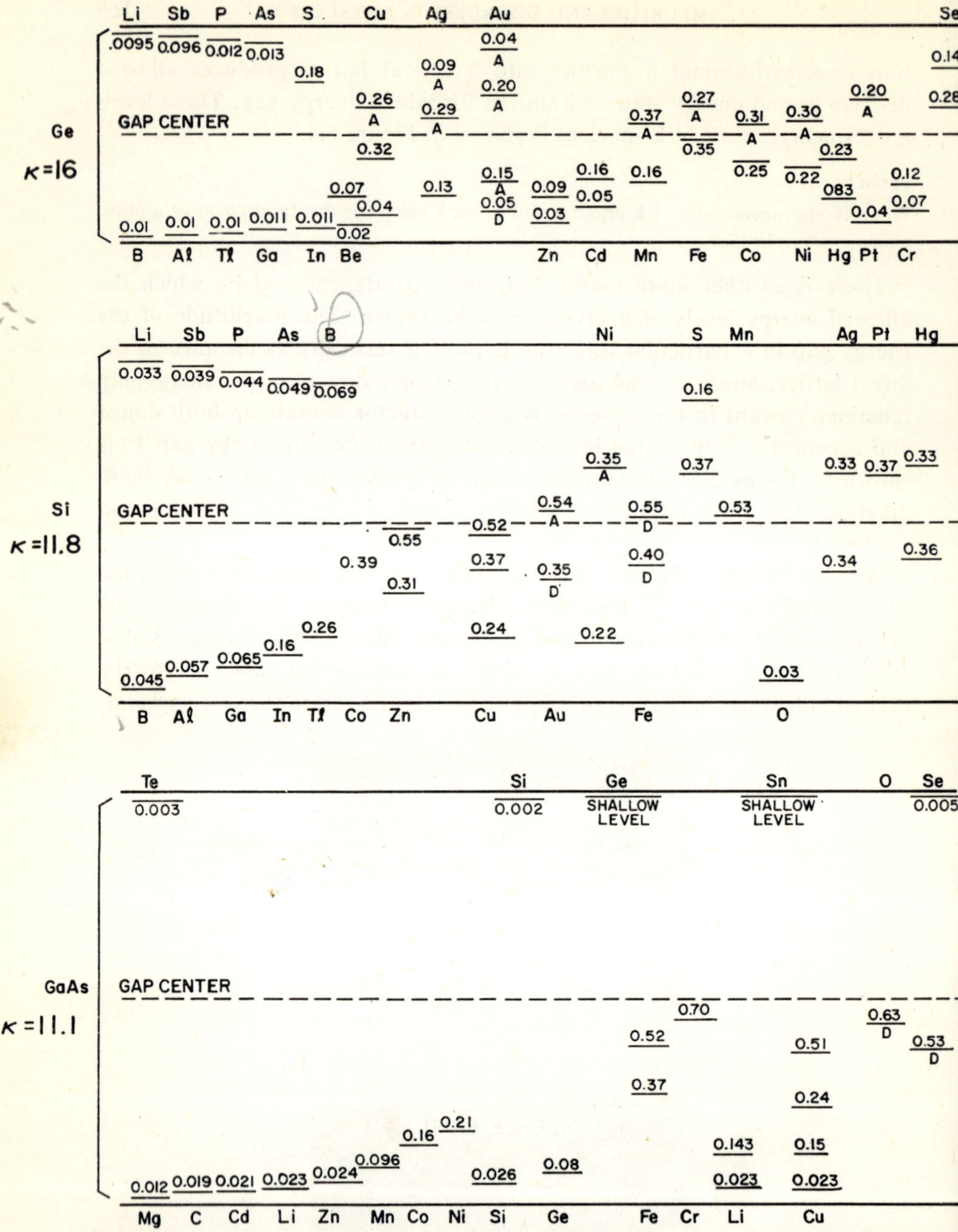

Figure 5.6 Impurity energy levels for various elements in Ge, Si and GaAs. The levels below the gap centers are measured from the top of the valence band and are acceptor levels unless indicated by *D* for donor level. The levels above the gap centers are measured from the bottom of the conduction band and are donor levels unless indicated by *A* for acceptor level [from S. M. Sze and J. C. Irvin, *Solid State Electronics*, **11**, 599, 1968].

5.2 IMPURITY ENERGY BANDS[1]

In the preceding discussion we have shown the existence of discrete bound states which arise from treating substitutional impurities in a hydrogen-like fashion. An interesting specific case of impurity-dominated materials is that of high substitutional impurity densities. We have learned from our treatment of the homopolar bond that the overlap of discrete atomic wave functions produces bands of allowed and unallowed energies due to the splitting of the degeneracy of the atomic wave functions. In a similar fashion the overlap of many impurity electronic wave functions may produce energy bands, in the case of small interimpurity distances, from the allowed discrete impurity energy states lying within the forbidden energy gap.

If we assume that the substitutional impurities in a material are homogeneously and regularly distributed and thus form what may be called an impurity sublattice, we can, by the imposition of the Bloch condition determine the extent of impurity banding as a function of interimpurity distance and thus of impurity concentration.

To make use of the Wigner-Seitz cellular method we shall set up Wigner-Seitz spheres of radius r_s given by

$$r_s = \left(\frac{3}{4\pi N}\right)^{1/3} \tag{5.2.1}$$

and where N is the density of impurity atoms. To determine the widths in energy of the impurity energy bands within the forbidden gap, we shall consider that the width of the bands is determined by the points $k = 0$ and $k = (\pi/a)$, where a is the lattice constant. Conditions (4.5.13), (4.5.15), (4.5.16), and (4.5.17) yield the following conditions on the solutions (5.1.1) to the radial wave equation

$$\left(\frac{1}{n} - \frac{l\varkappa a}{r_s}\right)\mathscr{F}\left(l+1-n,\ 2l+2,\ \frac{2r_s}{n\varkappa a}\right) = \frac{1}{n}\left[\frac{l+1-n}{l+1}\right]\mathscr{F}\left(l+2-n,\ 2l+3,\ \frac{2r_s}{n\varkappa a}\right) \tag{5.2.2}$$

and

$$\mathscr{F}\left(l+1-n,\ 2l+2,\ \frac{2r_s}{n\varkappa a}\right) = 0 \tag{5.2.3}$$

where

$$\mathrm{K}a = \frac{\hbar^2}{m^*e^2}$$

Equations (5.2.2) and (5.2.3) determine $\mathbf{k}$ and thus E as a function of r_s and thus of the impurity concentration. The width of the $1s$, $2s$ and $2p$ impurity energy bands within the energy gap are shown in Figure 5.7 as a function of $r_s/\varkappa a$. Thus we see that the discrete energy levels produced by the non-interacting impurities begin to widen and form bands as the number of impurities is increased. The impurity concentration at which the impurity band merges with the conduction band is determined by setting the impurity ionization energy equal to zero ($n = \infty$), yielding

$$(2l + 2)\left[\frac{8r_s}{\varkappa a}\right]^{-\frac{1}{2}} J_{2l+1}\left[\left(\frac{8r_s}{\varkappa a}\right)^{\frac{1}{2}}\right] = J_{2l}\left[\left(\frac{8r_s}{\varkappa a}\right)^{\frac{1}{2}}\right] \qquad (5.2.4a)$$

$$J_{2l+1}\left(\frac{8r_s}{\varkappa a}\right) = 0 \qquad (5.2.4b)$$

where J is the well known Bessel function.

The values of $r_s/\varkappa a$ where the $3s$, $3d$, $3p$ and $4f$ bands intersect the conduction band are also indicated in Figure 5.7.

If we take a hypothetical spherical Brillouin zone of radius k_B we see that

$$\frac{4}{3}\pi k_B^3 = N \qquad (5.2.5)$$

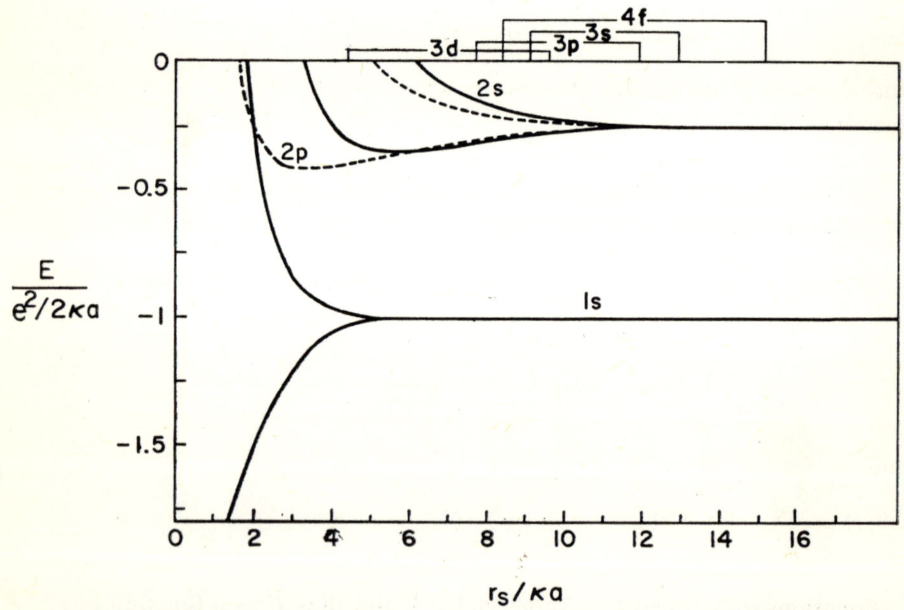

Figure 5.7 Width of impurity energy bands as a function of interatomic distance [from W. Baltensperger, *Phil. Mag.*, **44**, 1355, 1955].

We may define the width ΔE of our bands by

$$\Delta E = \frac{\hbar^2 k_B^2}{2m^+} \tag{5.2.6}$$

where m^+ is the effective mass of an electron in the impurity band. Thus we find for the width of the impurity energy band

$$\Delta E = \frac{m^{*3}}{m^+}\left(\frac{1}{2\varkappa^2 a e^2}\right)(k\hbar^2)^2\left(\frac{1}{.171}\right)\left[\left(\frac{4\pi}{3}\right)^{2/3} N^{2/3}\right] \tag{5.2.7}$$

where m^* is the mass of a conduction band electron.

It has been shown that inclusion of the actual random distribution of the impurities yields further broadening of the impurity energy bands. In addi-

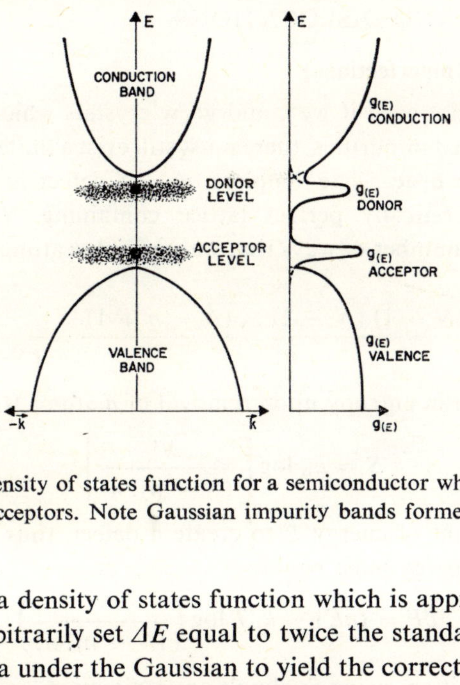

Figure 5.8 Total density of states function for a semiconductor which contains a great many donors and acceptors. Note Gaussian impurity bands formed within the energy gap.

tion, it produces a density of states function which is approximately Gaussian.[2] We can arbitrarily set ΔE equal to twice the standard deviation and normalize the area under the Gaussian to yield the correct number of states $2N$. We find the density of states g_{imp} is given by[3]

$$g_{\text{imp}}(E) = \frac{2\sqrt{2}\,N}{\Delta E\pi}\exp\left[-2\left(\frac{E}{\Delta E}\right)^2\right] \tag{5.2.8}$$

In addition to producing a Gaussian density of impurity states (Figure 5.8), the introduction of a large number of impurities perturbs the shape of the conduction and valence band edges. Bonch-Bruyevich has shown that, for large densities of impurities ($>10^{18}$cm^{-3}) the conduction band density of states function may be written[3]

$$g_{deg}(E) = \frac{(2m^*E)^{3/2}}{6\pi^{3/2}a\hbar^3}\exp[-x][I_{1/4}(x) + I_{-1/4}(x) + I_{3/4}(x) + I_{-3/4}(x)] \quad (5.2.9)$$

where

$$x = \frac{E^2}{8a^2} \qquad a = \frac{\pi^{7/12}\, e^{3/2}\, \hbar^{1/2}\, N^{5/12}}{2^{1/2}\, 3^{1/12}\, \varkappa^{3/4}\, m^{*1/4}}$$

the I's are Bessel functions with imaginary arguments and N is the impurity density — thus a large density of impurities causes the conduction band edge to 'tail' into the forbidden energy gap.

5.3 DEFECTS AND DISLOCATIONS

Defects (Lattice Imperfections)

Statistics show that even if we could grow crystals which are theoretically free from unwanted impurities, there must still exist a finite number of defects in the crystalline order. The simplest type of defect is a lattice vacancy. Consider a theoretically perfect lattice containing N indistinguishable atoms. The total number of ways that we can pick n atoms out of the system is

$$P = \frac{N(N-1)(N-2)\ldots(N-n+1)}{n!} = \frac{N!}{(N-n)!\,n!} \quad (5.3.1)$$

Thus the increase in entropy upon removal of n atoms from the crystal is

$$S = k_0 \log\left[\frac{N!}{(N-n)!\,n!}\right] \quad (5.3.2)$$

It takes an amount of energy E to create a defect, thus the change in the Helmholtz free energy must be

$$\partial F = (nE) - k_0 T \log\left(\frac{N!}{(N-n)!\,n!}\right) \quad (5.3.3)$$

which may be written, using Stirling's approximation

$$\partial F = (nE) - k_0 T\,[N \ln N - (N-n)\ln(N-n) - n \ln n] \quad (5.3.4)$$

In equilibrium, $\left(\frac{\partial F}{\partial n}\right)_{V,T} = 0$. Thus

$$\left(\frac{\partial F}{\partial n}\right)_{V,T} = 0 = E - k_0 T \ln\left(\frac{N-n}{n}\right) \quad (5.3.5)$$

which yields, for $N \gg n$,

$$n = Ne^{-E/k_0T} \tag{5.3.6}$$

Thus, for all finite temperatures, a number of lattice vacancies must exist. The above also holds for the number n of interstitial atoms.

Other types of defects are lattice vacancies in which an atom moves from a lattice site to a similar site (Schottky defects) or to an interstitial position (Frenkel defects) as shown in Figure 5.9. In the case where there are N lattice sites and N' interstitial positions, the increase in entropy due to the formation of a Frenkel pair is

$$S = k_0 \ln\left[\left\{\frac{N!}{(N-n)!\,n!}\right\}\left\{\frac{N'!}{(N'-n)!\,n!}\right\}\right] \tag{5.3.7}$$

which yields, for the case where $n \ll N, N'$

$$n = \sqrt{NN'}\, e^{-E/2k_0T} \tag{5.3.8}$$

where E is the energy necessary to create a Frenkel pair. Equation (5.3.8) also holds for the formation of Schottky defects.

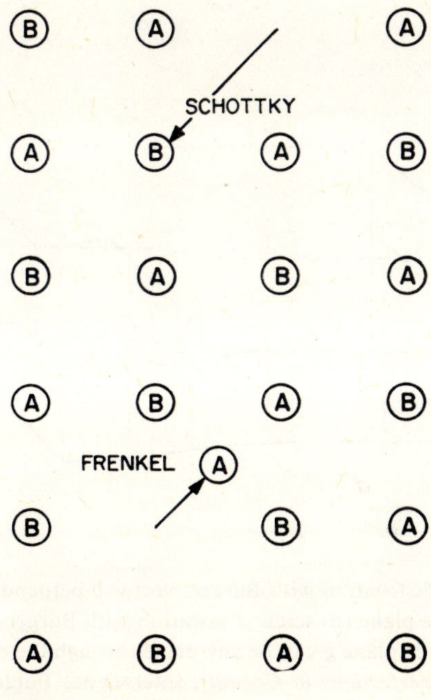

Figure 5.9 Schottky and Frenkel type defects.

Dislocations

A dislocation is a source of crystalline imperfection arising along a line or plane of atoms. Two simple types of dislocations, the edge and screw types are shown in Figures 5.10a and b respectively. An edge dislocation occurs when a section of crystal "slips" along a glide plane relative to another section. The edge dislocation axis *d* marks the boundary between the slipped and unslipped regions of the crystal. It is perpendicular to the direction of slip which is shown by the Burgers vector **b**. It may be thought of as arising from the termination of an extra plane of atoms as shown in Figure 5.11.

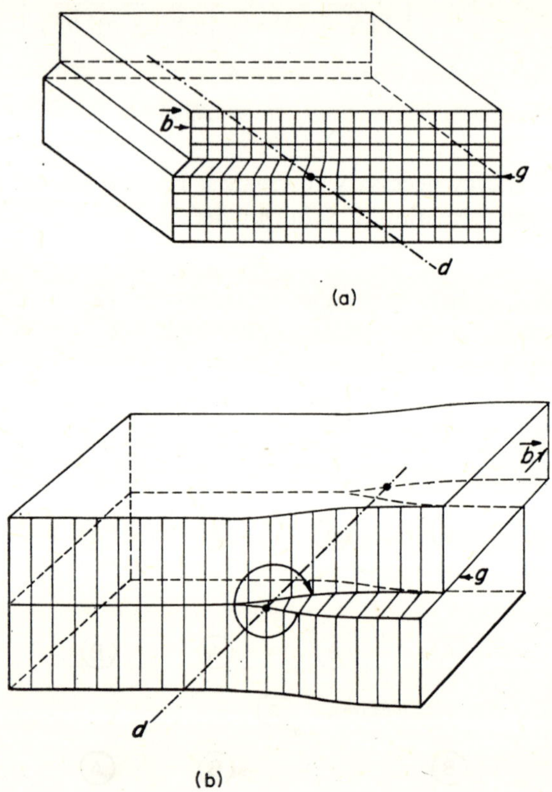

Figure 5.10 (a) Edge dislocation with Burgers vector **b** perpendicular to the dislocation line *d*. *g* marks the glide plane (b) screw dislocation with Burgers vector **b** parallel to the dislocation axis. The glide plane *g* can be any plane through the dislocation line *d*. [From H. G. Van Bueren, *Imperfections in Crystals*, Interscience Publishers, Inc., New York, 1960, p. 49.]

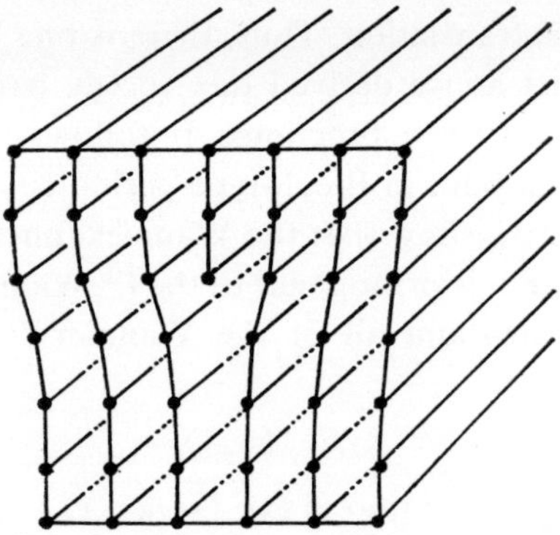

Figure 5.11 Edge dislocation caused by the termination of an extra plane of atoms.

A screw dislocation occurs when two sections of crystal slide relative to one another, resulting in a dislocation axis d which is parallel to the Burgers vector as is shown in Figure 5.10. In both the edge and screw type dislocations, the Burgers vector must be an integral number of lattice translations. Actual dislocations which are usually a mixture of the edge and screw types, can only terminate at the surface of a crystal, and can usually be observed by chemical etching. The dislocation density N_{Dis} is therefore given in cm^{-2} since the defects may be observed only at the surface.

5.4 WANNIER FUNCTIONS

In our treatment of the tight binding approximation we saw the utility of using, for insulating crystals especially, atomic orbital wave functions with very small overlap from site to site. In actual calculations, however, the atomic wave functions $\psi_k(\mathbf{r})$ are not necessarily orthogonal for different values of $\mathbf{k}$. Wannier has devised a type of localized wave function, built up from Bloch type wave functions[4], which has the advantage of orthogonality, and is independent of k within a given Brillouin zone. It is useful for the investigations of impurity wave functions.[5]

The Wannier function $\eta_l(\mathbf{r})$ for the lth energy band localized about the rth lattice site is constructed in the following way

$$\eta_l(\mathbf{r} - n\hat{a}) = \frac{1}{N^{\frac{1}{2}}} \sum_k \exp\left[-ikr\right] b_l(\mathbf{k}, \mathbf{r}) \tag{5.4.1}$$

where $b_l(\mathbf{k}, \mathbf{r})$ is a Bloch type wave function appropiate for our lattice for the lth energy band, and N is the number of atoms in our crystal. The vector $\hat{a}$

describes a unit lattice translation. Thus, there is one Wannier function for each energy band, just as we derived one energy band from each of our localized atomic orbital wave functions. It follows from (5.4.1) that the Wannier function has a built in Bloch-type periodicity.

It is a simple matter to show that the Wannier function for the jth lattice sity of the lth band, $\eta_l(\mathbf{r}_j)$, is orthogonal to the lth Wannier function for each of the other lattice sites and all of the Wannier functions for different bands.

We form

$$\eta_n^*(\mathbf{r}_j)\,\eta_l(\mathbf{r}_{j'}) = \left[\frac{1}{N}\sum_k\sum_{k'} \exp\,[i\mathbf{k}\cdot\mathbf{r}]\exp\{-i\mathbf{k}'\cdot\mathbf{r}_{j'}\}\, b_n^*(\mathbf{k},\mathbf{r}_j)\, b_n(\mathbf{k}'\mathbf{r}_{j'})\right] \tag{5.4.2}$$

But

$$b_n^*(\mathbf{k},\mathbf{r}_j)\, b_l(k',\mathbf{r}_{j'}) = \delta_{nl} \tag{5.4.3}$$

since $b_n(\mathbf{k},\mathbf{r}_j)$ and $b_n(\mathbf{k},\mathbf{r}_j)$ are eigenfunctions of the same translation operator Γ which correspond to different eigenvalues.

We find

$$\eta_n^*(\mathbf{r}_j)\,\eta_l(\mathbf{r}_{j'}) = \frac{1}{N}\sum_k\sum_{k'} \exp i[\mathbf{k}\cdot\mathbf{r}_j - \mathbf{k}'\cdot\mathbf{r}_{j'}]\,\delta_{n,\,l} \tag{5.4.4}$$

Thus the Wannier function for the lth band is orthogonal to that for the nth band.[6]

For Wannier functions of the same band, equation (5.4.2) becomes

$$\eta_l^{**}(\mathbf{r}_j)\,\eta_l(\mathbf{r}_{j'}) = \frac{1}{N}\sum_k\sum_{k'} \exp\,[i(\mathbf{k}\cdot\mathbf{r}_j - \mathbf{k}'\cdot\mathbf{r}_{j'})]\, b_l^*(\mathbf{k},\mathbf{r}_j)\, b_l(\mathbf{k}',\mathbf{r}_{j'}) \tag{5.4.5}$$

By the orthogonality of the Bloch functions for different values of **k**, we find

$$\eta_l^*(\mathbf{r}_j)\,\eta_l(r_{j'}) = \frac{1}{N}\sum_k\sum_{k'} \exp i[\mathbf{k}\cdot\mathbf{r}_j - \mathbf{k}'\cdot\mathbf{r}_{j'}]\,\delta_{k,\,k'}$$

$$= \frac{1}{N}\sum_k \exp i[\mathbf{k}(\mathbf{r}_j - \mathbf{r}_{j'})] = \delta_{j,\,j'} \tag{5.4.6}$$

which follows from the lattice sum rule. Thus Wannier functions for different lattice sites are orthogonal.

A Simple Example of a Wannier Function

We can get a good idea of the localized form of the Wannier function by considering a simple example. We shall consider the Wannier functions for a simple cubic lattice with basic vectors $\hat{a}_i$, $\hat{a}_j$ and $\hat{a}_k$. Under the assumption that a Bloch type wave function for a nearly free electron is a good approximation, we get for our Wannier function

$$\eta_l(\mathbf{r} - n\hat{a}) = \frac{1}{N^{\frac{1}{2}}} \sum_k \exp(-ikn)\, U_l(\mathbf{r})\, e^{i\mathbf{k}\cdot\mathbf{r}} \tag{5.4.7}$$

We can change the sum over **k** to an integral by the use of the densities of allowed values of **k**, which yields

$$\eta_l(\mathbf{r}) = \frac{VU(\mathbf{r})}{8\pi^3 N^{\frac{1}{2}}} \iiint e^{i\mathbf{k}\cdot\mathbf{r}} d\mathbf{k} \tag{5.4.8}$$

where V is the volume of the direct lattice. We integrate only over one Brillouin zone to determine the Wannier function for that zone, which yields

$$\eta_l(\mathbf{r}) = \frac{\sin\dfrac{\pi x}{a}\sin\dfrac{\pi y}{a}\sin\dfrac{\pi z}{a}}{\dfrac{\pi^3(xyz)}{a^{3/2}}} \tag{5.4.9}$$

A plot of the probability density $\eta_l^*(\mathbf{r}_j)\,\eta_l(\mathbf{r}_j)$ of this simple Wannier function will show that it is localized about, and has its maximum intensity at the rth lattice site.

PROBLEM

Show that the Wannier function (5.4.9) is normalized and graph $\eta_l^*(\mathbf{r}_j)\,\eta_l(\mathbf{r}_j)$.

5.5 A GENERAL THEOREM CONCERNING THE EQUATION OF MOTION OF AN ELECTRON IN A PERTURBED LATTICE

In Chapter 4 we saw the utility of relating the motion of an electron in a weakly perturbing lattice to the free electron state. The wave function for an electron in a metal was treated as if it were not much different than a free electron state. Slater has shown[7] that the state of an electron in a slowly perturbed periodic potential may, in analogous fashion, be related to the allowed electron states in the unperturbed periodic potential.

We consider that the solutions to the perturbed periodic potential for the nth band are $\Psi_n(\mathbf{k}, \mathbf{r})$ and that they satisfy the Schrödinger equation

$$\mathscr{H}\Psi_n(\mathbf{k}, \mathbf{r}) = E_k\Psi_n(\mathbf{k}, \mathbf{r}) \tag{5.5.1}$$

Again we set $\mathscr{H} = \mathscr{H}_0 + \mathscr{H}_1$ where $\mathscr{H}_0$ is the Hamiltonian for the unperturbed periodic lattice and $\mathscr{H}_1$ is a slowly varying perturbation. From the solutions $\psi_{0,n}(\mathbf{k}, \mathbf{r})$ to the unperturbed problem

$$\mathscr{H}_0\psi_{0,n}(\mathbf{k}, \mathbf{r}) = E_{0,n}\psi_{0,n}(\mathbf{k}, \mathbf{r}) \tag{5.5.2}$$

we form the Wannier function $\eta_n(\mathbf{r})$ for the nth band

$$\eta_n(\mathbf{r} - m\hat{a}) = \frac{1}{N^{1/2}} \sum_k \exp\left[-i\mathbf{k} \cdot m\hat{a}\right] \psi_{0,n}(\mathbf{k}, \mathbf{r}) \tag{5.5.3}$$

Multiplying by

$$\sum_m e^{i\mathbf{k}' \cdot m\hat{a}} \tag{5.5.4}$$

we find

$$\sum_m e^{i\mathbf{k}' \cdot m\hat{a}} \eta_n(\mathbf{r} - m\hat{a}) = \frac{1}{N^{1/2}} \sum_k \left[\sum_m e^{i[(\mathbf{k}' - \mathbf{k}) \cdot (m\hat{a})]}\right] \psi_{0,n}^{(\mathbf{k})} \tag{5.5.5}$$

and using the lattice sum rule one can transform the right hand side of (5.5.5) into

$$(N)^{1/2}\, \psi_{0,n}(\mathbf{k}, \mathbf{r}) \tag{5.5.6}$$

Thus

$$\psi_{0,n}(\mathbf{k}, \mathbf{r}) = \frac{1}{N^{1/2}} \sum_m e^{i\mathbf{k}' \cdot m\hat{a}} \eta_n(\mathbf{r} - m\hat{a}) \tag{5.5.7}$$

The solutions to the unperturbed lattice problem may therefore be generated by a linear combination of modulated Wannier functions. We now assume in a similar manner, that the solutions $\psi_n(\mathbf{k}, \mathbf{r})$ may also be expressed as a linear combination of Wannier functions

$$\psi_n(\mathbf{k}, \mathbf{r}) = \sum_m \xi_n(m\hat{a})\, \eta_n(\mathbf{r} - m\hat{a}) \tag{5.5.8}$$

where $\xi_n(m\hat{a})$ is a modulating function which must be determined.

Substitution of (5.5.8) into (5.5.1) yields

$$\mathscr{H}\left[\sum_m \xi_n(m\hat{a})\, \eta_n(\mathbf{r} - m\hat{a})\right] = E_n \sum_m \xi_n(m\hat{a})\, \eta_n(\mathbf{r} - m\hat{a}) \tag{5.5.9}$$

Multiplying by $\eta_n^*(\mathbf{r} - l\hat{a})$ and integrating over direct lattice space yields

$$\iiint d\mathbf{r} \sum_m \left[\eta_n^*(\mathbf{r} - l\hat{a}) \left\{\mathscr{H}_0 + \mathscr{H}_1 - E_n\right\} \xi_n(m\hat{a})\, \eta_n(\mathbf{r} - m\hat{a}) = \right. \tag{5.5.10}$$

If the perturbation $\mathscr{H}_1$ is sufficiently slowly varying over an atomic cell, we may write

$$0 = \sum_m \int\int\int d\mathbf{r}[\eta_n^*(\mathbf{r} - l\hat{a})\, \mathscr{H}_0 \xi_n(m\hat{a})\, \eta_n(\mathbf{r} - m\hat{a})] + \mathscr{H}_1(l\hat{a})\, \xi_n(l\hat{a}) - E_n \xi_n(l\hat{a}) \tag{5.5.11}$$

The term in brackets in (5.5.11) may be rewritten

$$\sum_m \sum_k \sum_{k'} \frac{1}{N} \exp\,[i(\mathbf{k} \cdot l\hat{a} - \mathbf{k}' \cdot m\hat{a})]\, \xi_n(m\hat{a}) \int_{\mathbf{r}} \psi_{o,\,n}^*(\mathbf{k}, \mathbf{r})\, \mathscr{H}_0 \psi_{0,\,n}(\mathbf{k}', \mathbf{r})\, d\mathbf{r} \tag{5.5.12}$$

From the orthogonality of the Bloch-type wave functions $\psi_{0,\,n}(\mathbf{k}, \mathbf{r})$ for different values of $\mathbf{k}$ we find that (5.5.12) becomes

$$\sum_k \sum_m \frac{1}{N} \exp i[\mathbf{k} \cdot (l\hat{a} - m\hat{a})]\, \xi_n(m\hat{a})\, E_0(\mathbf{k}) \tag{5.5.13}$$

where

$$\int\int\int d\mathbf{r} \psi_{0,\,n}(\mathbf{k}, \mathbf{r})\, \mathscr{H}_0 \psi_{0,\,n}(\mathbf{k}', \mathbf{r}) = E_0(\mathbf{k})\, \delta_{\mathbf{k},\,\mathbf{k}'} \tag{5.5.14}$$

We have found, quite generally, that the allowed energy of an electron in a purely periodic lattice is a periodic function of $\mathbf{k}$ with period π/a. Thus we may expand $E_0(\mathbf{k})$ in a Fourier series

$$E_0(\mathbf{k}) = \sum_m A(m\hat{a}) \exp\,[-i\mathbf{k} \cdot m\hat{a}] \tag{5.5.15}$$

where, of course, the Fourier coefficients $A(m\hat{a})$ are given by

$$A(m\hat{a}) = \sum_k \left(\frac{1}{N}\right) E_0(\mathbf{k}) \exp i(\mathbf{k} \cdot m\hat{a}) \tag{5.5.16}$$

Thus (5.5.13) becomes

$$\sum_s A(s\hat{a})\, \xi_n(l\hat{a} - s\hat{a}) \tag{5.5.17}$$

where we have made the substitution

$$l - m = s \tag{5.5.18}$$

We expand the modulating function $\xi_n(l\hat{a} - s\hat{a})$ about the lth lattice site

$$\begin{aligned} \xi_n(l\hat{a} - s\hat{a}) &= \xi_n(l\hat{a}) - m\hat{a} \cdot \nabla \xi_n(l\hat{a})\, [l\hat{a} - m\hat{a}] + \ldots \\ &\approx \xi_n(l\hat{a}) \exp - [l\hat{a} - m\hat{a}] \cdot \nabla \end{aligned} \tag{5.5.19}$$

Equation (5.5.17) becomes, using (5.5.19),

$$\sum_s A(s\hat{a}) \exp [(m\hat{a} - l\hat{a}) \cdot \mathbf{\nabla}]\, \xi_n(l\hat{a}) \tag{5.5.20}$$

Thus it seems that (5.5.20) tells us that we may substitute $i\mathbf{\nabla}$ for $\mathbf{k}$ in our expression for the unperturbed energy $E_0(\mathbf{k})$, which yields

$$[E_0(i\mathbf{\nabla}) + \mathscr{H}_1]\, \xi_n(\mathbf{r}) = E_n\xi_n(\mathbf{r}) \tag{5.5.21}$$

which is the equation that must be solved in order to determine the modulating functions $\xi_n(l\hat{a})$. Equation (5.5.21) is an analogue of Schrödinger's equation in which the kinetic energy operator is taken as the energy of the unperturbed problem $E_0(\mathbf{k})$ with $\mathbf{k}$ replaced by $-i\mathbf{\nabla}$. The potential is the perturbing potential $\mathscr{H}_1$.

Thus the problem of the allowed energy states of an electron in a perturbed lattice may be solved in terms of modulated solutions to the unperturbed problem.

If we assume that E has a simple form, e.g., the form for the spherical energy surface in the (100) direction of Si,

$$E_0(\mathbf{k}) = \frac{\hbar^2}{2}\left[\frac{k_x^2 + k_y^2}{m_t^*} + \frac{k_z^2}{m_l^*}\right] \tag{5.5.22}$$

we find

$$\left[-\frac{\hbar^2}{2m_t^*}\left(\frac{\partial^2}{\partial x^2} + \frac{\partial^2}{\partial y^2}\right) - \frac{\hbar^2}{2m_l^*}\frac{\partial^2}{\partial z^2} + \frac{e^2}{\varkappa r}\right]\xi_n(\mathbf{r}) = E_n\xi_n(\mathbf{r}) \tag{5.4.23}$$

where we have again chosen the spherical Coulomb potential weakened by the dielectric constant. 5.9.23 is the Schrödinger equation with ∇^2 reploced by

$$\frac{m}{m_t^*}\left(\frac{\partial^2}{\partial x^2} + \frac{\partial^2}{\partial y^2}\right) + \frac{m}{m_l^*}\frac{\partial^2}{\partial z^2} \tag{5.4.24}$$

5.6 DISORDERED MATERIALS

When disorder due to rapid cooling from the melt, fluctuations in composition or density becomes great enough to destroy the long range periodicity of a material, the density of allowed energy states takes on a peculiar structure. Examples of these types of materials are liquids, thin film covalently bound amorphous semiconductors such as Ge and Si and the compound chalcogenide glasses. In these materials only short range order remains. For example, in amorphous Ge, although the average number of nearest neighbors [the coordination number] remains the same (4) as in the crystalline

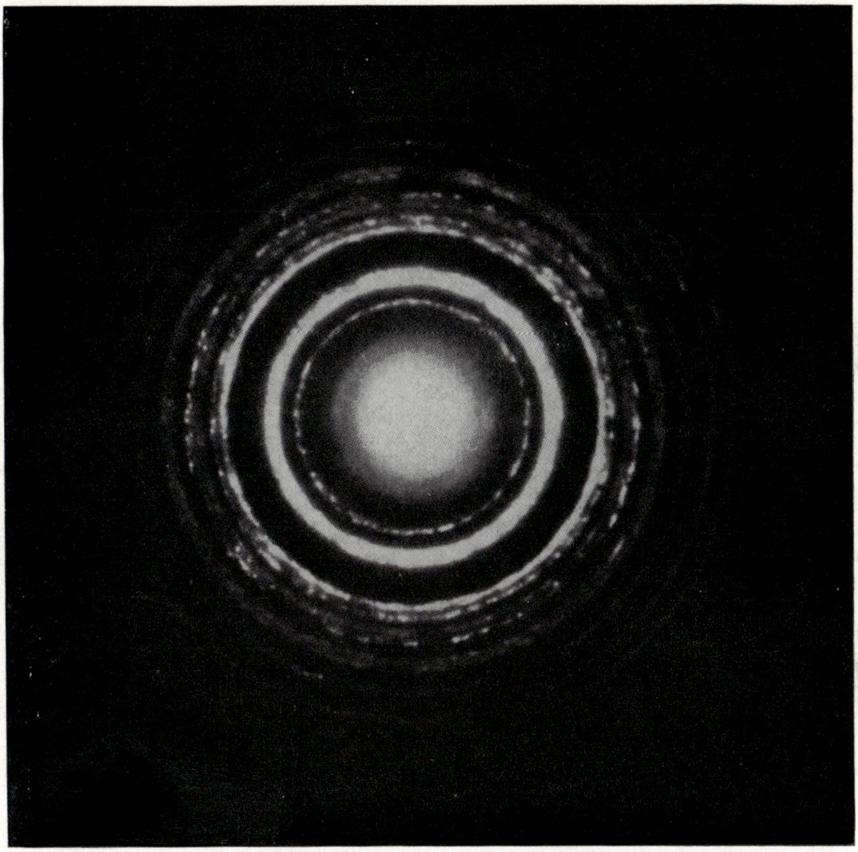

Figure 5.12 Electron diffraction pattern of thin film Amorphous $As_{30}Te_{48}Si_{12}Ge_{10}$ (From B. G. Bagley, and W. R. N. Nathover, Tnl. Non-Crystalline Solids, **2**, 161, 1970.)

form, the structure is translationally aperiodic. One may determine the structure of these materials by X-ray diffraction which shows a smeared pattern which is the result of the superposition of many diffraction patterns from randomly oriented microcrystals. Such a pattern is shown in Figure 5.12.

In order to solve for the allowed energy levels of such a disordered system, we must solve a Schrödinger equation of the form

$$\nabla^2\psi_{(\mathbf{r})} + \frac{2m}{\hbar^2}(E - V_{(\mathbf{r})})\,\psi_{(\mathbf{r})} = 0 \qquad (5.6.1)$$

where $V_{(r)}$, is an aperiodic atomic potential. A solution to (5.6.1) even in the one-dimensional case is very complicated and we shall not attempt it. Anderson,[8] in particular has treated the problem by superimposing a random potential $V_{(r)}$ on the solutions to a three dimensional crystalline array of atoms whose $1s$ states form an energy band of width J. If the variation $|\langle V\rangle^2|^{1/2}$ of the random potential is large enough with respect to the band width J, then, in addition to the continuous non-localized allowed states a number of allowed solutions to (5.6.1) may be localized in space and have the form

$$\psi_{(\mathbf{r})} = A\,e^{-\gamma(\mathbf{r}-\mathbf{r}_n)} \qquad (\text{real } \gamma) \qquad (5.6.2)$$

The Anderson criteria for the existence of these localized energy states is given by

$$|\langle V^2\rangle|^{1/2} > 10ZI\,e^{-\gamma(\mathbf{r}-\mathbf{r}_n)} \qquad (5.6.3)$$

where Z is the coordination number and $\mathbf{I}$ is an overlap integral. These localized states lie near the edges of the conduction and valence bands which are still separated by an amount of energy roughly equivalent to that required for breaking of covalent bonds. In an amorphous material these localized states extend well into the conduction and valence bands and may completely obliterate the energy gap. In fact, where previously we found an energy gap, we now have a "pseudogap," merely a minimum in the density of states function, as is shown in Figure 5.13.

Thus, in glassy materials, the short range order is enough to retain the functional forms for the conduction and valence bands and the destruction of the long range Bloch periodicity produces a continuum of localized energy states in the "pseudogap".

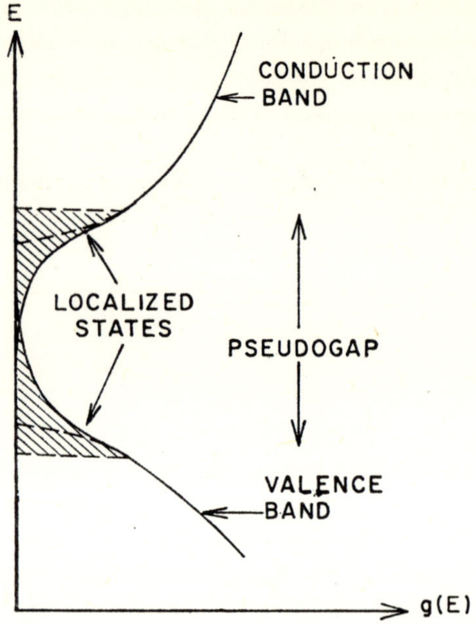

Figure 5.13 Density of states of an amorphous solid showing non-localized conduction and valance bands and obliteration of the energy gap by localized states (pseudogap).

References

1. This section follows W. Baltensperger, *Phil. Mag.* **44**, 1355, 1953.
2. H. M. James and A. S. Ginsbarg, *J. Phys. Chem.*, **57**, 840, 1953. T. P. Brody, *Jnl. App. Phys.*, **33**, 100, 1962. See also T. N. Morgan, *Phys. Rev.*, **139** (1965) A343.
3. V. L. Bonch-Bruyevich, *The Electronic Theory of Heavily Doped Semiconductors*, American Elsevier, New York, 1966.
4. G. H. Wannier, *Phys. Rev.*, **52**, 191, 1937.
5. G. F. Koster and J. C. Slater, *Phys. Rev.*, **95**, 1167, 1954.
6. This is strictly true only where the bands are not degenerate in energy.
7. J. C. Slater, *Phys. Rev.*, **76**, 1592, 1949.
8. P. W. Anderson, *Phys. Rev.*, **109**, 1498, 1958. See also N. F. Mott. *Rev. Mod. Phys.* **40**, (1968) 677. N. F. Mott, *Contemp. Phys.* **10** (1969) 125. M. H. Cohen, *Jnl. Non-Crys, Solids.* **4** (1970) 391.

CHAPTER 6

Statistics of Electrons in Solids

6.1 STATISTICS OF PURE INTRINSIC SEMICONDUCTORS AND INSULATORS

We are now in a position to investigate the processes by which the allowed energy bands which we have calculated for pure materials will be occupied by electrons. Since an electron must obey the Pauli exclusion principle, the distribution of electrons (and thus holes) amongst the allowed energy levels must obey Fermi-Dirac statistics.

It is easy to see that at $T = 0°K$, the Fermi-Dirac distribution function (2.2.63) predicts that all energy levels will be filled up to a maximum level μ_0. This is the Fermi level at 0°K and the three dimensional surface in $\mathbf{k}$ space formed by this level is what we have defined as the Fermi surface (Figure 4.17). As the temperature is increased, the Fermi surface smears due to the thermal excitation of electrons to higher energy levels. A look at Figure 2.3 will convince the reader that at moderate temperatures the spread of the Fermi-Dirac distribution function (approximately $2k_0T$) is much smaller than the total number of occupied levels below the Fermi level. Thus only the electrons at or close to the Fermi surface can influence the electrical and thermal properties of the material. As excitation occurs, the Fermi level μ behaves in a calculable manner.

We will now investigate the statistics governing the excitation of electrons across an energy gap from the valence to the conduction bands of pure semiconductors and insulators. In this case we may treat the few free electrons above the Fermi level as if they are a perfect gas of non-interacting Fermi-Dirac particles, and we may ignore the degeneracy factor g_i in $f_{FD}'(E_i)$ since the electrons have so many conduction band energy levels to choose from that they may be assumed to be free. We can take the simple spin degeneracy 2 into account by merely multiplying $g(E)$ by 2.

The sum of the Fermi-Dirac distribution function, which has a value 1 for filled energy levels and zero for unfilled ones, over all of the allowed energy levels in a solid is equal to the number of electrons n in the solid, i.e.,

$$n = \sum_i \left\{ \frac{1}{e^{\frac{E_i - \mu}{k_0 T}} + 1} \right\} \tag{6.1.1}$$

Assuming that very many closely spaced energy levels exist, we can convert the summation to an integral of the form

$$\int_0^\infty g(E) f(E) \, dE = n \tag{6.1.2}$$

As we have seen before, we can write a simple density of states function as proportioned to E^j where j need not be an integer. Thus we may write

$$n = A \int_0^\infty E^j f(E)\, dE \tag{6.1.3}$$

where A is an appropriate constant. This integral is one of a series of well known integrals called the Fermi-Dirac integrals. They are of the general form

$$\mathscr{F}_j[-\mu/k_0T] = \frac{1}{\Gamma_{(j+1)}} \int_0^\infty \frac{E^j dE}{\left(1 + e^{\frac{E-\mu}{k_0T}}\right)} \tag{6.1.4}$$

where $\Gamma_{(j+1)}$ is the gamma function defined by[1]

$$\Gamma_n = \int_0^\infty e^{-x} x^{n-1}\, dx \tag{6.1.5}$$

where $\Gamma_{(n+1)} = n\Gamma_{(n)}$. An excellent tabulation of these integrals may be found in Blakemore.[2] It can be shown that

$$\frac{\partial \mathscr{F}_j[\mu/k_0T]}{\partial[\mu/k_0T]} = \mathscr{F}_{(j-1)}[\mu/k_0T] \tag{6.1.6}$$

Integrals of type (6.1.4) show up quite often in semiconductor physics and so we shall study a number of approximate solutions. The first of these is the so-called "non-degenerate" or "classical" case where all energies we deal with are far from the Fermi level, $((E - \mu) > k_0T)$. Thus we may substitute the $f_{mB}(E)$ for $f_{FD}(E)$. For this case the number of electrons excited from the valence to conduction bands can be approximated by

$$n_i = \left(\frac{4m_e^*}{h^2}\right)^{3/2} \sqrt{2}\,\pi \int_{E_g}^\infty e^{\frac{\mu-E}{k_0T}} (E - E_g)^{1/2}\, dE \tag{6.1.7}$$

where n_i is called the intrinsic electron concentration. Although we have chosen our zero of energy at the valence band maximum, we can begin our integration at the minimum of the conduction band at $E = E_g$ since no allowed energy levels exist in the forbidden energy gap. Letting

$$x = \frac{E - E_g}{k_0T} \tag{6.1.8}$$

we get

$$n_i = \left(\frac{4m_e^*}{h^2}\right)^{3/2} \sqrt{2}\,\pi(k_0T)^{3/2}\, e^{\frac{\mu - Eg}{k_0T}} \int_0^\infty x^{1/2}\, e^{-x}\, dx \tag{6.1.9}$$

The integral on the right hand side of (6.1.9) may be shown to be

$$\int_0^\infty x^{1/2}\, e^{-ax}\, dx = \frac{1}{2a}\sqrt{\pi/a} \tag{6.1.10}$$

Thus we find that the number of electrons excited into the conduction band of a pure semiconductor or insulator is given by

$$n_i = 2\left(\frac{2\pi m_e^* k_0 T}{h^2}\right)^{3/2} e^{\frac{\mu - Eg}{k_0T}} \tag{6.1.11}$$

Or, in the usual notation

$$n_i = N_c\, e^{\frac{\mu - Eg}{k_0T}} \tag{6.1.12}$$

where

$$N_c = 2\left(\frac{2\pi m_e^* k_0 T}{h^2}\right)^{3/2} \tag{6.1.13}$$

Thus a knowledge of the position of the Fermi level and the energy gap will enable us to make a determination of the number of free electrons in the conduction band. In practice, however, one must make a further calculation in order to determine an expression for the position of the Fermi level as a function of temperature.

Of course, in a pure semiconductor or insulator, each electron excited across the energy gap leaves a positively charged vacancy behind in the valence band. These holes also contribute to conduction processes. If we denote the number of intrinsic holes in the valence band by p_i, we must have

$$n_i = p_i. \tag{6.1.14}$$

A knowledge of the temperature dependence of p_i will enable us to determine the position of the Fermi level as a function of temperature. The probability $f_{FD}(h)$ that an energy level E in the valence band is occupied by a hole (unoccupied by an electron) is

$$f_{FD}(h) = 1 - f_{FD}(E) \tag{6.1.15}$$

Thus the number of holes p_i in the valence band is given by

$$p_i = \int_{-\infty}^{0} f_{FD}(h)\, g(E)\, dE \tag{6.1.16}$$

where

$$g(E) = 2\pi\left(\frac{2m_h^*}{h^2}\right)^{3/2}(-E)^{1/2} \tag{6.1.17}$$

and where m_h^* is the effective mass of the holes in the valence band. The top of the valence band is again taken as $E = 0$, yielding

$$p_i = 2\left[\frac{2\pi m_h^* k_0 T}{h^2}\right]^{3/2} e^{-\mu/k_0T} \tag{6.1.18}$$

Using (6.1.14) and (6.1.18), we find

$$\mu = 3/4 k_0 T \ln\left[\frac{m_h^*}{m_e^*}\right] + \frac{Eg}{2} \tag{6.1.19}$$

Thus if $\left[\frac{m_h^*}{m_e^*}\right] = 1$, the Fermi level lies in the center of the forbidden energy gap in an intrinsic material. As the temperature is increased the Fermi level moves closer to the band with the smaller effective mass.

The equilibrium product of the number of electrons and holes is

$$n_i p_i = 4\left[\frac{2\pi k_0 T}{h^2}\right]^3 [m_e^* m_h^*]^{3/2} e^{-Eg/k_0T} \tag{6.1.20}$$

Thus

$$n_i = p_i = 2\left[\frac{2\pi k_0 T}{h^2}\right]^{3/2} [m_e^* m_h^*]^{3/4}\, e^{Eg/2k_0T} \tag{6.1.21}$$

i.e., the number of carriers thermally excited across the energy gap of a semiconductor or insulator varies with $e^{-Eg/2k_0T}$.

One should note that we have taken into account the contributions due only to a single conduction band minimum and a single valence band maximum. Where equivalent conduction band minima exist the density of conduction band states must be multiplied by M, the number of equivalent minima. Similarly for the density of valence band states.

Problem

For the case of two degenerate valence bands at $\mathbf{k} = 0$, determine the intrinsic hole concentration for the case where both bands may be treated independently as paraboloids of revolution with effective hole masses m_{h1}^* and m_{h2}^*.

Intrinsic Degeneracy

The condition of degeneracy implies that the distribution of allowed electron energies is very close to the Fermi level. This usually occurs in very impure semiconductors or metals. Our approximate determination of the Fermi level in terms of the effective masses, energy gap and temperature, (6.1.19), indicates that the condition of degeneracy may also arise at moderate temperatures if a number of conditions coincide to force the Fermi level to a position close to either the conduction or valence band edge. Two cases which may induce degeneracy according to (6.1.19) are: (1) $m_h^* \approx m_e^*$ in a material with a very small energy gap and (2) a large discrepancy in the effective hole and electron masses in a material with a small or moderate energy band gap. In either or both of the above cases, condition (6.1.19) may no longer be valid. The latter is the case for InSb. In this material the extreme disparity between the hole and electron effective masses near the valence band maximum and conduction band minimum, $m_h^*/m = 0.18$, $m_e^*/m = 0.013$[3] respectively, combines with a small energy gap ($E_g = 0.27$ eV at 0°K) to yield the degenerate condition.

PROBLEM

Calculate μ for InSb at 0° and 300°K. Assume that the variation in energy gap with temperature may be given by $(0.27 - 3.0 \times 10^{-4}\ T)$ eV.

In such cases as the above, one must resort to the use of other approximations (see section 6.6) or exact tabulations of the Fermi integrals.

PROBLEM

Determine the exact defining equation for the Fermi level of an intrinsic material in terms of the Fermi integrals.

6.2 STATISTICS OF IMPURITY INSULATORS AND SEMICONDUCTORS

At present, that part of solid state technology which deals with the purification of solids is quite advanced. However, residual unwanted impurities and defect levels remain, which we have investigated in Chapter 5. This necessitates consideration of to the effect of unwanted impurities on the electrical and thermal properties of solids. We have already treated the ionization energy of impurity atoms in a hydrogen-like approximation and found that, at low concentrations, impurities produce discrete bound states within the forbidden energy gap. At higher concentrations we determined that allowed impurity bands are formed in the energy gap. In this section we shall take up the problem of the statistical determination of the effect of these impurities on the number of charge carriers present in semiconductors and insulators.

In our treatment of the derivation of the Fermi-Dirac destribution function, we tacitly assumed that we would be attempting to determine the probability of occupation of a number of non-interacting, non-degenerate energy levels. In the conduction band of semiconductors and insulators, and in the free electron gas of a metal, the large number of allowed energy levels per unit energy allowed us to include the effect of energy level degeneracy by the simple mechanism of multiplying our density of states function $g(E)$ by a constant (2 for spin degeneracy). When one deals with the discrete impurity levels produced in the forbidden energy gap by a number of localized defect or impurity levels, the above procedure is no longer a valid one.

We must now utilize our generalization of the Fermi-Dirac distribution function which includes the effect of level multiplicity. In the case of a Group V donor impurity in a host lattice of Group IV atoms, for instance, its ionized state will allow it to accept an electron with a spin orientation either up or down. Once an electron is captured in one of the spin states, however, coulombic repulsion makes it highly improbable that the energy state with the reverse spin orientation will be occupied. Thus, for the occupation probability of an impurity or defect energy level E_l with degeneracy g_l we must utilize

$$f(E_l) = \frac{1}{1 + g_l^{-1}\left[e^{\frac{E_l-\mu}{k_0T}}\right]} \tag{6.2.1}$$

The most important case of multiplicity is the case of spin degeneracy. In this case the occupation probability of a donor impurity atom in which the spin degeneracy is 2 is

$$f(E_l) = \frac{1}{1 + \frac{1}{2}e^{\frac{E_l-\mu}{k_0T}}} \tag{6.2.2}$$

PROBLEM

Using (6.1.2) show that in the classical approximation ($E \gg \mu$) putting the degeneracy factor in the distribution function or in the density of states function yields the same result.

In the case of an acceptor impurity which, in the ionized state contains an extra electron, the entropy of the system becomes

$$S = -k_0 \ln \prod_{j=1}^{n} \left[\frac{L_j!\, g_j^{(L_j-n_j)}}{n_j!\,(L_j-n_j)!}\right] \tag{6.2.3}$$

since only those acceptors which are unionized can contribute extra permutations, e.g.; the ionizing electron must have spin opposite to that already occupying the acceptor.

By a procedure similar to the one utilized in Sect. 2.2, we find for the probability of occupancy of an acceptor atom by an electron

$$\underset{\text{acceptor}}{f(h)} = \frac{1}{1 + g^{+1} e^{\left(\frac{E-\mu}{k_0 T}\right)}}. \tag{6.2.4}$$

PROBLEM

Prove 6.2.4.

And, of course, the probability of acceptor occupancy by a hole $f(h)$ is

$$f(h) = 1 - f(E) = \frac{1}{1 + g^{-1} \exp\left(\frac{(\mu - E}{k_0 T}\right)}. \tag{6.2.5}$$

The degeneracy factor for acceptor states is usually taken as 4 when the acceptor state is derived from the fourfold degenerate valence band maximum as e.g. in Si.[4]

We may now proceed to determine the effect of donor and acceptor impurities on the number of free carriers and the Fermi level in semiconductors and insulators. Under conditions of equilibrium, our solid must be electrically neutral at all times. Thus, the number of positive charges, both bound and free, must equal the number of bound and free negative charges. Let us consider an insulating or semiconducting crystal which contains r different donor and s different acceptor type impurities. If we denote the density of the kth type donor species and the jth type acceptor species by N_{d_k} and N_{a_j}, respectively, and the density of ionized k^{th} type donor species, and j^{th} type acceptor species by $N_{d_k}^+$ and $N_{a_j}^-$, respectively, the condition of charge neutrality is

$$\sum_{j=1}^{s} pa_j + p_i + \sum_{k=1}^{r} N_{d_k}^+ = n_i + \sum_{j=1}^{s} N_{a_j}^- + \sum_{k=1}^{r} n_{d_k} \tag{6.2.6}$$

where n_{d_k} and p_{a_j} are the densities of electrons and holes in the conduction and valence bands due to ionization of k type donor and j type acceptor impurities, respectively.

Considering the case where each of the donor and acceptor impurities contributes a single energy level with degeneracy g_k and g_j, respectively, we find, using the top of the valence band as our zero in energy as shown in Figure 6.1

$$N_{d_k}^+ = N_{d_k}[1 - f_d(E_k)] = \frac{N_{d_k}}{1 + g_k \exp\left[\dfrac{-\phi_{d_k} + \mu_k}{k_0 T}\right]} \tag{6.2.7}$$

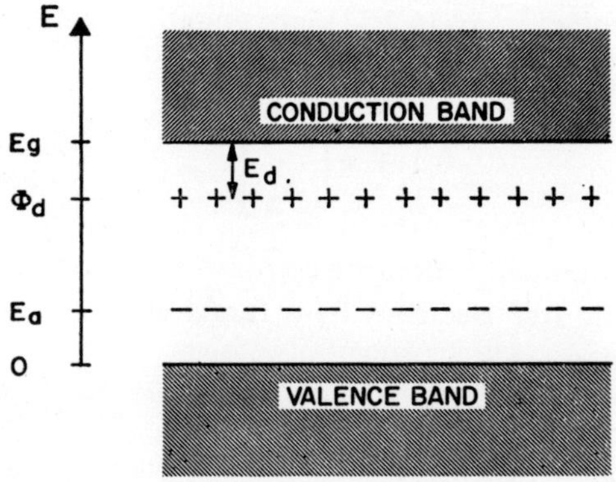

Figure 6.1 Energy level scheme for impurity semiconductors.

and

$$N_{a_j}^- = N_{a_j}[f_a(E_j)] = N_{a_j} \frac{1}{\left[1 + g_j \exp \dfrac{E_{a_j} - \mu_j}{k_0 T}\right]} \tag{6.2.8}$$

Thus (6.2.6) becomes

$$\sum_{j=1} p_{a_j} + p_i + \sum_{k=1} N_{d_k}\left[1 + g_k \exp \frac{\mu - \phi_{d_k}}{k_0 T}\right]^{-1}$$

$$= n_i + \sum_{k=1}^{r} n_{d_k} + \sum_{j=1}^{s} N_{a_j}\left[1 + g_j \exp\left(\frac{E_{a_j} - \mu_j}{k_0 T}\right)\right]^{-1} \tag{6.2.9}$$

Let us consider first the case of a strongly n-type crystal which contains a great many atoms of a single donor species and in which $N_a \approx 0$. The density of electrons in the conduction band contributed by this donor species is

$$n_d = N_a \left[\frac{1}{1 + g^{+1} \exp\left(\dfrac{\mu - \phi_d}{k_0 T}\right)}\right] \tag{6.2.10}$$

The number of intrinsic electrons in the conduction band which are thermally ionized from the valence band is given by (for the nondegenerate case)

$$n_i = N_c \exp\left(\frac{\mu - E_g}{k_0 T}\right) \tag{6.2.11}$$

Since the Fermi energies of identical particles must be equal, we may substitute the value for $e^{\mu/k_0 T}$ of (6.2.11) into equation (6.2.10), which yields

$$n_d = \frac{N_d}{1 + g\left[\dfrac{n_i}{N_c}\right] e^{\left[\frac{E_g - \phi_d}{k_0 T}\right]}} \tag{6.2.12}$$

which relates the number of donor electrons in the conduction band to the number of intrinsic free electrons by a function similar to a Fermi-Dirac distribution function.

PROBLEM

Show by considering the Fermi-Dirac distribution function that the Fermi level for a strongly n-type sample with a single donor species obeys the equations

$$\mu = k_0 T \ln\left[\frac{N_d - n_d}{n_d g}\right] + \phi_d \tag{6.2.13a}$$

$$\mu_0 \approx \lim_{\substack{T \to 0 \\ n_d \to 0}} \left[\ln\left(\frac{N_d - n_d}{n_d g}\right)\right] + \phi_d \tag{6.2.13b}$$

Since both T and n_d are zero at $T = 0$, (6.2.13b) is merely an approximation to show that at very low temperatures the Fermi level lies somewhere above the donor level. Thus, since the energy levels in the valence band lie below the Fermi energy, by far the great percentage of them should contain an electron.

Equation (6.2.6) may be written

$$n_i + n_d - p_i + N_d\left[1 + g \exp\left[\frac{\mu - \phi_d}{k_0 T}\right]\right]^{-1} \tag{6.2.14}$$

and we may now assume that the density of holes p_i in the valence band is small, and define the total number of electrons in the conduction band as $n = n_i + n_d$. (6.2.14) becomes

$$n = N_d\left[1 + g \exp\left[\frac{\mu - \phi_d}{k_0 T}\right]\right]^{-1} \tag{6.2.15}$$

Equation (6.2.12) is of course still valid as a determination of the total density of electrons in the conduction band provided we make a consistent determination of the position of the Fermi level. Thus we may write

$$n = N_c\, e^{\frac{\mu - E_g}{k_0 T}} \tag{6.2.16}$$

Using the result (6.2.13) and (6.2.16) we again equate the Fermi energy for similar particles

$$k_0 T \ln\left[\frac{n}{N_c}\right] + E_g = k_0 T \ln\left[\frac{N_d - n_d}{n_d g}\right] + \phi_d \tag{6.2.17}$$

which yields

$$\frac{n n_d g}{N_c(N_d - n_d)} = \exp\left[\frac{\phi_d - E_g}{k_0 T}\right] \tag{6.2.18}$$

Let us assume that we are at a temperature low enough that $n \approx n_d$ ($k_0 T \ll E_g$). In this case, (6.2.18) becomes

$$\frac{n^2 g}{N_c[N_d - n]} = \exp\left[\frac{\phi_d - E_g}{k_0 T}\right] \tag{6.2.19}$$

Solving for n we find

$$n = -$$

$$\frac{N_c \exp\left(\frac{\phi_d - E_g}{k_0 T}\right) \pm \sqrt{N_c^2 \exp\left(\frac{2\phi_d - 2E_g}{k_0 T}\right) + 4g N_c N_d \exp\left(\frac{\phi_d - E_g}{k_0 T)}\right)}}{2g} \tag{6.2.20}$$

Using the positive sign only, since n must be a positive quantity, we find the dependence of the density of conduction electrons on the temperature, effective mass, energy gap and donor ionization energy

$$n = \frac{N_c}{2g} \exp\frac{-E_d}{k_0 T}\left[-1 + \sqrt{\left(1 + 4g\frac{N_d}{N_c}\exp\frac{E_d}{k_0 T}\right)}\right] \tag{6.2.21}$$

(where we have taken E_d as a positive quantity as shown in Figure 6.1).

Using (6.2.16), we find for the Fermi energy

$$\mu = E_g - E_d + k_0 T \ln\left[\frac{1}{2g}\left(-1 + \sqrt{1 + 4g\frac{N_d}{N_c}\exp\frac{E_d}{k_0 T}}\right)\right] \tag{6.2.22}$$

At very low temperatures such that

$$4g \frac{N_d}{N_c} \exp \frac{E_d}{k_0 T} \gg 1 \tag{6.2.23}$$

the carrier concentration may be approximated by

$$n = \frac{\sqrt{N_c N_d}}{\sqrt{g}} \exp\left[-\frac{E_d}{2k_0 T} \right] \tag{6.2.24}$$

Thus, at low temperatures the density of free electrons in the conduction band of an n-type sample varies exponentially as $-E_d/2k_0T$ in a manner similar to our result that the number of intrinsic electrons depends exponentially on $E_g/2k_0T$.

At low temperatures, the Fermi level may be written [see (6.2.13)]

$$\mu = E_g - \frac{E_d}{2} + k_0 T \ln\left(\frac{N_d}{2N_c} \right) \tag{6.2.25}$$

Thus at zero degrees Kelvin the Fermi energy lies one-half of the way between the donor energy level and the bottom of the conduction band. As the temperature increases the Fermi level first rises very slightly due to the ionization of the donor levels and then falls to the center of the energy gap as the sample becomes intrinsic.

We can investigate the temperature dependence of the carrier concentration and the Fermi level at slightly higher temperatures. Under the conditions

$$4g \left(\frac{N_d}{N_c} \right) \exp \frac{E_d}{k_0 T} \ll 1 \tag{6.2.26}$$

and

$$k_0 T \ll E_g \tag{6.2.27}$$

we find for the carrier concentration

$$n = N_d \left[1 - 2g \frac{N_d}{N_c} \exp \frac{E_d}{k_0 T} \right] \tag{6.2.28}$$

where we have used the series approximation

$$(1 \pm x)^n = 1 \pm nx + n(n-1)\frac{x^2}{2!} + \cdots \tag{6.2.29}$$

Substitution of (6.2.16) into (6.2.28) yields

$$\mu = E_g + k_0 T \ln\left[\frac{N_d}{N_c}\right] + k_0 T \ln\left[1 - 2g\frac{N_d}{N_c}\exp\left(\frac{E_d}{k_0 T}\right)\right] \tag{6.2.30}$$

Using

$$\ln(1 + x) = x - \tfrac{1}{2}x^2 + \tfrac{1}{3}x^3 + \cdots \tag{6.2.31}$$

we find for the Fermi level at intermediate temperatures

$$\mu = E_g + k_0 T \ln\left(\frac{N_d}{N_c}\right) - k_0 T\, 2g\left(\frac{N_d}{N_c}\right)\exp\left(\frac{E_d}{k_0 T}\right) \tag{6.2.32}$$

As the temperature is increased, the last term in (6.2.32) becomes dominant and the Fermi level drops from its position at 0°K toward the center of the energy gap. When it drops below the donor energy level, essentially all of the donor atoms will be ionized.

PROBLEM

Derive expressions for the carrier concentration and Fermi level of a strongly p-type semiconductor ($N_d = 0$). [Expressions analogous to (6.2.21), (6.2.22), (6.2.28) and (6.2.32).

6.3 THE INFLUENCE OF EXCITED STATES OF IMPURITIES AND DEFECTS[5]

A consequence of our hydrogen-like model for impurities in a lattice is that excited states of the impurity atom exist. Thus each impurity may introduce a number of discrete energy levels ϕ_j into the energy gap (see Figure 6.2). These excited levels are taken into account quite simply by the following method: We consider a sample with a density N_d of similar donor atoms and define the density of donors which have electrons trapped in the ith excited level by N_d^i.

$$N_d^i = N_d\left[\frac{1}{1 + g_i^{-1}\exp\dfrac{\phi_i - \mu}{k_0 T}}\right] - \sum_{j \neq i} N_d^j\left[\frac{1}{1 + g_i^{-1}\exp\left(\dfrac{\phi_i - \mu}{k_0 T}\right)}\right] \tag{6.3.1}$$

Thus the total density of available i levels is equal to the total density of donors minus the density of donors which have electrons in other levels than the ith level. This is equal to the effective number of i levels occupied because again coulombic repulsion allows only one electron per atom. But

$$N_d - \sum_{j \neq i} N_d^j = N_d + N_d^i - \sum_j N_d^j \tag{6.3.2}$$

where

$$\sum_j N_d^j = N_d^0 \tag{6.3.3}$$

Thus 6.3.2 may be written

$$N_d - \sum_{j \neq i} N_d^j = (N_d - N_d^0) + N_d^i = N_d^+ + N_d^i \tag{6.3.4}$$

Substituting (6.3.4) into (6.3.1) yields the ratio between the number of electrons in the ith energy level to the total number of ionizid donors

$$\frac{N_d^i}{N_d^+} = \frac{1}{g_i^{-1} \exp\left(\dfrac{\phi_i - \mu}{k_0 t}\right)} \tag{6.3.5}$$

The total number of neutral donors is given by $\sum_i N_d^i$. Thus

$$N_d^0 = N_d^+ \sum_i g_i^{+1} \exp\left(\frac{\mu - \phi_i}{k_0 T}\right) \tag{6.3.6}$$

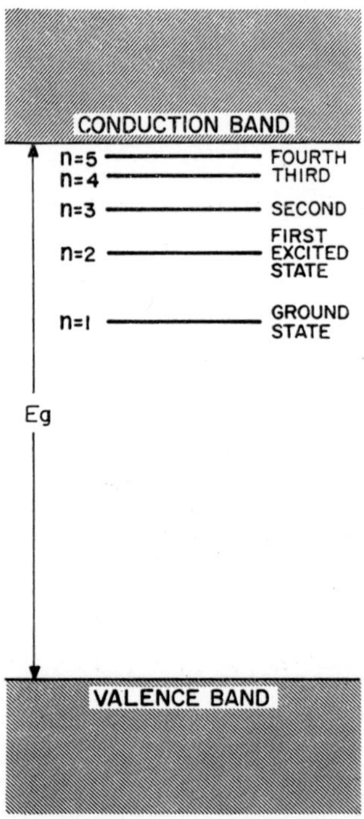

Figure 6.2 Excited states of a hydrogen-like impurity within the energy gap of a semiconductor.

Using

$$N_d^+ = N_d - N_d^0 \tag{6.3.7}$$

(6.3.6) may be written

$$N_d^+ = \frac{N_d}{1 + \sum_i \left[g_i^{+\prime} \exp\left(\frac{-\phi_i + \mu}{k_0 T} \right) \right]} \tag{6.3.8}$$

We should note that we need to consider only the first few excited states of (6.3.8) since the higher states will form part of the conduction band. The effect of the inclusion of the excited states on a graph of $\ln(n)$ vs $1/T$ for a compensated semiconductor is shown in Figure 6.3. Thus we have found the

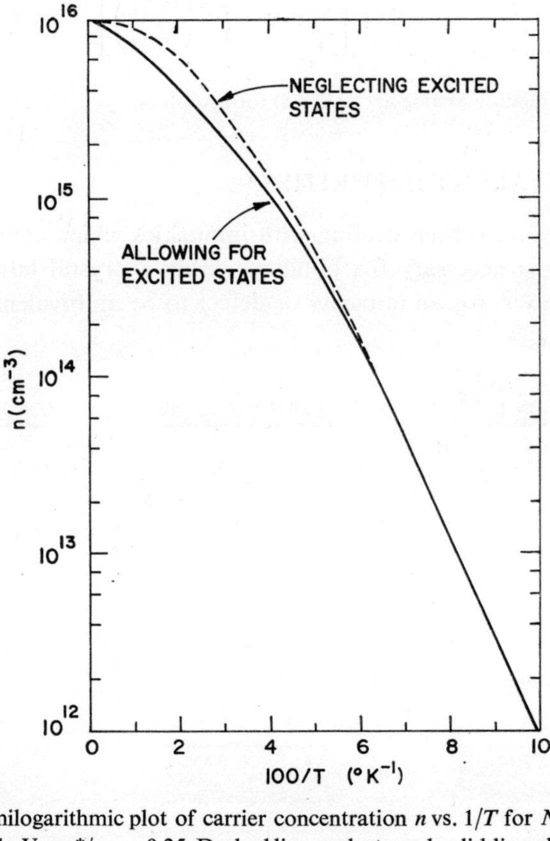

Figure 6.3 Semilogarithmic plot of carrier concentration n vs. $1/T$ for $N_d = 1.1 \times 10^{16}$ cm^{-3}. $E_d = 0.01$ eV, $m_e^*/m = 0.25$. Dashed line neglects and solid line allows for excited states. [This sample also contains 10^{15} cm^{-3} acceptor impurities. See section 6.5] (from J. S. Blakemore, *Semiconductor Statistics*, Pergamon Press, New York, 1962, Page 145).

dependence of the density of ionized donors on the total density of donor impurities each of which has a set of excited energy states of degeneracy g_i. Note that the quantity

$$\frac{1}{1 + \sum_i g_i \exp\left(\frac{-\phi_i + \mu}{k_0 T}\right)} \tag{6.3.9}$$

acts as if it were

$$1 - f_{\varepsilon_i}(E) \tag{6.3.10}$$

where the subscript ε_i indicates a type of distribution function valid for a set of energy levels i.e.,

$$f_{\varepsilon_i}(E) = \frac{1}{1 + \left[\sum_i g_i \exp\left(\frac{\mu - \phi_i}{k_0 T}\right)\right]^{-1}} \tag{6.3.11}$$

PROBLEM

Determine the equation analogous to (6.3.8) for acceptors.

6.4 MULTIVALENT IMPURITIES[6]

Up to now we have been dealing with impurities which contain one more electron than is necessary for bonding within a crystal lattice. It is also possible, however, for an impurity or defect to be multivalent n- or p-type.

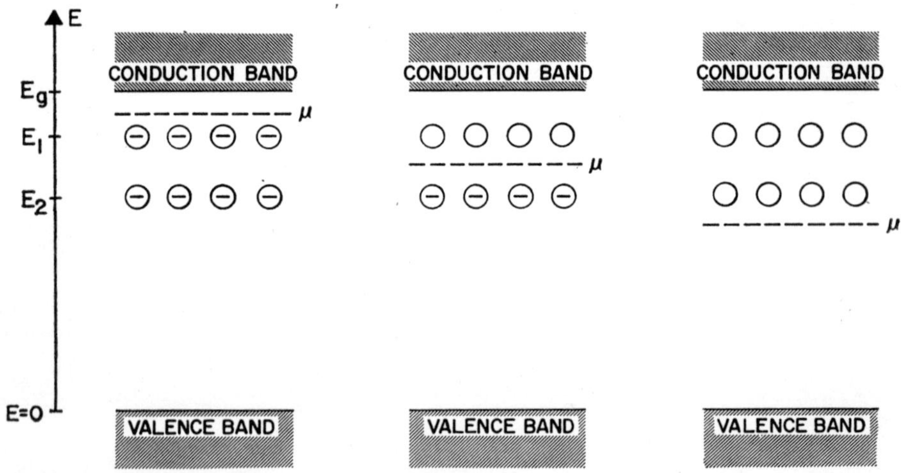

Figure 6.4 (a) Neutral, (b) singly ionized and (c) doubly ionized helium-like donor impurities. Degree of ionization depends on the position of the Fermi level.

For example, the substitution of a Group VI impurity such as oxygen in a group IV lattice produces a site with two more electrons than necessary for covalent bonding in the lattice. In this case, the first electron may be ionized with energy E_1 and the second with energy E_2 as is depicted in Figure 6.5. As is shown in Figure 6.4, the donor will be singly or doubly ionized depending on the position of the Fermi level. When the Fermi level is above the first ionization energy E_1, all N cm^{-3} impurities are unionized and neutrally charged. When the temperature increases and the Fermi level moves down between E_1 and E_2, the impurities become singly ionized $(+1)$. As μ moves down still farther, past E_2, the impurities become doubly ionized and charged $2+$. In this case it is neither valid to treat the shallow level as an excited

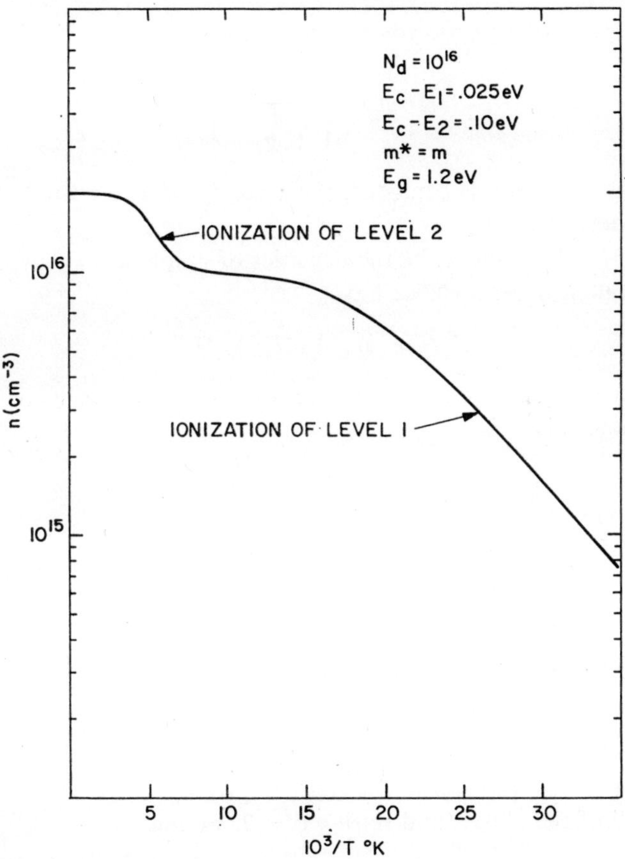

Figure 6.5 Predicted carrier concentration n vs. inverse temperature for 10^{16} cm^{-3} helium-like impurities with energy levels $E_c = E_1 = 0.025$ eV and $E_c - E_2 = 0.1$ eV.

state of the singly ionized impurity nor is it valid to treat the levels as independent since the number of electrons in one level is dependent on the number of electrons in the other levels. In this section we will show, how one can treat complicated systems by use of simple Fermi functions, if one counts states correctly.

Let us examine the simple case of N cm^{-3} helium-like impurity atoms of the type shown in Figure 6.4. In its doubly charged state, filling of the level E_2 follows the usual occupation statistics for a g times degenerate donor level with

$$f_{FD}(E_1) = \frac{1}{1 + g_2^{-1}\, e^{\frac{E_2-\mu}{k_0 T}}} \tag{6.4.1}$$

However, filling of E_1 proceeds according to (6.2.4)

$$f_{FD}(E_2) = \frac{1}{1 + g_2\, e^{\frac{E_2-\mu}{k_0 T}}} \tag{6.4.2}$$

Once the first electron is captured, the second has a restricted spectrum of allowable degeneracies.

Letting M_0, M_1 and M_2 be the densities of unionized, singly and doubly ionized divalent impurities, we have

$$N = M_0 + M_1 + M_2 \tag{6.4.3}$$

The total density of electrons n in the conduction band which come from the divalent donors is

$$n = M_1 + 2M_2 \tag{6.4.4}$$

The densities of available states at E_1 and E_2, N_1 and N_2, respectively, are

$$N_1 = M_0 + M_1 \tag{6.4.5a}$$

$$N_2 = M_1 + M_2 \tag{6.4.5b}$$

The densities of electrons n_1 and n_2 in levels E_1 and E_2 respectively are

$$n_1 = M_0 = N_1 f_{FD}(E_1) \tag{6.4.6a}$$

$$n_2 = M_1 \tag{6.4.6b}$$

Utilizing all of the above, and setting $g = 2$, we find

$$n = \frac{N[2 - f(E_2)]\,[1 - f(E_1)]}{1 - f(E_1) + f(E_1)f(E_2)} \tag{6.4.7}$$

n is, of course, also given by (6.1.11). (6.4.7) may be written in the following form

$$x^3 + 2\alpha x^2 - \alpha(1 - \beta)\,x - \alpha\beta = 0 \qquad (6.4.8)$$

where

$$x = \frac{N_c}{2N}\exp\left[\frac{\mu - E_g}{k_0 T}\right] = \frac{n}{2N} \qquad (6.4.9)$$

$$\alpha = \frac{N_c}{2N}\exp\left[\frac{E_1 - E_g}{k_0 T}\right] \qquad (6.4.10)$$

$$\beta = \frac{N_c}{2N}\exp\left[\frac{E_2 - E_g}{k_0 T}\right] \qquad (6.4.11)$$

A plot of n vs $10^3/T$ predicted by (6.4.8) is shown in Figure 6.5. (6.4.8) has relatively simple forms in the limit of very high and very low temperatures. At very low temperatures where $n \ll N$, level E_1 begins to ionize according to

$$n = \frac{2N}{1 + (1 + \alpha^{-1})^{1/2}} \qquad (6.4.12)$$

At very high temperatures, (6.4.8) becomes

$$n = \frac{4N\beta}{(1 - \beta)\left[\pm\left\{1 + \dfrac{8\beta}{(1 - \beta)^2}\right\}^{1/2} - 1\right]} \qquad (6.4.13)$$

At very low temperatures (6.4.8) yields the temperature dependence of the number of ionized electrons from level E_1

$$n = \sqrt{2NN_c}\,\exp\left[\frac{E_1 - E_g}{2k_0 T}\right] \qquad (6.4.14)$$

[see (6.2.24)].

At slightly higher temperatures level E_1 becomes exhausted according to

$$n = N\left[1 - \left(\frac{N}{2N_c}\right)\exp\left(\frac{E_g - E_1}{k_0 T}\right)\right] \qquad (6.4.15)$$

(6.4.13) predicts that ionization of E_2 proceeds with twice the exponent as is usual for hydrogen-like impurities

$$n = N + \frac{N_c}{2}\exp\frac{E_2 - E_g}{k_0 T} \qquad (6.4.16)$$

and exhausts at higher temperatures according to

$$n = 2N\left[1 - \frac{2N}{N_c}\exp\frac{E_g - E_2}{k_0T}\right] \tag{6.4.17}$$

It is possible also for an impurity or defect to be amphoteric, i.e., to act as either a donor or acceptor, depending on the position of the Fermi level. An example is gold, which introduces 3 acceptor and one donor level in Ge, as shown in Figure 5.6.

6.5 COMPENSATED MATERIALS

We have derived the formulas which predict the temperature dependence of the number of carriers in intrinsic, strongly n-type and strongly p-type materials. In most cases, however, it is not valid to assume that a semiconductor is dominated by only one type of impurity. One must take into account the presence of both donor and acceptor impurities when making actual calculations. The presence of both n- and p-type impurities in a material is called *compensation*. In this section, we shall investigate the statistics which goven the number of free carriers in a partially compensated semiconductor.

Let us assume that conduction in our crystal is dominated by the densities N_d and N_a of donor and acceptor impurity atoms, respectively. In this case, the charge neutrality condition is

$$n_i + n_d + N_a^- = p_i + p_a + N_d^+ \tag{6.5.1}$$

The introduction of substantive amounts of both donor and acceptor atoms will, as could be expected, shift the Fermi level from its intrinsic position. In the approximation that our system remains non-degenerate (the Fermi level is a few times k_0T from a band edge within the forbidden energy gap), we may again assume that the number of (intrinsic) holes in the valence band is equal zero, (6.5.1) may be written

$$n = N_d\left[\frac{1}{1 + g_d\, e^{\left(\frac{\mu - \phi_d}{k_0T}\right)}}\right] - N_a\left[\frac{1}{1 + g_a\dfrac{E_a - \mu}{k_0T}}\right] \tag{6.5.2}$$

Let us investigate first the case where $N_d > N_a$. As we can expect from our earlier work, the Fermi level will be somewhat closer to the donor impurity level than to the acceptor level, and thus the density of ionized acceptors N_a will be approximately equal to the total density of acceptor atoms N_a. Thus

$$n = N_d \left[\frac{1}{1 + g_d e^{\frac{\mu - \phi_d}{k_0 T}}} \right] - N_a \tag{6.5.3}$$

which may be solved to yield

$$\frac{n(n + N_a)}{N_d - n - N_a} = \frac{N_c}{g_d} \exp\left[\frac{-E_g + \phi_d}{k_0 T} \right] = \frac{N_c}{g_d} \exp\left(\frac{-E_d}{k_0 T} \right) \tag{6.5.4}$$

PROBLEM

Find the solution corresponding to (6.5.4) which includes the first two excited states of a donor atom.

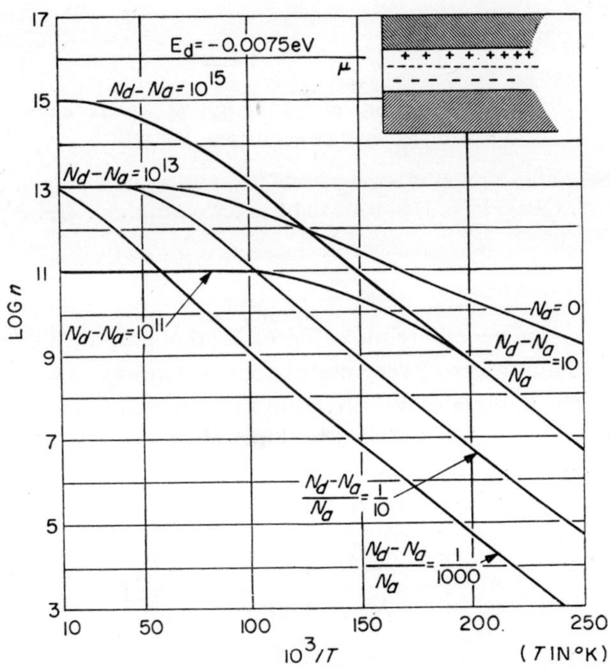

Figure 6.6a Natural logarithm of the free carrier concentration in an n-type compensated semiconductor for different degrees of compensation. At high temperatures n is determined by $(N_d - N_a)$ but at low temperatures it depends on $(N_d - N_a)/N_a$. (From E. H. Putley, *The Hall Effect and Related Phenomena*, Butterworths, London, 1960, page 126.)

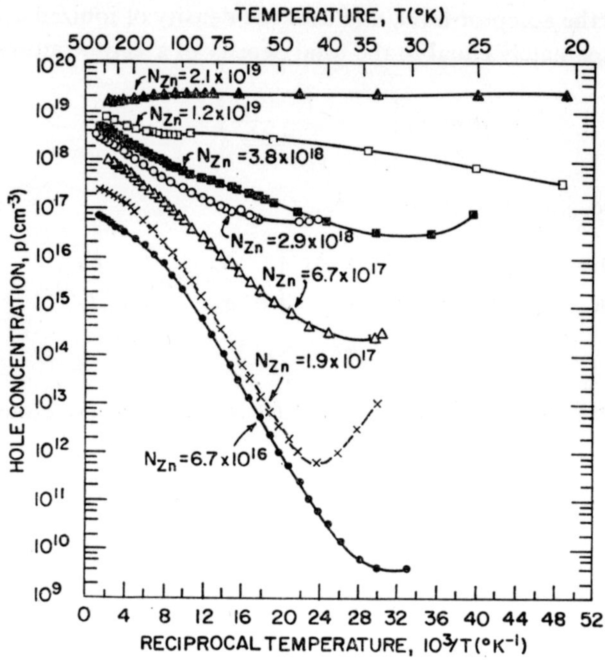

Figure 6.6b Natural logarithm of experimental free carrier concentration in zinc doped GaP (from H. C. Casey, Jr, F. Hermanis and K. B. Wolfstirn, *Jnl. App. Phys.*, **40**, 2495, 1969).

To solve (6.5.4) in general requires a great deal of graphical labor. We can, however, make a number of very useful approximations. As we go to lower temperatures, the number of free electrons in the conduction band decreases. In the temperature range $k_0T < E_d$, the density of conduction electrons may become much smaller than the difference in the densities of donor and acceptor impurities. In this case (6.5.4) approximates to

$$n = \left[\frac{N_d - N_a}{N_a}\right]\frac{N_c}{g_d}\exp\left(\frac{-E_d}{k_0T}\right) \tag{6.5.5}$$

PROBLEM

Show for the case of moderate temperatures ($k_0T \approx E_d$) $N_a \ll n \ll N_d$ that the number of free electrons in the conduction band again depends exponentially on $E_d/2k_0T$, i.e.

$$n = (N_d - N_a)^{1/2}\left(\frac{N_c}{g_d}\right)^{1/2}\exp\left(\frac{-E_d}{2k_0T}\right) \tag{6.5.6}$$

At higher temperatures in the range

$$E_g - \phi_d \ll k_0T \ll E_g \tag{6.5.7}$$

(6.5.5) reduces to

$$n = N_d - N_a \tag{6.5.8}$$

Figure 6.6a shows a plot of ln (n) vs $1/T$ for different values of $(N_d - N_a)$. Figure 6.6b shows the free carrier concentration as a function of reciprocal temperature for a number of p-type GaP samples. Note that as the concentration increases, the overlap of impurity wave functions produces a smaller ionization energy (section 5.1) which results in a smaller slope.

PROBLEM

Derive the appropriate equations for a compensated p-type semiconductor, analogous to (6.5.5). What is the high temperature intercept? How does the number of holes in the valence band at low temperatures depend on E_a, the acceptor ionization energy?

6.6 DEGENERATE STATISTICS

For the case where very many impurities or defects have been introduced into the crystal lattice, pushing the Fermi level to within k_0T of either the conduction or valence bands, we may no longer use the classical approximation. This is the case which must be used for highly doped semiconductors and metals. We shall now investigate an approximate solution for integrals of the form (6.1.4) (in particular $\mathscr{F}_{1/2}(\mu/k_0T)$ which arises in the parabolic band approximation). Degenerate systems may be treated quite simply by making use of the following approximation[7]

$$n = \int_0^\infty f(E)\,g(E)\,dE = -\int_0^\infty G(E)\frac{\partial f(E)}{\partial E}\,dE$$

$$\approx \int_0^\infty g(E)\,dE + \frac{\pi^2(k_0T)^2}{6}\left(\frac{\partial g(E)}{\partial E}\right)\bigg|_{E=\mu_0} + \frac{7\pi^4}{360}(k_0T)^4\frac{\partial^3 g(E)}{\partial E^3}\bigg|_{E=\mu_0} \tag{6.6.1}$$

where $f(E)$ is the Fermi distribution function and

$$G(E) = \int_0^E g(E')\,dE' \tag{6.6.2}$$

Let us prove (6.6.1) by considering the integral

$$I = \int_0^\infty \frac{g(E)\,dE}{e^{\frac{E-\mu}{k_0T}} + 1} \tag{6.6.3}$$

Integration by parts yields

$$I = f(E)\,G(E)|_0^\infty - \int_0^\alpha G(E)\frac{\partial f(E)}{\partial E}\,dE \tag{6.6.4}$$

The Fermi distribution function approaches 0 and 1 at $E = \infty$ and $E = 0$, respectively. Thus, if the density of states function $g(E)$ approaches ∞ slower than exponentially, and approaches zero for $E = 0$, the first term on the right of (6.6.4) will tend to zero yielding

$$I = -\int_0^\infty \frac{\partial f(E)}{\partial E}\,G(E)\,dE \tag{6.6.5}$$

(Note that our parabolic density of states function satisfies these requirements). A plot of the derivative of the Fermi-Dirac distribution function with respect to energy (Figure 6.7) shows that it has an appreciable value only for energies near the Fermi level. Thus the integrand in (6.1.11) has an appreciable value only for $E = \mu$. We may therefore expand $G(E)$ in a Taylor series about μ.

$$G(E) = \sum_{n=0}^\infty \frac{(E-\mu)^n}{n!}\left.\frac{\partial^{(n-1)}g(E)}{\partial E^{(n-1)}}\right|_{E=\mu} \tag{6.6.6}$$

It is easily verified that

$$\frac{\partial f(E)}{\partial E} = -\frac{1}{k_0T}\left[\frac{e^{\frac{E-\mu}{k_0T}}}{\left(e^{\frac{E-\mu}{k_0T}}+1\right)^2}\right] \tag{6.6.7}$$

which in turn may be written

$$\frac{1}{k_0T}\left[\sum_{m=1}^\infty (-1)^m\,m\,e^{-m\left(\frac{E-\mu}{k_0T}\right)}\right] \tag{6.6.8}$$

Substitution of (6.6.6) and (6.6.8) into (6.6.5) yields

$$\begin{aligned} I = &-\int_0 g(E)\,dE\int_0 dE\left.\frac{\partial f(E)}{\partial E}\right|_{E=\mu} \\ &-\frac{1}{k_0T}\int\sum_{n=1}^\infty\sum_{m=1}^\infty(-1)^m\,m\,e^{-m\left(\frac{E-\mu}{k_0T}\right)}\frac{(E-\mu)^n}{n!}\left.\frac{\partial^{(n-1)}g(E)}{\partial E^{(n-1)}}\right|_{E=\mu}dE \end{aligned} \tag{6.6.9}$$

Noting that

$$\int_0^\infty dE \frac{\partial f(E)}{\partial E} = 1 \tag{6.6.10}$$

we find

$$I = G(E) - \frac{1}{k_0 T} \int_0^\infty \sum_{n=0}^{\infty} \sum_{m=0}^{\infty} (-1)^m \, m \, e^{\frac{-m(E-\mu)}{k_0 T}} \frac{(E-\mu)^n}{n!} \frac{\partial^{(n-1)} g(E)}{\partial E^{(n-1)}}\Bigg|_{E=\mu} dE \tag{6.6.11}$$

Changing to the new variable

$$x = \frac{E - \mu}{k_0 T} \tag{6.6.12}$$

we get

$$I = G(E) - \sum_{n-1}^{\infty} \sum_{m=1}^{\infty} \frac{\partial^{(n-1)} g(E)}{\partial E^{(n-1)}}\Bigg|_{E=\mu} \frac{(k_0 T)^n (-1)^m m}{n!} \int_{-\infty}^{\infty} x^n e^{-mx} \, dx \tag{6.6.13}$$

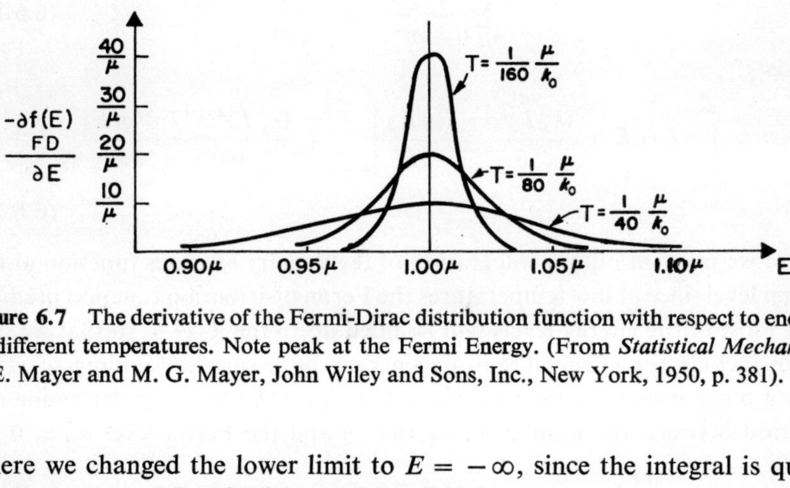

Figure 6.7 The derivative of the Fermi-Dirac distribution function with respect to energy at different temperatures. Note peak at the Fermi Energy. (From *Statistical Mechanics*, J. E. Mayer and M. G. Mayer, John Wiley and Sons, Inc., New York, 1950, p. 381).

where we changed the lower limit to $E = -\infty$, since the integral is quite small at $x = -\mu/k_0 T$ for low temperatures.

The integrand is asymetrical in x, thus the integral is zero for odd n. For even values of n, we may write

$$I = G(E) - 2 \sum_{m=1}^{\infty} \sum_{n=2,4\ldots}^{\infty} \frac{\partial^n g(E)}{\partial^n (E)}\Bigg|_{E=\mu} \frac{m (k_0 T)^n (-1)^m}{n!} \int_0^\infty x^n e^{-mx} \, dx \tag{6.6.14}$$

We note that

$$\int_0^\infty x^n e^{-mx}\,dx = \frac{n!}{(m)^{n+1}} \tag{6.6.15}$$

Thus (6.6.14) becomes

$$I = G(E) - 2\sum_{m=1}^{\infty}\sum_{n=2}^{\infty}\left[\frac{\partial^n g(E)}{\partial E^n}\bigg|_{E=\mu}\frac{(k_0T)^n(-1)^m\,m}{n!}\left(\frac{n!}{m^{n+1}}\right)\right] \tag{6.6.16}$$

Expanding the first two terms in n we find

$$I = G(E) - 2\sum_{m=1}^{\infty}\left[\frac{\partial^2 g(E)}{\partial E^2}\frac{(k_0T^2)(-1)^m}{m^2} + \frac{\partial^4 g(E)}{\partial E^4}\frac{(k_0T)^4(-1)^m}{m^4} + \cdots\right] \tag{6.6.17}$$

Utilizing

$$\sum_{m=1}^{\infty}\frac{(-1)^m}{m^2} = \frac{\pi^2}{12} \tag{6.6.18}$$

and

$$\sum_{m=1}^{\infty}\frac{(-1)^m}{m^4} = \frac{-7\pi^4}{120} \tag{6.6.19}$$

yields

$$I = n = \int_0^\mu g(E)\,dE + \frac{(k_0T)^2\pi^2}{6}\frac{\partial g(E)}{\partial E}\bigg|_{E=\mu} + \frac{(k_0T)^4\pi^4 7}{360}\frac{\partial^3 g(E)}{\partial E^3}\bigg|_{E=\mu} + \cdots \tag{6.6.20}$$

where we may cut off our integration of the density of states function at the Fermi level since at low temperatures the Fermi distribution function predicts that most of the energy levels will be filled up to the level μ, so that we can again include the effect of spin degeneracy by a simple doubling of $g(E)$. Using our parabolic density of states function $g(E)$, we may determine the relation between the number of carriers n and the Fermi level μ_0 at 0°K.

$$n = \int_0^{\mu_0}\frac{1}{2\pi^2}\left(\frac{2m^*}{\hbar^2}\right)^{3/2}E^{1/2}\,dE = \frac{2}{6\pi^2}\left(\frac{2m^*}{\hbar^2}\right)^{3/2}\mu_0^{3/2} \tag{6.6.21}$$

The radius μ_0 of an approximate Fermi sphere is thus

$$\mu_0 = n^{2/3}(3\pi^2)^{2/3}\left(\frac{\hbar^2}{2m^*}\right) \tag{6.6.22}$$

To second order in the temperature (in the parabolic band approximation) the change in the radius of the Fermi sphere with temperature may be determined approximately by making use of the first two terms of (6.6.20). Thus

$$n \approx \frac{1}{3\pi^2}\left(\frac{2m^*}{\hbar^2}\right)^{3/2}(3n\pi^2)\left(\frac{\hbar^2}{2m^*}\right)^{3/2} + \frac{(k_0T)^2}{12}\left(\frac{2m^*}{\hbar^2}\right)^{3/2}\frac{\mu^{-1/2}}{2} \tag{6.6.23}$$

which yields the change in the Fermi level with temperature. Which results in the "smearing out" of the surface of the Fermi sphere due to the excitation of electrons to higher allowed energy levels of our crystal.

References

1. *Handbook of Mathematical Functions*, Ed. by M. Abramowitz and I. A. Stegun. National Bureau of Standards Applied Mathematics Series, 7055, May 1968.
2. J. S. Blakemore, *Semiconductor Statistics*, Pergamon Press, New York, 1962, p. 346.
3. G. Dresselhaus, A. Kip, C. Kittel and G. Wagoner, *Phys. Rev.*, **98**, 556, 1955.
4. W. Kohn, *Solid State Physics*, Ed. by F. Seitz and D. Turnbull, Vol. 5, Academic Press, New York, 1957.
5. This section follows J. S. Blakemore, *Semiconductor Statistics*, Pergamon Press, New York, 1962, p. 140.
6. This section follows C. H. Champness, *Proc. Phys. Soc.* Lon. ***B*69**, 1335, 1956. For a more general treatment, see W. Shockley and J. T. Last, *Phys. Rev.* **107** (1957) 392.
7. A. Sommerfeld, *Z. Physik*, **47**, 1, 1928.

CHAPTER 7

Dynamics of Crystal Lattices

7.1 BINDING FORCE AND BINDING ENERGY

Binding Forces

In Chapter 4, we discussed the formation of a stable covalent bond. We can gain greater insight, however, into the bonding forces which hold crystals together by considering an ionic crystal. The total force F_T which holds a crystal together is the sum of an attractive force F_A derived from a spherically symmetric inverse r coulombic potential U_A.

$$F_A = -\frac{\partial U_A}{\partial r} = -\frac{\partial}{\partial r}\left[\frac{-e^2 z_1 z_2}{\varkappa r}\right] = \frac{-e^2 z_1 z_2}{\varkappa r^2} \tag{7.1.1}$$

and a repulsive force F_R which is probably due to the shielded ion-ion coulombic repulsion. Not much can be said about this repulsive force except that it can be expected to drop off far more rapidly than the unshielded coulomb forces. Therefore, we will assume that it too is derivable from an inverse r law

$$F_R = -\frac{\partial U_R}{\partial r} = -\frac{\partial}{\partial r}\left[\frac{B}{r^n}\right] = \frac{nB}{r^{(n+1)}} \tag{7.1.2}$$

where n, the Born exponent, is unknown. We will now show that the repulsive term F_R is very short ranged and therefore may be neglected in calculations of the crystal binding energy.

The potentials U_A and U_R combine, as shown in Figure 7.1, to produce a total potential U_T which contains a stable equilibrium separation a, which is, of course, the lattice constant. At this separation, the total force on the atoms is equal to zero.

$$F_T = -\left.\frac{\partial U_T}{\partial r}\right|_{r=a} = -\frac{e^2 z_1 z_2}{\varkappa a^2} + \frac{nB}{a^{n+1}} = 0 \tag{7.1.3}$$

Thus

$$a = \left[\frac{nB\varkappa}{e^2 z_1 z_2}\right]^{\frac{1}{n-1}} \tag{7.1.4}$$

Up to now we have been considering crystal lattices as if the lattice sites were rigidly held by interatomic forces. We know, however, that solids expand and contract, for example with pressure and heat, and this implies that the interatomic forces are not rigid, but allow for at least a small deviation from their equilibrium positions. Let us therefore assume a small adiabatic, reversible contraction or dilation of our lattice and expand the

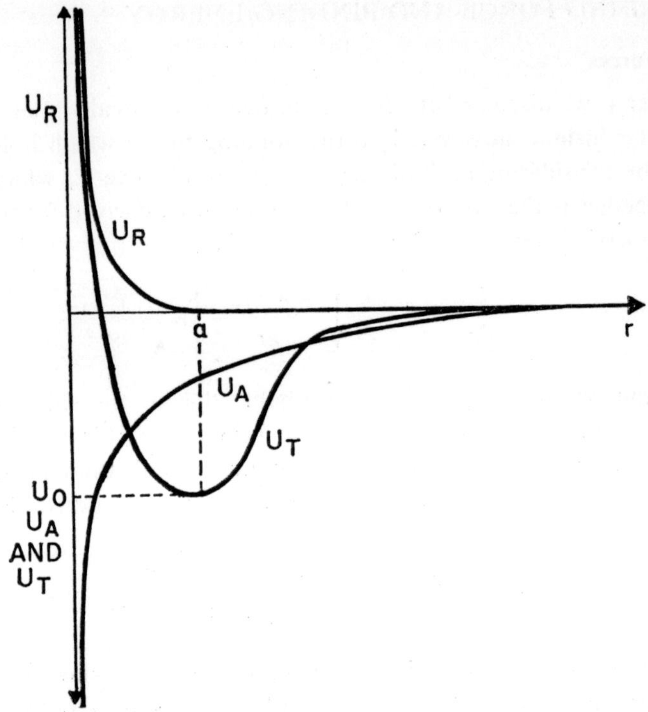

Figure 7.1 Addition of the short range repulsive potential U_R and the attractive potential U_A to produce the total crystal binding potential U_T. The lattice constant is a.

total binding potential energy in terms of a Taylor series about the equilibrium volume V_0, i.e.,

$$U_T \approx U_0 + \Delta V\left(\frac{\partial U_T}{\partial V}\right)_{V_0} + \frac{(\Delta V)^2}{2}\left(\frac{\partial^2 U_T}{\partial V^2}\right)_{V_0} + \ldots \tag{7.1.5}$$

In equilibrium $\left(\frac{\partial U_T}{\partial V}\right)_{V_0} = 0$. Thus the change in the internal energy upon dilation is

$$U_T - U_0 = \frac{(\Delta V)^2}{2}\left(\frac{\partial^2 U_T}{\partial V^2}\right)_{V_0} \tag{7.1.6}$$

For small changes in volume, the bulk compressibility may be written

$$\beta = -\frac{\Delta V}{V_0 P} \tag{7.1.7}$$

where we have assumed that the initial pressure is equal to zero. Thus the change in energy during adiabatic dilation is also given by

$$U_T - U_0 = -\int_{V_0}^{V} P\,dV = \frac{(\Delta V)^2}{2\beta V_0} \tag{7.1.8}$$

Equating (7.1.6) and (7.1.8) yields

$$\beta = \frac{1}{V_0}\left(\frac{\partial^2 U_T}{\partial V^2}\right)_{V_0}^{-1} \tag{7.1.9}$$

Thus the bulk compressibility is inversely proportional to the curvature of the total internal potential energy curve near its equilibrium value.

The volume of a simple cubic crystal is given by ca^3 where c is a constant. Thus we find

$$\frac{\partial^2 U_T}{\partial V^2} = \frac{1}{9c^2a^2}\left[-\frac{4e^2z_1z_2}{\varkappa a^5} + \frac{n(n+3)\,B}{a^{n+4}}\right] \tag{7.1.10}$$

Using our value of B from (7.1.4) yields

$$n = 1 + \frac{9c\varkappa a^4}{\beta e^2 z_1 z_2} \tag{7.1.11}$$

The bulk compressibility is primarily a function of the valence electrons of an atom. Thus it is an oscillating function of the atomic number as is shown in Figure 7.2. The equilibrium separation increases quite strongly with atomic number, thus its fourth power dominates over z_1 and z_2 in the denominator of (7.1.11). Thus it is usually found that the Born exponent increases with increasing atomic number as is shown in Table 7.1. The large values of n thus confirm that the repulsive potential is very short ranged.

Table 7.1 Values of the Born Exponent n. (from L. Pauling, *Nature of the Chemical Bond*, Cornell University Press, Ithaca, N.Y., 1960, p. 509)

Ion	n
He	5
Ne	7
Ar, Cu^+	9
Kr, Ag^+	10
Xe, Au^+	12

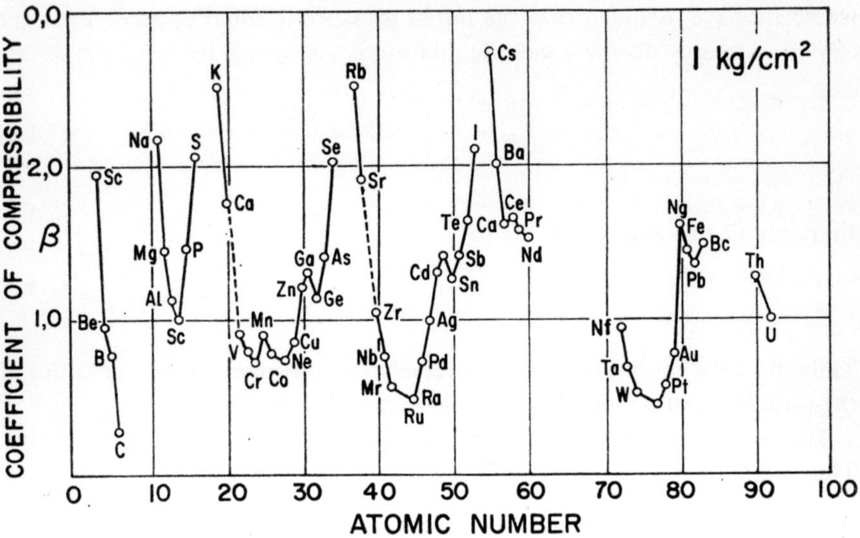

Figure 7.2 Variation of the bulk compressibility β with the atomic number (from L. F. Vereshchagin, *Nature*, **6**, 3, 1957).

Neglecting repulsion terms we may write the total crystalline energy as

$$U = \sum_{i \neq j} \sum \frac{z_i z_j e^2}{\varkappa r_{ij}} \tag{7.1.12}$$

We introduce the notation $r_{ij} = m_{ij} a$ where m_{ij} is a dimensionless parameter

$$U = \frac{N e^2}{a} \sum_{i \neq j} \frac{z_i z_j}{\varkappa m_{ij}} \tag{7.1.13}$$

We may define

$$\sum_{i \neq j} \frac{(\pm)}{m_{ij}} = A \tag{7.1.14}$$

which is the Madelung constant which for simple structures may be easily evaluated. We will evaluate A for a linear lattice of Na^+ and Cl^- atoms as is shown in Figure 7.3. Choosing a negative ion as a reference site we find

$$A = 2[1 - \tfrac{1}{2} + \tfrac{1}{3} - \tfrac{1}{4} + \tfrac{1}{5} - \tfrac{1}{6} + \cdots] \tag{7.1.15}$$

Using the converging series

$$\ln(1 + x) = x - \frac{x^2}{2} + \frac{x^3}{3} - \frac{x^4}{4} - \cdots \tag{7.1.16}$$

we find

$$A = 2 \ln 2 \tag{7.1.17}$$

For most real structures the convergence of (7.1.14) is quite slow and computer methods must be used (although various approximate mathematical schemes have been worked out[1,2]). The Madelung constants for a few structures are shown in Table 7.2.

Table 7.2 The Madelung constant A for various structures

Structure	A
NaCl	1.7476
CsCl	1.7627
Zincblende	1.6381
Wurzite	1.6413

7.2 VIBRATIONS OF THE INFINITE ONE-DIMENSIONAL MONATOMIC LATTICE

For simplicity let us examine in one dimension the binding force which arises from the binding potential (7.1.5)

$$F(r) = -\frac{\partial U_T}{\partial r} = -\Delta r\left[\left(\frac{\partial^2 U_T}{\partial r^2}\right)_{r_0}\right] - (\Delta r)^2\left[\frac{1}{2}\left(\frac{\partial^3 U_T}{\partial r^3}\right)_{r_0}\right] + \cdots \tag{7.2.1}$$

The first term of (7.2.1) constitutes the harmonic approximation in which an atom is treated as if it were bound to its neighbors by forces which increase linearly with the deviation from equilibrium (Hooke's law). The spring constant K is given by the quantity in brackets.

If we treat the interatomic forces in the harmonic approximation, and assume that the atom is bound strongly only to its nearest neighbor, we can gain insight into the magnitude of the elastic atomic vibration. Let us consider the results of our investigation of the simple quantum mechanical oscillator in Chapter 1. We found that the allowed energies of vibration of the simple oscillator were $(n + 1)\,\hbar\omega$. Thus even at 1°K ($k_0 T = 8.6 \times 10^{-5}$ eV) lattice vibrations of the frequency 4×10^{10} sec^{-1} may take place. Even at low temperatures the lattice ions vibrate frantically and thus the concept of a rigid lattice is not a realistic one, even at low temperatures.

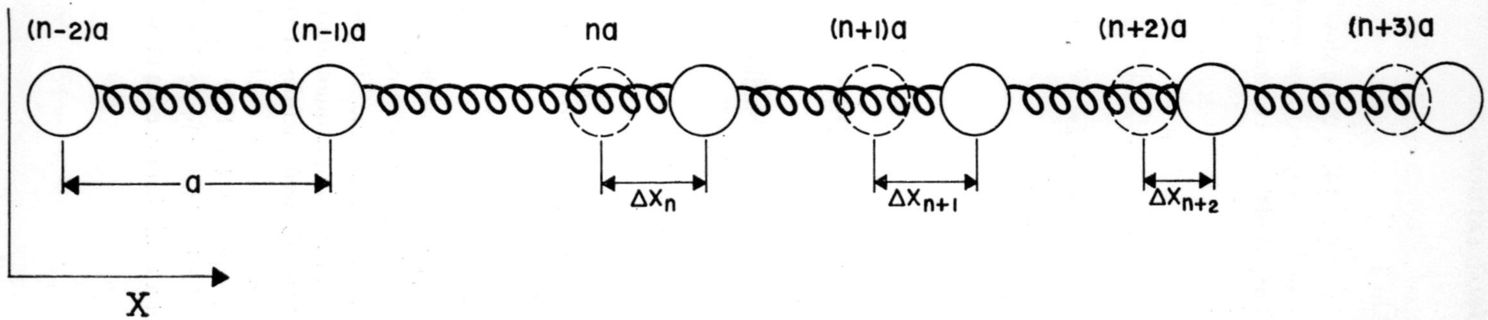

Figure 7.4 Linear monatomic lattice.

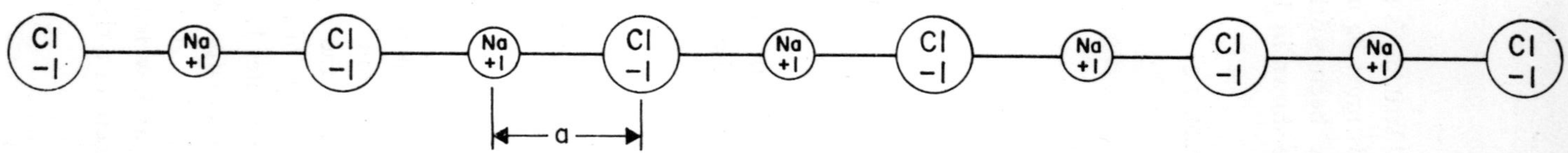

Figure 7.3 Linear Na^+ Cl^- lattice.

Although the harmonic approximation serves to demonstrate many of the features associated with the crystalline lattices, we shall find that we must include anharmonic terms proportional to $(\Delta r)^3$, etc., in our force equation to properly account for phenomena such as thermal expansion.

In the present section we shall investigate the effect of the constraints imposed by lattice continuity on the allowed frequencies of vibration of a number of simple quantum mechanical harmonic oscillators. We shall find that the constraint of lattice continuity has the effect of limiting the spectrum of allowed frequencies of vibration. Thus the student will realize that the determination of the allowed frequencies of vibration of a lattice is treated as a perturbation on the allowed frequencies of vibration of a collection of independent harmonic oscillators, in much the same way as we determined the allowed energy of an electron travelling in the lattice by considering the periodic potential of the ions as a perturbation on the energy of a free electron.

Consider an infinite one-dimensional lattice of similar atoms each of mass m, whose equilibrium lattice spacing is "a" as shown in Figure 7.4. We shall treat only nearest neighbor interactions.

If we examine the arbitrary vibration of an atom in the lattice we find that vibration may be expressed as the result of three independent harmonic vibrations of amplitudes Δx, Δy and Δz, with effective spring constants K_x, K_y and K_z, respectively. The two directions of vibration perpendicular to an arbitrarily chosen reference direction [×] are called *transverse* directions. The remaining vibration in the x direction is the *longitudinal* vibration.

The above allows us to reduce what would have been an extremely complicated three-dimensional problem to three simple independent one-dimensional problems. Whether we choose to investigate the longitudinal or transverse directions in no way inhibits our solution, for the solutions must be equivalent by a simple 90° rotation of our crystalline axes and substitution of the appropriate spring constants. Let us then examine the longitudinal vibrations of an atom along the x direction. To facilitate writing the equations of motion, we shall number the lattice sites as shown in Figure 7.4 and denote the longitudinal displacement from equilibrium of the nth atom as Δx_n. The equation of motion of the nth atom may be written

$$F_n = -K_x[\Delta x_{n+1} - \Delta x_n] - K_x[\Delta x_{n-1} - \Delta x_n] \tag{7.2.2}$$

where K_x is the interaction constant and each term in brackets represents the longitudinal change in equilibrium separation between adjacent atoms each of which generates a force which tends to restore the atom to its original position. Thus we find

$$F_n = m\,\ddot{\Delta x}_n = -K_x[\Delta x_{n+1} + \Delta x_{n-1} - 2\Delta x_n] \qquad (7.2.3)$$

Of course, since we are dealing with an infinite number of atoms along the x direction, we must in principle solve an infinite number of equations of the form (7.2.3). However, the atoms are identical and thus we can from (7.2.3) itself determine the allowed frequencies of vibration of the nth atom (and thus of the whole lattice.)

We are dealing with only the harmonic terms of nearest neighbor interactions and therefore expect the solution of (7.2.3) to be of the form

$$\Delta x_n = A\, e^{i(\omega t + kna)} \qquad (7.2.4)$$

where ω is the angular frequency of longitudinal vibration.

Substitution of (7.2.4) in (7.2.3) yields

$$-m\omega^2\, e^{i(kna)} = -K_x[e^{i(n+1)ka} + e^{i(n-1)ka} - 2e^{i(kna)}] \qquad (7.2.5)$$

where we have used

$$\frac{d}{dt}(\Delta x_n) = i\omega\, \Delta x_n$$

$$\frac{d^2}{dt^2}(\Delta x_n) = -\omega^2\, \Delta x_n \qquad (7.2.6)$$

Dividing (7.2.5) by $-e^{ikna}/m$ yields

$$\omega^2 = \frac{K_x}{m}[e^{ika} + e^{-ika} - 2] \qquad (7.2.7)$$

Making the substitution

$$[e^{ika} + e^{-ika} - 2] = \left[-e^{-\frac{ika}{2}} + e^{\frac{ika}{2}}\right]^2 = 4\sin^2\frac{ka}{2} \qquad (7.2.8)$$

we find for the allowed frequencies of vibration

$$\omega = \pm 2\sqrt{\frac{K_x}{m}}\sin\frac{ka}{2} \qquad (7.2.9)$$

Plotting ω vs $\mathbf{k}$ we find that the allowed frequencies of longitudinal vibration are periodic in $\mathbf{k}$ with a period of $2\pi/a$ (Figure 7.5a). Again we have found that the constraints imposed by lattice continuity have produced a periodicity in the allowed energy of our system. In the present case, as was the case for electron energy, all of the pertinent information concerning the allowed frequencies of vibration is contained in the first Brillouin zone of our extended zone scheme, and thus we may again make use of our reduced zone scheme to obtain our information. The reduced zone for a linear lattice of lattice constant a is determined as before, i.e., by a translation through $2n$ reciprocal lattice vectors, where n is an integer. The reduced scheme is shown in Figure 7.5b. This curve is the so-called acoustical longitudinal branch of the vibrational spectrum, for the vibrational frequencies are of the order of the frequencies of sound waves in the crystal.

Figure 7.6 includes the two branches of the curve which would have resulted if we had also considered the two perpendicular transverse oscillations of the lattice. The two additional ω vs $\mathbf{k}$ curves are called the acoustical transverse branches of the vibrational spectrum. We shall see that when we deal with a lattice which has more than one type of atom another branch (the optical branch) appears at higher frequencies. Note that we have again found that there is symmetry about $\mathbf{k} = 0$, as we have for the allowed energy of an electron in a lattice.

It is quite simple to demonstrate that for small values of $\mathbf{k}$, the velocity of propagation of energy $\left\{\frac{\partial \omega}{\partial k}\right\}$ along the lattice is a constant.

For small values of $\mathbf{k}$, (7.2.9) may be written

$$\omega = \pm \sqrt{\frac{K_x}{m}}\, ka \tag{7.2.10}$$

The velocity of propagation of energy along the lattice becomes

$$\frac{\partial \omega}{\partial k} = \pm a \sqrt{\frac{K_x}{m}} \tag{7.2.11}$$

Thus for small values of ka the lattice acts as if it were a non-dispersive medium for both the transverse and the longitudinal waves.

Assuming a non-isotropic unit cell in which the forces tending to restore each atom to equilibrium are different for the longitudinal and the two transverse modes, we find that the largest frequencies of vibration are (Figure 7.6)

$$\omega_{l(\max)} = 2\sqrt{\frac{K_x}{m}}$$

$$\omega_{t1(\max)} = 2\sqrt{\frac{K_y}{m}}$$

$$\omega_{t2(\max)} = 2\sqrt{\frac{K_z}{m}} \tag{7.2.12}$$

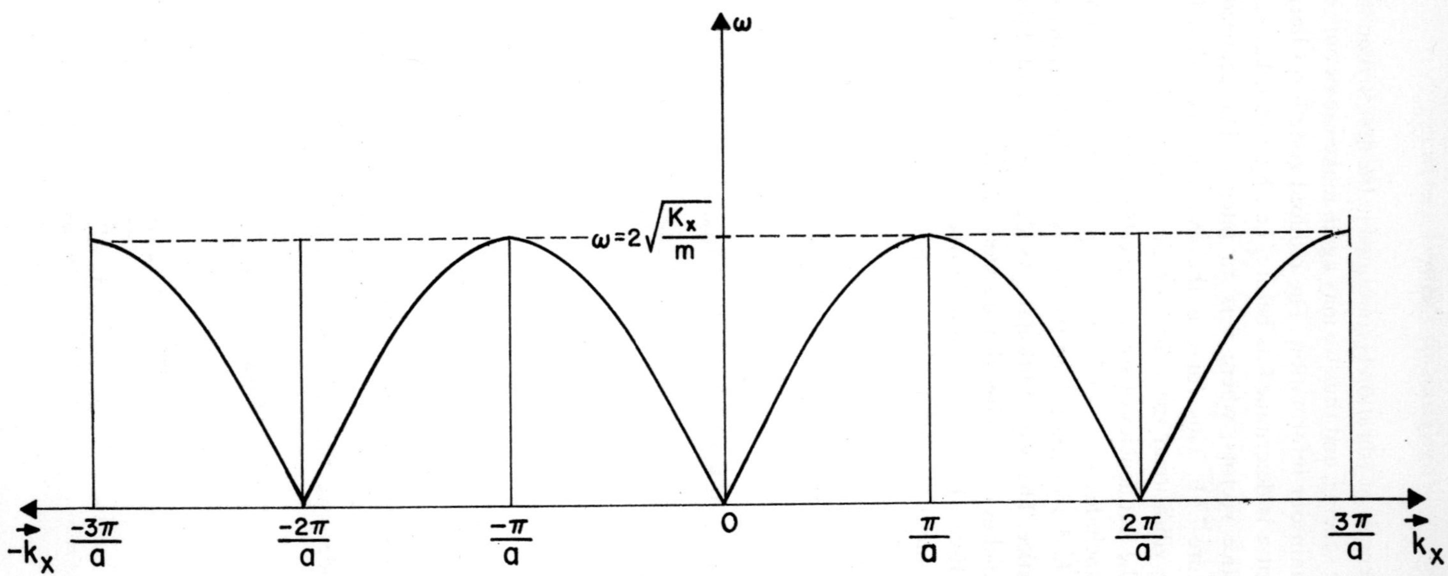

Figure 7.5a

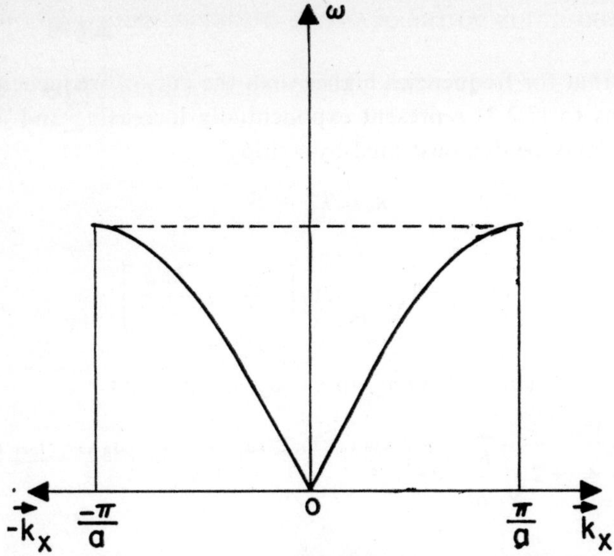

Figure 7.5b

Figure 7.5 (a) Extended and (b) reduced representations of the allowed frequency *vs* ***k*** curves for a one-dimensional monatomic lattice with lattice constant *a*.

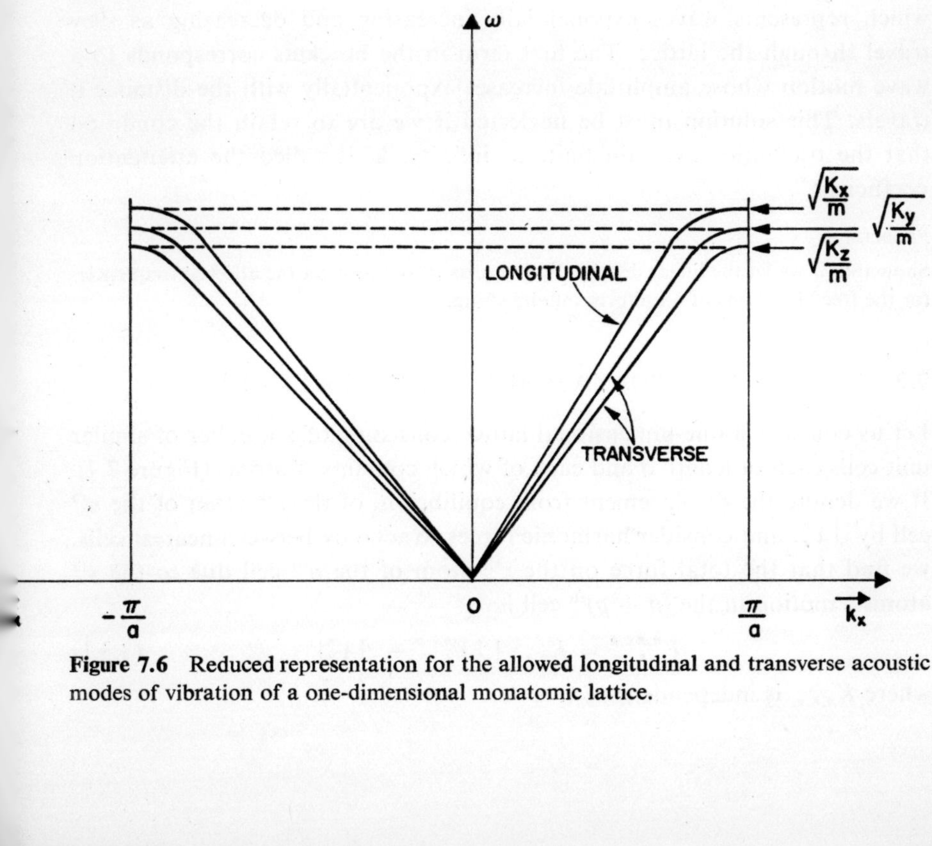

Figure 7.6 Reduced representation for the allowed longitudinal and transverse acoustic modes of vibration of a one-dimensional monatomic lattice.

We note that for frequencies higher than the cut-off frequencies (7.2.12), the solutions to (7.2.3) represent exponentially increasing and attenuating waves. This may be demonstrated by writing

$$k = k_0 + ik' \tag{7.2.13}$$

Thus from (7.2.9)

$$\omega = \pm 2\sqrt{\frac{K_x}{m}} \sin\left[\frac{k_0 a}{2} + \frac{ik'a}{2}\right] \tag{7.2.14}$$

Utilizing

$$\sin(x + y) = \sin x \cos y + \cos x \sin y \tag{7.2.15}$$

we find

$$\omega = \pm 2\sqrt{\frac{K_x}{m}} \left\{ e^{\frac{k'a}{2}} \left(\frac{\sin ka}{2\ 2} + \frac{i\cos ka}{2\ 2}\right) + e^{\frac{-k'a}{2}} \left(\frac{\sin ka}{2\ 2} - \frac{i\cos ka}{2\ 2}\right) \right\} \tag{7.2.16}$$

which may be written

$$\omega = \pm 2\sqrt{\frac{K_x}{m}} \left[f(\mathbf{k})\, e^{+\frac{k'a}{2}} + g(\mathbf{k})\, e^{-\frac{k'a}{2}} \right] \tag{7.2.17}$$

which represents waves exponentially increasing and decreasing as they travel through the lattice. The first term in the brackets corresponds to a wave motion whose amplitude increases exponentially with the distance it travels. This solution must be neglected if we are to retain the condition that the oscillations remain finite at infinity. $\mathbf{k}'$ is called the attenuation coefficient.

PROBLEM

Show that if we let the linear density $M/a = \varrho$, as $a \to 0$, we get the allowed frequencies for the free vibrations of a uniform infinite string.

7.3 THE INFINITE POLYATOMIC LINEAR LATTICE[3]

Let us consider a one-dimensional lattice consisting of a number of similar unit cells each of length d and each of which contains N atoms (Figure 7.7). If we denote the displacement from equilibrium of the rth atom of the nth cell by ΔX_r^n, and consider harmonic forces to act only between nearest cells, we find that the total force on the rth atom of the nth cell due to the sth atoms's motion in the $(n + p)^{\text{th}}$ cell is

$$F_{r,s}^{n,n+p} = K_{p,r,s}\left[\Delta X_s^{n+p} - \Delta X_r^n\right] \tag{7.3.1}$$

where $K_{p,r,s,}$ is independent of n.

The force on the sth atom in the $(n + p)^{\text{th}}$ cell due to the motion of the rth atom in the nth cell is

$$F_{s,r}^{n+p,n} = K_{p,s,r}[\Delta X_r^n - \Delta X_s^{n+p}] \tag{7.3.2}$$

Note that

$$F_{s,r}^{n+p,n} = -F_{r,s}^{n,n+p} \tag{7.3.3}$$

Thus

$$K_{r,s} = K_{s,r} \tag{7.3.4}$$

The total force acting on the rth particle in the nth cell is

$$F_r^n = \sum_{\substack{p \\ \text{over all} \\ \text{cells}}} \sum_{\substack{s \\ \text{over all atoms} \\ \text{in the} \\ (n+p)\text{th cell}}} K_{r,s}[\Delta X_s^{n+d} - \Delta X_r^n] \tag{7.3.5}$$

Introducing the harmonic approximation where x is the distance from the origin of the cell to the origin of the lattice, i.e., $x = nd$.

$$\Delta X_r^n = A_r^n e^{-i[knd - \omega t]} \tag{7.3.6}$$

Substitution of (7.3.6) into (7.3.5) yields

$$F_r^n = \sum_p \sum_s K_{r,s}[A_s^{n+p} e^{-ikpd} - A_r^n] e^{-i[knd - \omega \tau]}$$

$$M_r \frac{\partial^2(\Delta X_r^n)}{\partial t^2} = -M_r \omega^2 A_r^n e^{-iknd} = -M_r \omega^2 \Delta X_r^n \tag{7.3.7}$$

Letting

$$D_{r,s} = \sum_p K_{r,s} e^{-ikpd} \quad (r \neq s) \tag{7.3.8}$$

and

$$D_{r,r} = -\sum_p \sum_s K_{r,s} + \sum_p K_{r,r} e^{-ikpd} \tag{7.3.9}$$

By (7.3.7)

$$-\omega^2 A_r^n M_r = \sum_{s \neq r} \{\sum_p K_{r,s} A_s^{n+p} e^{-ikpd}\} + \sum_p K_{r,r} A_r^{n+p} e^{-ikpd} - \sum_{s,p} K_{r,s} A_r^n \tag{7.3.10}$$

Thus we must solve the set of equations

$$\sum_s [D_{r,s} + M_r \omega^2 = \delta_{r,s}] A_s^{n+p} = 0 \tag{7.3.11}$$

in order to determine the allowed frequencies of vibration of a polyatomic lattice.

Let us apply (7.3.11) to the simple case of a linear diatomic lattice in the approximation of only nearest neighbor interactions. Let the masses of the two atoms be m and M as shown in Figure 7.8. The interatomic distance a corresponds to one-half of the length of the unit cell. For simplicity we shall number our cells such that our origin of coordinates is located at the nth cell. Considering the nearest neighbor interactions which affect an arbitrary atom by mass M, we find

$$\begin{Bmatrix} K_{-1,r,s} = K \\ K_{+1,r\ s} = 0 \end{Bmatrix} \tag{7.3.12}$$

Examining the nearest neighbor interactions which affect the atom of mass m in the same cell as above we find

$$\begin{Bmatrix} K_{-1,r,s} = 0 \\ K_{+1,r,s} = K \end{Bmatrix} \tag{7.3.13}$$

thus we find

$$\begin{aligned} D_{1,1} &= -2K \\ D_{1,2} &= K(1 + e^{ikd}) \\ D_{2,1} &= K(1 + e^{-ikd}) \\ D_{2,2} &= -2K \end{aligned} \tag{7.3.14}$$

Thus, (7.3.11) becomes, in matrix form

$$\begin{vmatrix} (-2K + M\omega^2) & K(1 + e^{ikd}) \\ K(1 + e^{-ikd}) & (-2K + m\omega^2) \end{vmatrix} = 0 \tag{7.3.15}$$

which may be solved for ω^2 to yield

$$\omega^2 = K\left[\frac{1}{m} + \frac{1}{M}\right] \pm K\left[\left(\frac{1}{m} + \frac{1}{M}\right)^2 - \frac{4\sin^2 ka}{mM}\right]^{1/2} \tag{7.3.16}$$

Examining the negative square root we find for small k

$$\omega_1^2 = \frac{2Kk^2a^2}{M} \tag{7.3.17}$$

ω_1^2 reaches a maximum

$$\omega_1^2 = \frac{2K}{M} \tag{7.3.18}$$

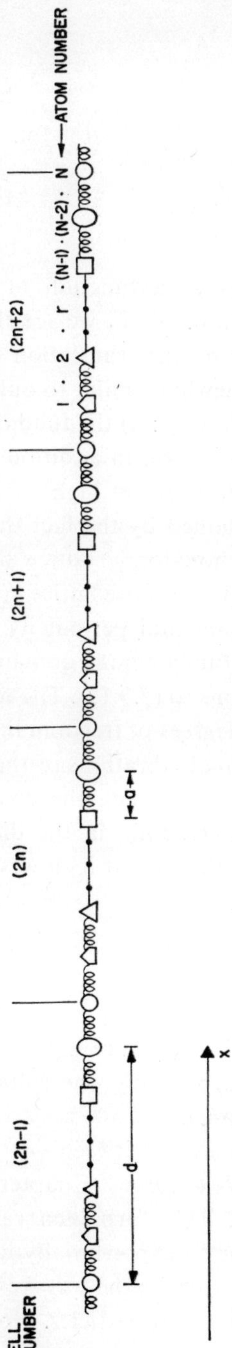

Figure 7.7 The infinite one-dimensional polyatomic lattice with unit cell length *d*.

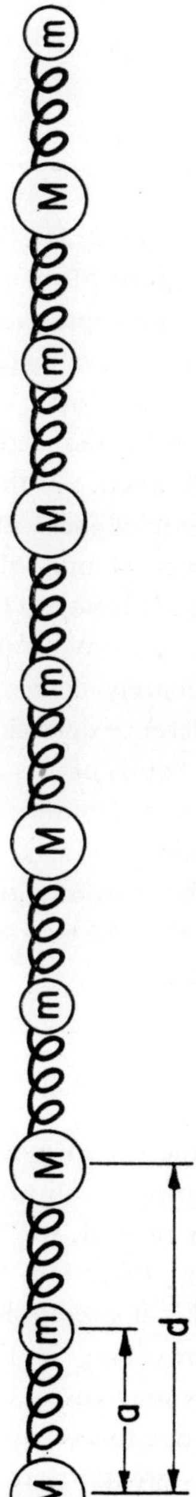

Figure 7.8 The infinite one-dimensional diatomic lattice.

for $k = \pi/2a$. The positive square root yields

$$\omega_2^2 = 2K\left[\frac{1}{m} + \frac{1}{M}\right] \tag{7.3.19}$$

for $\mathbf{k} = 0$

$$\omega_2^2 = \frac{2K}{m} \tag{7.3.20}$$

for $k = \pi/2a(m \ll M)$. The allowed frequencies as a function of $\mathbf{k}$ are shown in Figure 7.9 in the extended representation. Again we see that we can reduce the representation to a period (π/a) by the translation of any portion of the curve by $(2n\pi)/a$. The graph is somewhat similar to our result for a linear monatomic chain, except for the following: a) the fundamental period has been reduced from π/a to $\pi/2a$, b) we have, in addition to the acoustical branch, another branch of allowed vibrations at higher frequencies. The first of these differences is readily explained by the fact that a is the fundamental unit of length by which the monatomic lattice may be constructed. It is not so for the linear diatomic lattice whose lattice constant is $2a$. Thus we have found again that the fundamental periodicity of the lattice intimately determines the width of the fundamental domain. The second difference derives directly from the solutions to (7.3.11). The number of allowed bands of ω is equal to the number of degrees of freedom in which the lattice may vibrate. The case of the longitudinal vibrations of the monatomic chain produced one allowed band of ω vs $\mathbf{k}$, i.e., the acoustical branch. The introduction of another degree of freedom, in the diatomic chain introduced a thin second band shifted from the first at its lowest point by $\Delta\omega$ where

$$\Delta\omega = \sqrt{\frac{2K}{m}} - \sqrt{\frac{2K}{M}}. \tag{7.3.21}$$

The introduction of more degrees of freedom will introduce higher optical branches into the allowed vibrational spectrum, but only one solution to (7.3.11), in general, will produce the acoustic branch.

Note that there is a forbidden band of frequencies between $\omega_{Ac_{max}}$ and $\omega_{opt_{min}}$. This is a complete analogue to our E vs $\mathbf{k}$ curves (Chapter 4) for particles travelling in the crystalline potential. This forbidden range of frequencies may easily be shown again to be due to exponentially damped vibrations due to a complex k value in a manner similar to the way we treated frequencies greater than cutoff for a lattice of identical particles. Note too,

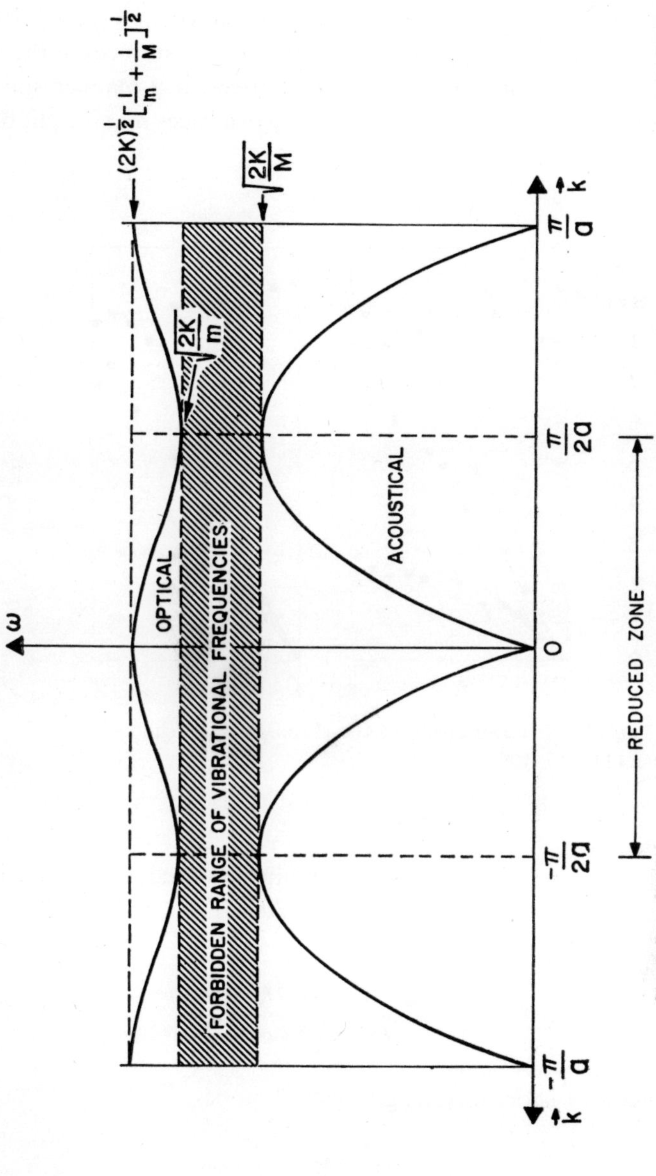

Figure 7.9 Acoustical and optical branches of the vibrational spectrum of the one-dimensional diatomic lattice (extended representation).

that if conditions are right, the frequency curves may intersect at some point in **k** space.

The vibrational spectrum of germanium is shown in Figure 7.10. One can plot for easy visualization the constant frequency surfaces in the fundamental Brillouin zone of various types of lattices in a manner similar to the way we did for the constant electron energy surfaces in our tight binding model.

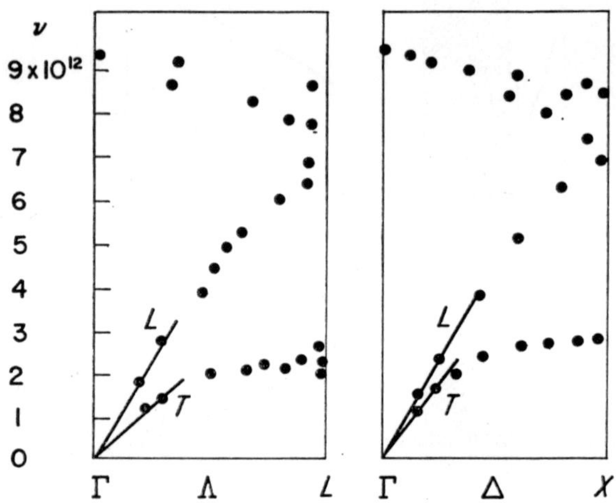

Figure 7.10 The vibrational spectrum of Ge. (From B. N. Brockhouse and P. K. Iyengar, *Phys. Rev.*, **111**, 747, 1958).

One can get a good picture of the high and low frequency vibrations of the lattice by solving for the amplitude ratio A_2/A_1 from (7.3.11). This yields

$$\frac{A_2}{A_1} = \frac{M\omega^2 - 2K}{K(2\cos^2 ka)} \tag{7.3.21}$$

which produces for short wavelength optical modes

$$\left(\frac{A_2}{A_1}\right)_{\text{opt}} = \frac{\dfrac{M}{m} - 1}{1 - \sin^2 ka} \tag{7.3.22}$$

Since the optical branch does not enclose a large width of frequencies, we can set $\mathbf{k} = 0$ to find that

$$\left(\frac{A_2}{A_1}\right)_{\text{opt}} = \frac{M}{m} \tag{7.3.23}$$

Thus, when the lattice vibrates in its optical mode the ratio of the displacement of the atoms is proportional to the ratio of their masses. This is pictured in Figure 7.11 a. In the low frequency acoustical mode we find that

$$A_1 \approx A_2 \tag{7.3.24}$$

which is what we would expect for the propagation of long wavelength lattice waves. This is shown in Figure 7.11 b.

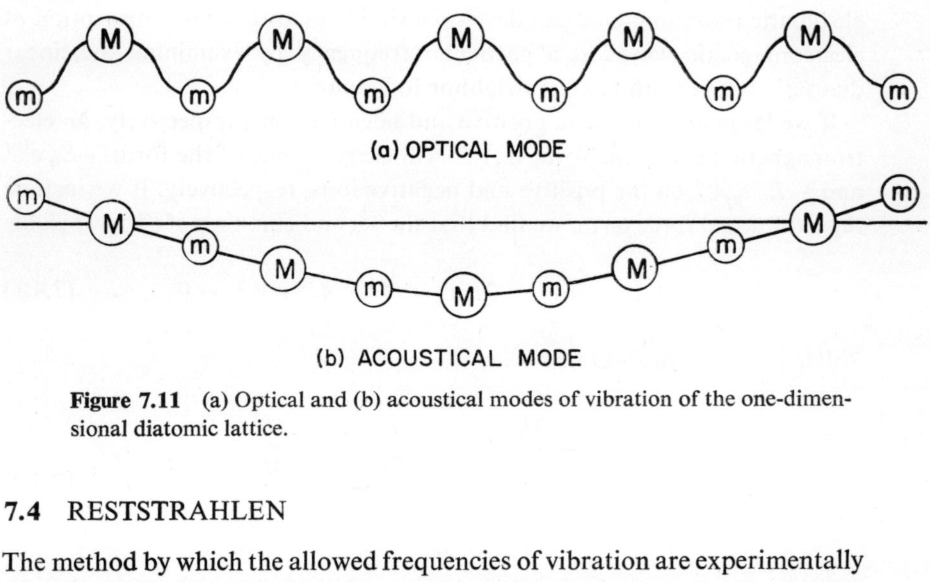

Figure 7.11 (a) Optical and (b) acoustical modes of vibration of the one-dimensional diatomic lattice.

7.4 RESTSTRAHLEN

The method by which the allowed frequencies of vibration are experimentally determined is to observe those frequencies which the lattice strongly absorbs. Although crystals will absorb all those frequencies which correspond to allowed frequencies of vibration, a particularly strong absorption is found for infrared frequencies in the range (20–150 microns). This corresponds to the optical branch of ionically bound crystals.

The linear diatomic lattice which we have studied is a good one-dimensional representation of an ionically bound lattice, e.g., NaCl. We have seen that the optical frequencies of vibration of such a lattice are such that they do not change much within the Brillouin zone. Thus let us assume for the moment that optical vibrations take place at a single frequency throughout the Brillouin zone.

(7.11 a) illustrates that the high frequency optical vibrations of a diatomic lattice are such that the amplitudes of vibration of alternating atoms form the maxima and minima of the lattice displacement waves. The shortest wavelength that the lattice can sustain is $2a$ in length. In this mode, the positively charged ions move in a direction opposite from the negative ions. Thus one would expect that besides the strong absorption of energy at all frequencies which excite the lattice vibrations, one would find a strong reradiation from the crystal when one forces it to oscillate at optical frequencies high enough to induce reinforced radiation from the oppositely oscillating positive and negative ions.

These ***residual*** rays or ***Reststrahlen*** have been utilized to produce monochromatic radiation. One can develop a simple picture for the absorption of electromagnetic waves at a particular frequency by examining the linear diatomic lattice with nearest neighbor interactions.

If we let m and M be the positive and negative ions, respectively, an electromagnetic field of the form $E_0 e^{i\omega t}$ will exert a force of the form $+E_0 e^{i\omega t}$ and $-E_0 e^{+i\omega t}$ on the positive and negative ions, respectively. If we include this additional force term, we find that the second equation of (7.3.11) yields

$$2K \cos kaA_1 - 2KA_2 + M\omega^2 A_2 + e E = 0 \tag{7.4.1}$$

Solving for A_2, one obtains

$$A_2 = \frac{-\dfrac{2K}{M} A_1 \cos ka}{\omega^2 - \dfrac{2K}{M}} - \frac{\dfrac{eE}{M}}{\omega^2 - \dfrac{2K}{M}} \tag{7.4.2}$$

Since the optical vibrational band is not wide, let us assume that in the optical mode the vibrational frequency of the atoms is given by $\omega_0 = 2k/M$, i.e., its value at $k = \pi/2d$.

Thus

$$A_2 = \frac{-\dfrac{e E}{M}}{\omega - \omega_0^2} \tag{7.4.3}$$

Thus the amplitude of vibration in the optical mode goes through a maximum when the forcing frequency ω reaches the natural frequency ω_0 of optical vibrations. In addition to the large increase in absorption at $\omega = \omega_0$, one also finds a reradiation of electromagnetic energy due to the harmonic accelerations of the positively and negatively charged ions.

7.5 SPECIFIC HEAT OF CRYSTALS

The study of the chronological development of the theory of the specific heat of solids is an extremely interesting and rewarding undertaking. In a nutshell it demonstrates the evolution of our concept of what a solid is and the triumphs of the old quantum theory over classical concepts in explaining observed specific heat phenomena. The calculation of the exact specific heat for even simple specific three dimensional lattices is an extremely long and involved procedure. Thus we shall endeavor only to show the evolution of the broad and general concepts. We shall first discuss the classical theory of specific heats.

The Classical Theory of Specific Heat

In 1819 P. L. Dulong and A. T. Petit reported that the experimental values of the specific heat c_p at high temperatures were the same for all solid substances, approximately 6 cal/deg.[4] Although this result was unexpected, it was explained by L. Boltzmann in quite simple terms.[5] The average kinetic energy of a classical particle in thermal motion can be shown to be $k_0T/2$ per degree of freedom. Boltzmann argued that one may consider that the forces which bind one atom to its neighbor are, to a first approximation, harmonic in nature. The average kinetic energy $\frac{1}{2}mv^2$ for a harmonic oscillator is equal to the average potential energy $\frac{1}{2}Kx^2$. Thus, for N classical three dimensional harmonic oscillators the total energy U associated with thermal motion is

$$U = 3(2)\left(\frac{k_0T}{2}\right)N = 3Nk_0T \tag{7.5.1}$$

Consequently the specific heat of a single oscillator is

$$c_V = \left(\frac{\partial U}{\partial T}\right)_V = 3Nk_0 = 3R \tag{7.5.2}$$

This yields the value $c_v = 5.96$ cal/deg which is in good agreement with the experimental value found by Dulong and Petit. It was later found experimentally by G. N. Lewis that the value c_v rather than c_p agreed at high temperatures with the theoretical value $3R$.[6]

Quantum Theory of Specific Heat

In the meantime, subsequent investigations had shown that the specific heats of many solids decreased with decreasing temperature approximately as T^3. This decrease in the specific heat at low temperatures was first recognized by Einstein as a breakdown of the classical description of the kinetics of thermal motion.[7] Einstein retained the concept of harmonic binding forces between atoms but postulated further that these oscillators had to obey the Planck hypothesis, i.e., that the allowed energy states of the ith oscilllator had to be of the form[8]

$$E_i = (n + 1/2)\, h\nu_i \quad n = 1, 2 \ldots \tag{7.5.3}$$

where ν_i is the frequency of the ith oscillator. The elementary excitations $h\nu_i$ of the harmonic oscillators which make up our crystal are given the name *phonons*. They bear the same relation to the lattice waves as do photons to light waves.

As we saw previously Bose-Einstein statistics derive from placing an assemblage of indistinguishable particles amongst a set of energy states allowing any number of particles per state. Thus if the thermal energy of the lattice is thought to be distributed in a set of energy levels of the form (7.5.3), we see that we may have any number of phonons per energy level and, of course, these phonons are indistinguishable. Thus phonons are Bose-Einstein type particles, and the distribution function for phonons in the ith mode of vibration is given by

$$f(E_T) = n_{\mathbf{k},j} = \frac{1}{\exp\left[\dfrac{h\nu(\mathbf{k},j)}{k_0 T}\right] - 1} \tag{7.5.4}$$

where we have used the zero point energy $(h\nu)/2$ as the origin of energy. (7.5.4) gives the average number of phonons in the mode of vibration of a harmonic oscillator which is specified by the variables $\mathbf{k}$ and j.

The average thermal energy of this mode of vibration is

$$U_{(\mathbf{k},j)} = h\nu_{(\mathbf{k},j)} n_{(\mathbf{k},j)} \tag{7.5.5}$$

Thus, the $(\mathbf{k}, j)^{\text{th}}$ mode of vibration of an assemblage of harmonic oscillators has the energy

$$U_{(\mathbf{k},j)} = \sum_{\substack{\text{over all}\\ \text{modes}\\ \text{of}\\ \text{vibration}}} h\nu_{i(\mathbf{k},j)} n_{i(\mathbf{k},j)}$$

$$= \sum_{\mathbf{k},j} \left[\frac{h\nu_{(\mathbf{k},j)}}{\exp\left(\dfrac{h\nu_{(\mathbf{k},j)}}{k_0 T}\right) - 1} \right] \tag{7.5.6}$$

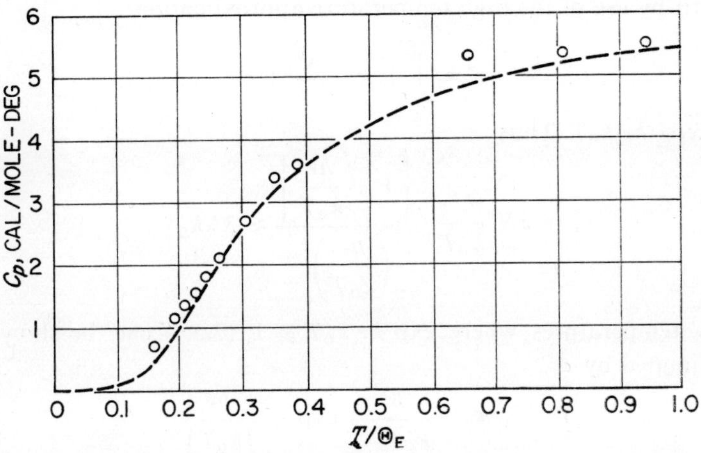

Figure 7.12 Experimental values for the heat capacity of diamond compared with the predicted results of the Einstein model ($\vartheta_E = 1320$ °K). (From A. Einstein, *Ann. Physik*, **22**, 180, 1907.)

Einstein made the simplifying assumption that at any specific temperature all of the oscillators (lattice sites) vibrate in the same mode of oscillation. Thus (7.5.6) may be written

$$U = \sum_{\substack{i\\ \text{all lattice}\\ \text{sites}}} \frac{h\nu_i}{\exp\left[\dfrac{h\nu_i}{k_0 T}\right] - 1} \tag{7.5.7}$$

and thus

$$U = \frac{3N' h\nu}{\exp \dfrac{h\nu}{k_0 T} - 1} \tag{7.5.8}$$

where $N' = mN$ and N is the number of lattice sites. We have treated each of the harmonic oscillators independently and assumed that each mode of lattice vibration could be broken up into three independent components. It is a simple matter to now find the specific heat c_v which is given by

$$c_v = \left(\frac{\partial U}{\partial T}\right)_v = \frac{3N'h\nu\left(\dfrac{h\nu}{k_0T^2}\right)\exp\left(\dfrac{h\nu}{k_0T}\right)}{\left[\exp\left(\dfrac{h\nu}{k_0T}\right) - 1\right]^2} \tag{7.5.9}$$

Expression (7.5.9) can be shown to agree with the classical law of Dulong and Petit by use of the high temperature approximation

$$e^x \approx 1 + x + \frac{x^2}{2} + \cdots \tag{7.5.10}$$

where $x = h\nu/k_0T$. Thus

$$c_v = 3N'\frac{h^2\nu^2}{k_0T^2}\frac{\left[1 + \dfrac{h\nu}{k_0T}\right]}{\left(\dfrac{h\nu}{k_0T}\right)^2} \approx 3Nk_0 \tag{7.5.11}$$

At low temperatures where $\exp h\nu/k_0T \gg 1$, (7.5.9) may be shown to be approximated by

$$c_V = \frac{3N'h^2\nu^2}{k_0T^2}\exp - \left\{\frac{h\nu}{k_0T}\right\} \tag{7.5.12}$$

Thus, at low temperatures, the Einstein picture of a solid as composed of many independent harmonic oscillators indeed predicts a c_v decreasing exponentially at low temperatures. The Einstein results for particular solids are usually discussed in terms of the mode of vibration ν_E which gives the best fit results for the observed specific heat data for that solid. This is done conveniently by defining the so-called Einstein temperature θ_E by

$$h\nu_E = k_0\theta_E \tag{7.5.13}$$

which gives for (7.5.9)

$$c_v = 3N'k_0\left[\frac{\theta_E^2}{T^2}\right]\frac{\exp(\theta e/T)}{\left[\exp\left(\dfrac{\theta e}{T}\right) - 1\right]} \tag{7.5.14}$$

Although the Einstein theory gives qualitative agreement with the experimentally observed decrease in specific heat with decrease in temperature, quantitatively its exponential decrease is too rapidly varying to account for the observed T^3 decrease in c_v (see Figure 7.14).

It was Peter Debye[9] who recognized that the Einstein assumption that the lattice sites acted like independent harmonic oscillators was an over-simplified picture which did not take into account the long wavelength vibrations of the lattice, which are particularly important at low temperatures. In these low frequency modes many lattice sites must vibrate coherently and thus may not be assumed to be independent of one another (see Figure 7.11 b). Thus we must include the fact that the lattice may be vibrating in a combination of any of a distribution of allowed frequencies. At very low temperatures the long wavelength, low frequency acoustic modes of oscillation were assumed to be more important than the high energy optical modes in which the atoms could be assumed to vibrate independently.

We will perform a calculation similar to Debye's. Debye assumed that at low temperatures the number of long wavelength acoustic modes in which the crystal could vibrate was limited to $3N$, the number of degrees of freedom of the crystal. The cutting off of the number of modes of vibration at $3N$ is a convenient device which allows for a large number of allowed modes, but which circumvents the blowing up of the lattice energy if one were to do the calculation in a countinuous medium with an infinite number of allowed modes of vibration. Further, he assumed that the solid would, when vibrating in these long wavelength modes, appear to be essentially isotropic, since the wavelength in these modes is so much larger than the interatomic spacings that the differences in spacing with direction in the crystal could be neglected. Assuming that we are dealing with a three dimensional Bravais lattice, j will include two transverse and one longitudinal acoustic mode. Assuming that the crystal is isotropic, these modes will be identical. Thus (7.5.6) becomes

$$U = 3 \sum_k \frac{h\nu_k}{\exp\left(\dfrac{h\nu_k}{k_0 T}\right) - 1} \tag{7.5.15}$$

We can convert (7.5.15) to an integration over $\mathbf{k}$ space by use of an approximate density $G(\mathbf{k})\, d\mathbf{k}$ of allowed frequencies of vibration

$$U = 3 \int_{\mathbf{k}} \frac{h\nu_k G_\nu(\mathbf{k})\, d\mathbf{k}}{\exp\left(\dfrac{h\nu_k}{k_0 T}\right) - 1} \tag{7.5.16}$$

Since we will be dealing with long wavelength (low **k**) modes of vibration, it is undoubtedly a valid approximation to go back to our long wavelength approximation for the dispersion relation for a lattice of similar particles (7.2.10)

$$\nu = \frac{ka}{2\pi}\sqrt{\frac{K}{m}} \tag{7.5.17}$$

At long wavelengths, ν varies linearly with **k** and each point in **k** space represents also an allowed frequency ν. Thus the density $G_\nu =$ (**k**) is the same as the density of allowed **k** values in **k** space. We find

$$G_\nu(\mathbf{k})\, d\mathbf{k} = \frac{V}{8\pi^3}\, d\mathbf{k} \tag{7.5.18}$$

Assuming again that the number $3N$ of allowed modes is large enough that a spherical distribution of k values is not a bad approximation, we get

$$U = 3\int_{k_0}^{k_{max}} \frac{h\nu_k \dfrac{V}{8\pi^3} 4\pi k^2\, dk}{\exp\left(\dfrac{h\nu_k}{k_0 T}\right) - 1} \tag{7.5.19}$$

Using

$$\nu = \frac{kS}{2\pi} \tag{7.5.20}$$

where S is the velocity of sound we get

$$U = \frac{3hSV}{4\pi^3}\int_{k_0}^{k_{max}} \frac{k^3\, dk}{\exp\left(\dfrac{hkS}{2\pi k_0 T}\right) - 1} \tag{7.5.21}$$

To find k_{max}, we set the total number of allowed frequencies equal to $3N$. For simplicity, however we may assume that we are dealing with an isotropic solid in which $\nu_l = \nu_t$. Thus we find

$$\begin{aligned} N &= \int_0^{k_{max}} G_\nu(\mathbf{k})\, d\mathbf{k} \\ &= \int_0^{k_{max}} \frac{V}{2\pi^2} k^2\, dk \end{aligned} \tag{7.5.22}$$

Thus

$$k_{max} = \left[\frac{16\pi^2 N}{V}\right]^{1/3} \tag{7.5.23}$$

Setting $(hSk)/(2\pi k_0 T) = x$ we find

$$\hat{a}k = dx\left[\frac{2\pi k_0 T}{hS}\right] \tag{7.5.24}$$

(7.5.21) becomes

$$U = \frac{12\pi V[k_0 T]^4}{h^3 S^3}\int_0^{x_{max}} \frac{x^3\, dx}{[e^x - 1]} \tag{7.5.25}$$

The number of modes $3N$ is so large that we may approximate x_{max} by infinity. Utilizing

$$\int_0^{\infty} \frac{x^3\, dx}{e^x - 1} = \frac{\pi^4}{15} \tag{7.5.26}$$

we find

$$U = \frac{4V\pi^5 k_0^4 T^4}{5h^3 S^3} \tag{7.5.27}$$

and the specific heat c_v becomes

$$c_v = \frac{16V\pi^5 k_0^4 T^3}{5h^3 S^3} \tag{7.5.28}$$

Thus we see that the Debye theory predicts the observed T^3 dependence of c_v. It is common to define a *Debye Temperature* θ_0 as

$$\theta_D = \frac{h\nu_{max}}{k_0} = \frac{hS}{2\pi k_0}\left(\frac{6\pi^2 N}{V}\right)^{1/3} \tag{7.5.29}$$

Thus we may write

$$c_v = \frac{12Nk_0\pi^4}{5}\left(\frac{T}{\theta_D}\right)^3 \tag{7.5.30}$$

See Figure 7.13 for a plot of Debye results. Table 7.3 lists the Debye temperature for some typical semiconductors. In truth however, the Debye temperature for any material is not a constant, but varies with temperature as is shown for GaAs in Figure 7.14. The decrease at high temperatures is due to anharmonic effects, while the minimum at low temperatures is due to coupling of very long wavelength acoustic modes.

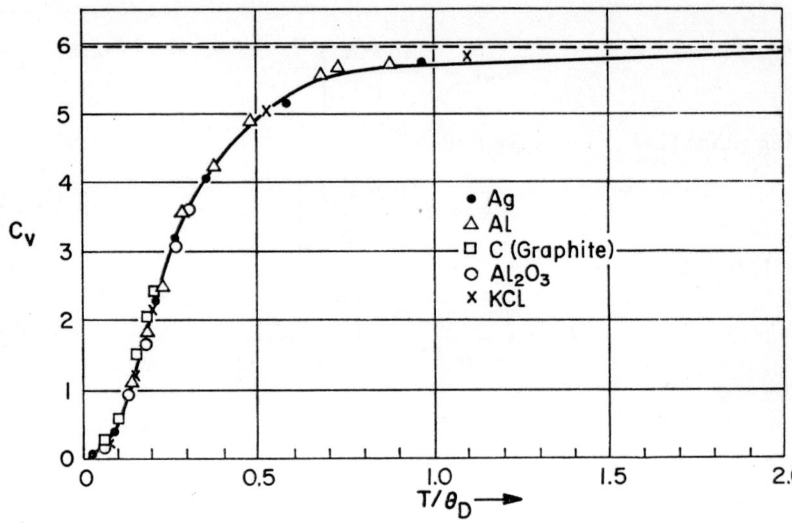

Figure 7.13 Comparison of the Debye specific heat curve and the observed specific heat of a number of simple substances. (From F. Seitz, *The Modern Theory of Solids*, McGraw-Hill Book, Co., Inc., New York, 1940, p. 109.)

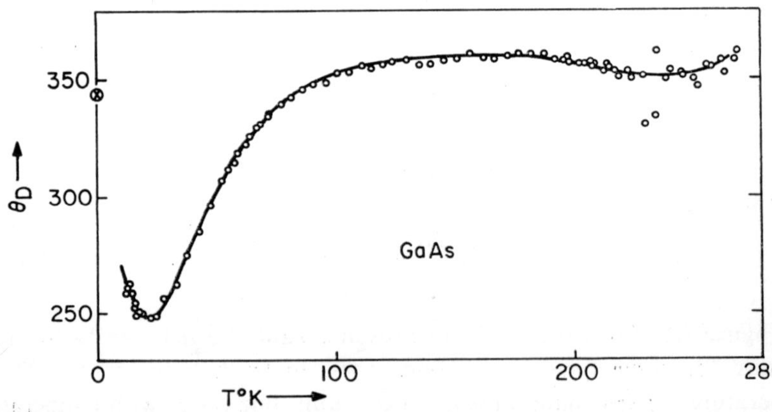

Figure 7.14 Variation of ϑ_D with temperature for GaAs. (From U. Piesberger, *Semiconductors and Semimetals*, Ed. by R. K. Willardson and A. C. Beer, Vol. 2, Academic Press, New York, 1966, p. 56.)

Table 7.3 The high temperature Debye temperature for various semiconductors (from, P. W. Sparks and C. A. Swenson, *Phys. Rev.*, **163**, 779, 1967)

Material	θ_D °K
Si	645
Ge	374
GaAs	345
GaSb	266
InAs	234
InSb	205

Specific Heat of Electrons

In metals particularly, a derivation of the specific heat of a solid in the low temperature region must include the specific heat associated with conduction electrons. If we treat the remaining conduction electrons in a metal at low temperatures as a gas of free particles, as we did before, the energy of the electrons in the crystal is given by

$$\int g(E)f(E)\,dE = E \tag{7.5.31}$$

[compare equation (6.6.1) of the Sommerfeld theory.] Again using the Sommerfeld approximation we calculate

$$\mathbf{E} = \frac{3}{5}N\mu_0\left[1 + \frac{5\pi^2}{12}\left(\frac{k_0T^2}{\mu_0}\right) + \cdots\right] \tag{7.5.32}$$

Thus the electronic specific heat decreases linearly with temperature

$$\underset{\text{electronic}}{c_v} = \left(\frac{\partial \bar{E}}{\partial T}\right)_V = \frac{3R}{2}\,\frac{\pi^2k_0T}{3\mu_0} \tag{7.5.33}$$

7.6 THERMAL EXPANSION

In this section we shall calculate the coefficient of thermal expansion $\varkappa_0$ for a solid in the "quasi-Harmonic" approximation and show its dependence on the anharmonic terms of (7.1.5). [It can be shown that the purely harmonic crystal in the nearest neighbor approximation does not exist except under very special conditions.[10]] The average value of an atom's displacement from equilibrium remains zero in the harmonic approximation regardless of how frantically the atom vibrates about its equilibrium position, i.e.

$$\langle x \rangle = \frac{\int_{-\infty}^{+\infty} x\, e^{-Ax^2/k_0T}}{\int_{-\infty}^{+\infty} e^{-Ax^2/k_0T}} = 0 \qquad (7.6.1)$$

where Ax^2 represents the harmonic potential. This is because the total potential U_T [Figure 7.1] is symmetric about the equilibrium separation $\boldsymbol{a}$.

PROBLEM:
Prove equation 7.6.1.

Let us however calculate the thermal expansion coefficient $\varkappa_0$ for a group of harmonic oscillators. The partition function may be written

$$z_i = \sum_j e^{-\left(j+\frac{1}{2}\right)\frac{h\nu_i}{k_0T}} = e^{\frac{h\nu_i}{2k_0T}} \sum_j \left[e^{-\frac{h\nu_i}{k_0T}}\right]^j \qquad (7.6.2)$$

which may be written

$$z_i = \frac{e^{-\frac{h\nu_i}{2k_0T}}}{1 - e^{\frac{h\nu_i}{k_0T}}} = \left[2 \sinh \frac{h\nu_i}{2k_0T}\right]^{-1}. \qquad (7.6.3)$$

We may write the total partition function for the system of oscillators as

$$Z = \prod_i z_i = \prod_i \left[2 \sinh \frac{h\nu_i}{2k_0T}\right]^{-1} \qquad (7.6.4)$$

and we find for the Helmholtz free energy

$$F = -k_0T \ln Z = k_0T \sum_i \ln 2 \sinh \left[\frac{h\nu_i}{2k_0T}\right] \qquad (7.6.5)$$

The total pressure is given by

$$P = -\left(\frac{\partial F}{\partial V}\right)_{T,V} = k_0T \sum_i \coth\left(\frac{h\nu_i}{2k_0T}\right) \frac{h}{2k_0T} \left[\frac{\partial \nu_i}{\partial V}\right]$$

$$= -\sum_i \left[\frac{1}{2} h\nu_i + \frac{h\nu_i}{e^{\frac{h\nu_i}{k_0T}} - 1}\right] \frac{\partial \ln \nu_i}{\partial V} \qquad (7.6.6)$$

Let us assume that we are in the classical region of energy and that all oscillators are vibrating with the same frequency. We shall pick this frequency as ν_{max} (7.5.29). Thus we find

$$P = -\frac{\gamma}{V_0} U \tag{7.6.7}$$

where $\frac{\partial \ln \nu_i}{\partial V} = -\frac{\gamma}{V_0}$, and γ is the Gruneisen coefficient. From (7.1.6) and (7.1.9), we find

$$P = \frac{\Delta V}{\beta V_0} = \frac{U\gamma}{V_0} \tag{7.6.8}$$

Differentiating with respect to temperature

$$\frac{\varkappa_0}{\beta} = \frac{\partial U}{\partial T}\left(\frac{\gamma}{V_0}\right) \tag{7.6.9}$$

where we've assumed in the classical limit that the frequency of vibration is independent of the temperature. Thus we find for the thermal expansion coefficient

$$\varkappa_0 = \frac{\gamma \beta c_V}{V_0} \tag{7.6.10}$$

$\varkappa_0$ is thus proportional to the specific heat. Note that the Gruneisen coefficient is inversely dependent on the bulk compressibility β and thus is a function of the curvature terms in our binding potential [see (7.1.9)].

Let us examine the Gruneisen coefficient which may be written

$$-\frac{\partial \ln \nu}{\partial \ln V} = \gamma = -\frac{\partial \ln \nu}{\partial \ln \varkappa_0}\left[\frac{\partial \ln \varkappa_0}{\partial \ln \nu}\right]\left[\frac{\partial \ln \nu}{\partial V}\right] \tag{7.6.11}$$

For a harmonic oscillator

$$\frac{\partial \ln \nu}{\partial \ln \varkappa_0} = \frac{1}{2} \tag{7.6.12}$$

and

$$\ln \varkappa_0 = \ln\left(\frac{e\, z_1 z_2}{\varkappa_0 n \beta}\right) + (n-1)\ln \nu \tag{7.6.13}$$

$$\frac{\partial \ln \varkappa_0}{\partial \ln \nu} = n - 1 \tag{7.6.14}$$

We may write $Na^3 = V$. Thus

$$\frac{\partial \ln a}{\partial V} = \frac{1}{3} \tag{7.6.15}$$

which yields

$$\gamma = \left[\frac{n-1}{6}\right] \tag{7.6.16}$$

Therefore the Gruneisen coefficient for an ideally harmonic lattice is constant and does not depend on the type of lattice. One actually has to deal with anharmonic interatomic forces to yield a meaningful γ.[11] Thus the Gruneisen coefficient is a convenient parameter by which one can characterize the anharmonicity of interatomic forces for different crystals. In actual fact, the Gruneisen coefficient is a function of temperature, and many substances display negative γ's in certain ranges of temperature, as can be seen from Figure 7.15.

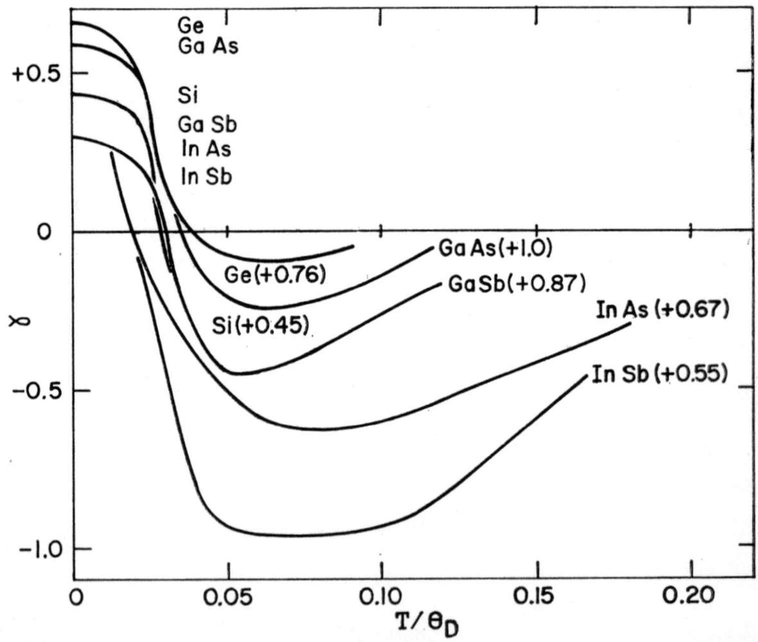

Figure 7.15 The Gruneisen coefficient γ as a function of temperature for various semiconductors. The values in parenthesis are estimated room temperature values of γ. (From P. W. Sparks and C. A. Swenson, *Phys. Rev.*, **163**, 779, 1967.)

References

1. P. P. Ewald, *Ann. d. Physik*, **64**, 253, 1921.
2. H. M. Evjen, *Phys. Rev.*, **39**, 675, 1932.
3. This section follows L. Brillouin, *Wave Propagation in Periodic Structures*, used with permission of McGraw Hill Book Co., Inc., N. Y. 1946, p. 65.
4. P. L. Dulong and A. T. Petit, *Ann. Chem. Phys.*, **10**, 395, 1919.
5. L. Boltzmann, *SitzG. KGL. Akad. Wiss. Wien*, **63:2**, 679, 1871.
6. G. N. Lewis, *J. Am. Chem. Soc.*, **29**, 1165, 1516, 1907.
7. A. Einstein, *Ann. Physik*, (4) **22**, 180, 1907; **22**, 800, 1906; **34**, 170, 1911.
8. The factor of $^1/_2$ was not used by Einstein but was a later development of the theory.
9. P. Debye, *Ann. Physik*, **39**, 789, 1912.
10. T. H. K. Barron, *Annals of Physics*, **1**, 77, 1957.
11. *Ibid.*

CHAPTER 8

Electron Interactions in Solids

8.1 THE ELECTRON-PHONON INTERACTION

Interaction of Electrons with Acoustical Vibrations (The Deformation Potential)

The collision of an electron with the vibrating lattice is the major interaction which serves to foul up the theory we have built up. In this section we will investigate this interaction and derive the selection rules which must be obeyed if such an interaction is to take place.

The transition probability per unit time of a system from an initial state $\langle\gamma_i|$ to a final state $|\gamma_f\rangle$ is given by (1.4.18). Thus it remains to find the form of the electron-phonon Hamiltonian. Let us write the displacement from equilibrium of the ith lattice site as $\Delta\mathbf{r}_i$. The ionic potential may be expanded as

$$U(\mathbf{r}) \approx U(\mathbf{r}) - \sum_i (\Delta\mathbf{r}_i) \cdot \nabla U(\mathbf{r}_i) + \cdots \tag{8.1.1}$$

The sum term is the perturbing potential produced by the deviation of our lattice sites from their equilibrium positions. We may expand $\Delta\mathbf{r}_i$ in terms of the normal coordinates $\mathbf{q}_\varkappa$.

$$\Delta\mathbf{r}_i = \sum_\varkappa q_\varkappa \hat{e}_\varkappa e^{i\varkappa\cdot\mathbf{r}_i} \tag{8.1.2}$$

where we have indicated the direction of the normal coordinate by the unit polarization vector $\hat{e}_\varkappa$.

To utilize our perturbation methods we must construct a state for our electron-lattice system. We will characterize the lattice by the state $|n_\varkappa\rangle$ which denotes that state of the lattice which contains n phonons of the wave vector $\varkappa$ (the number representation). We choose $|\gamma_\mathbf{k}\rangle$. for the electronic state. Thus we may write the total state $|N_{\mathbf{k},\varkappa}\rangle$ of the system as the product of electron and lattice states.

$$|N_{\mathbf{k},\varkappa}\rangle = |n_\varkappa\rangle\,|\gamma_\mathbf{k}\rangle \tag{8.1.3}$$

Neglecting the time dependence, we find from (1.4.18)

$$P_{\mathbf{k},\mathbf{k}'} = \frac{1}{\hbar^2}\left|\langle N_{\mathbf{k},\varkappa}|\sum_\varkappa\sum_{\mathbf{r}_i} q_\varkappa e^{i\varkappa\cdot\mathbf{r}_i}\,\hat{e}_\varkappa \cdot \nabla\mathrm{U}(\mathbf{r}_i)\,|N_{\mathbf{k}',\varkappa}\rangle\right| \tag{8.1.4}$$

We may write the normal coordinates $q_\varkappa$ in terms of our creation and annhilation operators β^* and β, respectively.

$$q_\varkappa = \sqrt{\frac{\hbar}{2m\omega_\varkappa}}(\beta^*_\varkappa + \beta_\varkappa) \tag{8.1.5}$$

We have found in section 1.3 that β^* and β act only between states which differ by one phonon. Thus we are led to

$$P_{\mathbf{k},\mathbf{k}'} = \frac{1}{2m\omega_\varkappa}\left|\begin{matrix}\{\langle n_\varkappa|\,\beta^*_\varkappa + \beta_\varkappa\,|n_{\varkappa-1}\rangle\, W_{\mathbf{k},\mathbf{k}'}\}\\ \{\langle n_\varkappa|\,\beta^*_\varkappa + \beta_\varkappa\,|n_{\varkappa+1}\rangle\, W_{\mathbf{k},\mathbf{k}'}\}\end{matrix}\right|^2 \tag{8.1.6}$$

where the upper term corresponds to phonon annhilation and the lower to phonon creation, and

$$W_{\mathbf{k},\mathbf{k}'} = \sum_{\mathbf{r}_i} e^{i\varkappa\cdot\mathbf{k}_i}\,\langle\gamma_\mathbf{k}\,|\hat{e}_\varkappa\cdot\nabla\mathbf{U}(\mathbf{r}_i)\,|\gamma_{\mathbf{k}'}\rangle \tag{8.1.7}$$

Since we usually know the wave vector of the lattice mode from which we scatter, we have dropped the sum over lattice modes. Retaining it introduces unnecessary complications.

We have

$$\langle n_\varkappa|\,\beta^*_\varkappa + \beta_\varkappa\,|n_{\varkappa-1}\rangle = 2n_\varkappa^{\frac{1}{2}} \tag{8.1.7a}$$

$$\langle n_\varkappa|\,\beta^*_\varkappa + \beta_\varkappa\,|n_{\varkappa+1}\rangle = 2(n_{\varkappa+1})^{\frac{1}{2}} \tag{8.1.7b}$$

which yields

$$P_{\mathbf{k},\mathbf{k}'} = \frac{2}{m\omega_\varkappa}\begin{Bmatrix} n_\varkappa & |W_{\mathbf{k},\mathbf{k}'}|^2 \\ (n_{\varkappa+1}) & |W_{\mathbf{k},\mathbf{k}'}|^2\end{Bmatrix} \tag{8.1.8}$$

Thus the transition probability is proportional to the number of phonons in the lattice state from which you are scattering, as we would intuitively predict.

Writing $|\gamma_\mathbf{k}\rangle$ as

$$\sum_{\mathbf{r}_i} e^{i\mathbf{k}\cdot\mathbf{r}}\psi_{\mathbf{k}(\mathbf{r}_i)} \tag{8.1.9}$$

we find

$$W_{\mathbf{k},\mathbf{k}'} = \sum_{\mathbf{r}_i} e^{i(\mathbf{k}'-\mathbf{k}+\varkappa)\cdot\mathbf{r}_i}\int_C \psi^*_\mathbf{k}(\mathbf{r})\,\hat{e}_x\cdot\nabla\mathbf{U}(r)\,\psi_{\mathbf{k}'}(\mathbf{r})\,d\mathbf{r} \tag{8.1.10}$$

where the integral extends over a unit cell. The sum in (8.1.10) is, of course, a lattice sum and has the value

$$N\,\delta_{(\mathbf{k}'-\mathbf{k}+\varkappa,\mathbf{G})} \tag{8.1.11}$$

where $\mathbf{G}$ is a reciprocal lattice vector. Thus the electron lattice interaction can take place only if the initial and final momentum vectors of the scattered electron equal the momentum $\varkappa$ of a phonon of the lattice mode of vibration (normal process), or $\varkappa + 2\pi\mathbf{G}$ (Umklapp process). This is called phonon emission or phonon absorption by the lattice. The normal and Umklapp processes are shown in Figure 8.1a and b. It should be noted that an Umklapp process in the reduced zone scheme (8.1b) may be regarded as a normal process in the extended zone scheme (Figure 8.1c).

It is seen that the integrand of (8.1.10) yields an appreciable contribution only when the lattice momentum vector $\boldsymbol{\varkappa}$ is parallel to the displacement.

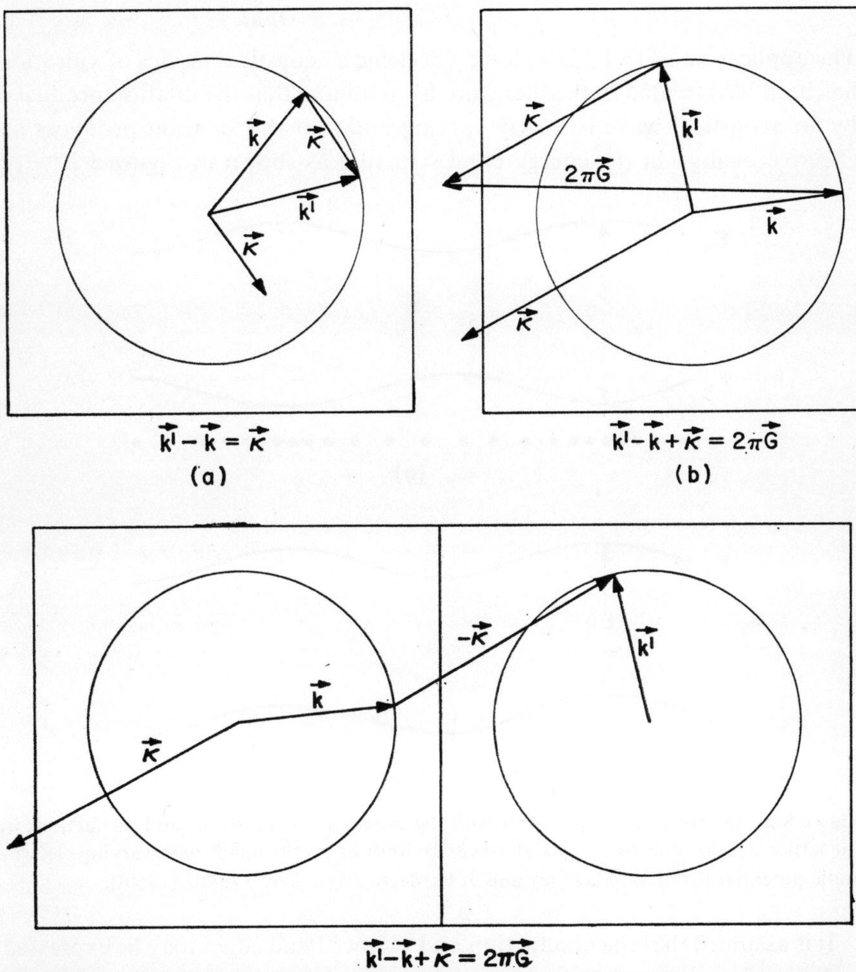

Figure 8.1 (a) Normal Scattering. The electron's momentum changes from $\mathbf{k}$ to $\mathbf{k}'$ through the emission of a phonon of wave vector $\boldsymbol{\varkappa}$. (b) Unklapp scattering in the reduced zone scheme. (c) Umklapp process in the extended zone scheme. In both (b) and (c) $\mathbf{k} - \mathbf{k}' + \boldsymbol{\varkappa} = 2\pi\mathbf{G}$. The square is the boundary of the Brillouin zone and the electron is scattered from one point to another on a spherical energy surface.

Thus an electron will interact most strongly with the longitudinal phonons. Inserting the time dependence, we find

$$P_{\mathbf{k},\mathbf{k}'} = \frac{2\hbar}{m\omega_{\varkappa}} \begin{Bmatrix} n_{\varkappa} & |\omega_{\mathbf{k},\mathbf{k}'}|^2 \\ (n_{\varkappa+1}) & |\omega_{\mathbf{k},\mathbf{k}'}|^2 \end{Bmatrix} \frac{\sin^2 \dfrac{t}{\hbar} \begin{Bmatrix} E_{\mathbf{k}'} - E_{\mathbf{k}} - \hbar\omega_{\varkappa} \\ E_{\mathbf{k}'} - E_{\mathbf{k}} + \hbar\omega_{\varkappa} \end{Bmatrix}}{\begin{Bmatrix} (E_{\mathbf{k}'} - E_{\mathbf{k}} - \hbar\omega_{\varkappa})^2 \\ (E_{\mathbf{k}'} - E_{\mathbf{k}} + \hbar\omega_{\varkappa})^2 \end{Bmatrix}} \tag{8.1.12}$$

The application of (8.1.12) to long wavelength acoustical modes of vibration has been well treated in the literature by assuming that the dilation produced by an acoustical wave is slowly varying and that this dilation produces an effective change in the energy band structure as shown in Figure 8.2.[1, 2, 3]

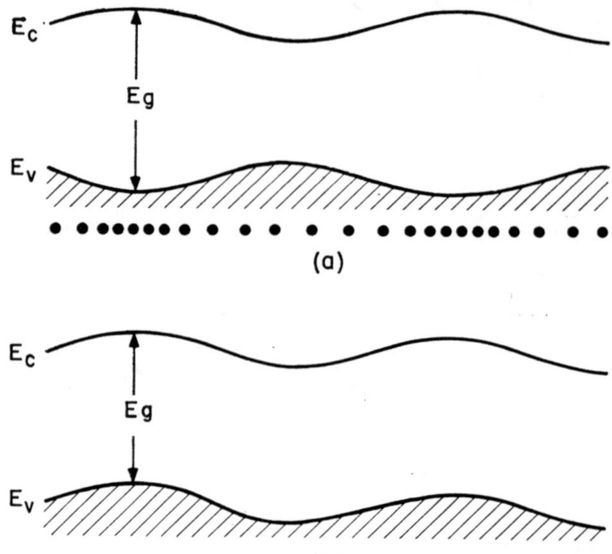

Figure 8.2 (a) Shift of energy bands with the alternate compressions and rarifactions in the lattice density due to acoustical wave. (b) Shift of energy bands with varying electrostatic potential [from W. Shockley and J. Bardeen, *Phys. Rev.* **77**, 407 (1950)].

It is assumed that the conduction and valence band edges may be expressed as linear functions of the tensor strain $\varepsilon_{i,j}$ produced by a long wavelength mode.

$$\begin{aligned} E_C &= E_{CO} + \sum_{i,j} B^C_{i,j}\varepsilon_{i,j} \\ E_V &= E_{VO} + \sum_{i,j} B^V_{i,j}\varepsilon_{i,j} \end{aligned} \tag{8.1.13}$$

where E_{CO} and E_{VO} are the conduction and valence band edges in the absence of strain. For a cubic crystal (8.1.13) reduces to

$$\begin{aligned} E_C &= E_{CO} + E_{IC}\,\delta V \\ E_V &= E_{VO} + E_{IV}\,\delta V \end{aligned} \tag{8.1.14}$$

where δV is the lattice dilation. The shift in the energy gap with strain is given by

$$E_C - E_V = (E_{CO} - E_{VO}) + (E_{IC} - E_{IV})\,\delta V \tag{8.1.15}$$

E_{1C} and E_{1V} are called the deformation potential constants for each band. Thus we may define a deformation potential dU which can be considered to be the perturbation which produces scattering by acoustical phonons.

Now we may examine $W_{\mathbf{k},\mathbf{k}'}$ in the long wavelength limit. We may write $|\nabla \mathbf{U}| \approx \frac{dU}{dV}\cdot$ In terms of a small lattice dilation $\delta V = \frac{dV}{-V}$ we have

$$dU = \left[\frac{dU}{dV}V\right]\delta V \tag{8.1.17}$$

Thus we may define $\frac{dU}{dV}V$ as a deformation potential E_1. When we express the dilation in terms of our Fourier analyzed displacement vectors, we find

$$W_{\mathbf{k},\mathbf{k}'} = \frac{E_1 i\varkappa}{N}\int_C \psi_{\mathbf{k}}^* \psi_{\mathbf{k}'}\,d\mathbf{r} \tag{8.1.17}$$

The probability of interaction becomes

$$P_{\mathbf{k},\mathbf{k}'} = \frac{2\hbar\varkappa^2 E_1^2}{m\omega_\varkappa N^2}\begin{Bmatrix}(n_\varkappa)\\(n_{\varkappa+1})\end{Bmatrix}\frac{\sin^2\dfrac{t}{\hbar}\begin{Bmatrix}E_{\mathbf{k}'} - E_{\mathbf{k}} - \hbar\omega_\varkappa\\ E_{\mathbf{k}'} - E_{\mathbf{k}} + \hbar\omega_\varkappa\end{Bmatrix}}{\begin{Bmatrix}(E_{\mathbf{k}'} - E_{\mathbf{k}} - \hbar\omega_\varkappa)^2\\(E_{\mathbf{k}'} - E_{\mathbf{k}} + \hbar\omega_\varkappa)^2\end{Bmatrix}} \tag{8.1.18}$$

Electron Interactions with the Optical Modes in Heteropolar Crystals

In polar crystals such as NaCl or the covalent III-V compounds the low frequency dielectric constant will show a resonance in the infrared region due to the polar modes of vibration. The high frequency dielectric constant will show a resonance in the optical or ultraviolet which corresponds to the oscillations of the electronic charge clouds surrounding the ions. We will examine the interaction between electrons and the polar modes below the frequency at which the optical resonance occurs.

The total polarization field **P** may be written in terms of its high and low frequency components $\mathbf{P}_\infty$ and $\mathbf{P}_0$, respectively

$$\mathbf{P} = \mathbf{P}_\infty + \mathbf{P}_0 \tag{8.1.19}$$

If we assume that the frequencies of the optical modes of vibration of the lattice are much smaller than the frequency of oscillation of the electron, the lattice will not be able to follow the electron oscillations, and will see only a static polarization field. Thus we will deal with $\mathbf{P}_0 \cdot \mathbf{P}$ is related to the electric displacement vector by

$$\mathbf{D} = \mathbf{E} + 4\pi\mathbf{P} \tag{8.1.20}$$

For static and dynamic fields, respectively

$$\begin{aligned} P &= P_0 \qquad D = \varkappa_0 E \\ P &= P_\infty \qquad D = \varkappa_\infty E \end{aligned} \tag{8.1.21}$$

Thus we find for $\mathbf{P}_0$

$$\mathbf{P}_0 = \frac{D}{4\pi\varkappa} \tag{8.1.22}$$

where

$$\frac{1}{\varkappa} = \left[\frac{1}{\varkappa_0} - \frac{1}{\varkappa_\infty}\right] \tag{8.1.23}$$

$\varkappa$ is the effective dielectric constant for lattice polarizability.

The polarization field is proportional to the real displacement of the atoms from their equilibrium positions

$$\mathbf{P}_0 = A \sum_{\varkappa} q_\varkappa \, \hat{e}_\varkappa \, e^{i\varkappa \cdot \mathbf{r}_i} \tag{8.1.24}$$

where A is a constant which may be determined from the following condition on the energy of our vibrating lattice in the field of an electron

$$m\omega_\varkappa^2 (\nabla \mathbf{r}_i)^2 = \mathbf{D} \cdot \mathbf{P}_0 \tag{8.1.25}$$

We find

$$\omega_\varkappa^2 = \frac{4\pi\varkappa}{m} \tag{8.1.26}$$

Thus we may take $\mathbf{P}_0$ equal to (8.1.24) if we make use of (8.1.26), i.e.,

$$\mathbf{P}_0 = \sqrt{\frac{\hbar\omega_\varkappa}{8\pi\varkappa}} \sum_{\varkappa} \hat{e}_\varkappa \, (\beta^* + \beta) \, e^{i\varkappa \cdot \mathbf{r}} \tag{8.1.27}$$

We may define a polarization potential φ_P by

$$\mathbf{P}_0 = \frac{\nabla\varphi_P}{4\pi} \tag{8.1.28}$$

which yields

$$\varphi_P = \frac{4\pi}{i}\sum_{\varkappa}\sqrt{\frac{\hbar\omega_{\varkappa}}{8\pi K^2\varkappa}}(\beta^* + \beta)\, e^{i\varkappa\cdot\mathbf{r}} \tag{8.1.29}$$

Thus the probability of interaction becomes, using our previous procedures,

$$P_{\mathbf{k},\mathbf{k}'} = \frac{1}{\hbar^2}|\langle N_{\mathbf{k},\varkappa}|\, e\varphi_P\, |N_{\mathbf{k}',\varkappa}\rangle|^2 = \frac{8\omega_{\varkappa}\, e^2\pi}{\hbar K^2\varkappa N}\begin{Bmatrix}(n_{\varkappa}) \\ (n_{\varkappa+1})\end{Bmatrix} \tag{8.1.30}$$

Note that we have neglected the polarization produced by the electronic charge on it surroundings. As the electron travels through the crystal, its localized polarization field will travel with it.[4,5] The electron and its associated polarization cloud are called a "Polaron." The polaron effective mass is usually larger than that of the free electron and may be written

$$m_p^* = m_e^*\left[1 + \frac{\alpha}{6}\right] \quad \alpha < \tfrac{1}{2} \tag{8.1.31}$$

$$m_p^* = 0.02 m_e^*\alpha^4 \quad \alpha \gg 1 \tag{8.1.32}$$

where α is a dimensionless measure of the strength of the electron-lattice interaction and is given by[6]

$$\alpha = \frac{e^2}{\varkappa}\sqrt{\frac{m_e^*}{2\omega_{\varkappa}\hbar^3}} \tag{8.1.33}$$

8.2 ELECTRON-PHOTON INTERACTIONS (DIRECT TRANSITIONS)

An important interaction within a crystal is that due to the interaction of an electron with a photon of energy $\hbar\omega_{\varkappa}$, where $\varkappa$ is the wave vector of the photon. This interaction is the basis for much of the interpretation of the results of optical absorption experiments in direct band gap materials. Let us derive the selection rules and their consequences for such an interaction.

The Hamiltonian for a particle of charge e located in an electromagnetic field is given by

$$\mathcal{H} = \frac{1}{2m}(\mathbf{p} - e\,\mathbf{A})^2 + V(\mathbf{r}) \tag{8.2.1}$$

where we have replaced the momentum operator p by $\mathbf{p} - e\mathbf{A}$. $\mathbf{A}$ is the vector potential which may be written

$$\mathbf{A} = A\,\hat{e}_{\varkappa}\,e^{i\varkappa\cdot\mathbf{r}} \tag{8.2.2}$$

where $\hat{e}_{\varkappa}$ is the unit polarization vector. Neglecting the terms in A^2 which are small we find

$$\mathscr{H} = -\frac{\hbar^2\nabla^2}{2m} + V(\mathbf{r}) - \frac{i\,e\hbar}{m}\mathbf{A}\cdot\nabla \tag{8.2.3}$$

We may trat the third term in $\mathscr{H}$ as a perturbation. Thus the probability of interaction between an electron and a photon is

$$P_{\mathbf{k},\mathbf{k}'} = \left| \frac{i\,e\hbar\mathbf{A}}{m_e^*} \sum_{\mathbf{r}_i} e^{i(\mathbf{k}'-\mathbf{k}+\varkappa)\cdot\mathbf{r}_i}\,\hat{e}_{\varkappa}\cdot\left[\int_C \psi_{\mathbf{k}}^*\,\nabla\,\psi_{\mathbf{k}'}\,d\mathbf{r} + i\mathbf{k}\int_C \psi_{\mathbf{k}}^*\psi_{\mathbf{k}'}\,d\mathbf{r}\right]\right| \tag{8.2.4}$$

As before we find that $P_{k,k'}$ is equal zero unless $\mathbf{k}' - \mathbf{k} + \varkappa$ is zero or $2\pi\mathbf{G}$. Thus we have again met the requirement for momentum conservation. The wave vevtor $\varkappa$ of incident light is very much smaller than $\mathbf{k}$ or $\mathbf{k}'$. Thus the momentum conservation condition may be written $\mathbf{k} \approx \mathbf{k}'$, i.e., in order for an electron to interact with a photon, the wave vector of the electron may not change. This requirement restricts the absorption of photon energy by an electron to those interactions resulting in a "direct" or "vertical" transition, as shown in Figure 8.3. (Intraband transitions of this type called, free-carrier absorption, may not take place.)[8]

In general the total probability is given by

$$P = \int_{-\infty}^{\infty} P_{\mathbf{k},\mathbf{k}'}[g_v(\mathbf{k})\,f_v(\mathbf{k})]\,[g_c(\mathbf{k}')\,\{1 - f_c(\mathbf{k}')\}]\,dk \tag{8.2.5}$$

where the first term in brackets gives the total number of occupied states within the valence band and the second describes the number of available states in the conduction band (of course, the density of states functions must be multiplied by the factor M, the number of equivalent minima).

We will calculate the probability of a direct transition from a particular $\mathbf{k}$ value in the valence band to an energy level located in the conduction band. Thus we shall set $g_v(\mathbf{k}) = 1$. Further assume that the conduction band minima is broad and flat and take the density of final states as $V/(4\pi^3)$ (including spin). We will make our calculation for sufficiently low temperatures that the valence band is filled and the conduction band empty, and

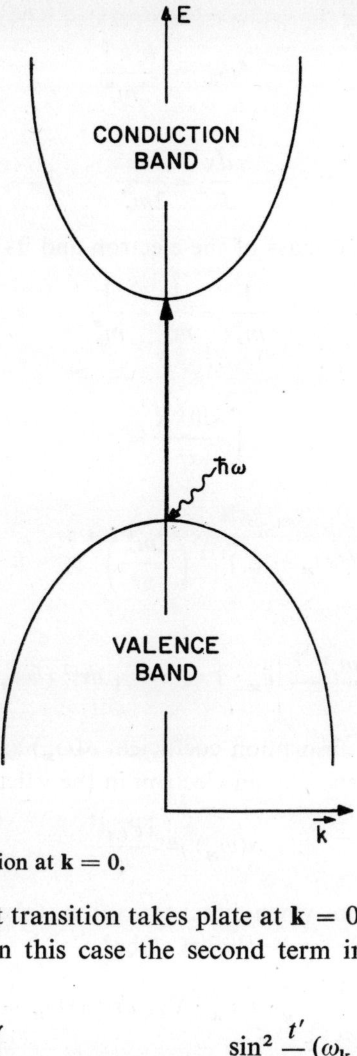

Figure 8.3 Direct transition at **k** = 0.

assume that our direct transition takes plate at $\mathbf{k} = 0$ (sometimes called the allowed transition). In this case the second term in (8.2.4) goes to zero yielding

$$P = \frac{-e^2A^2\nabla}{\hbar^2 m^{*2}(4\pi^3)} \int_{E_c}^{\alpha} \left\{ |\hat{e}_{\varkappa} \cdot \int_c \psi_{\mathbf{k}}^* \nabla \psi_{\mathbf{k}'}\, dr|^2 \, \frac{\sin^2 \frac{t'}{2} (\omega_{\mathbf{k}} - \omega_{\mathbf{k}} - \omega_{\varkappa})\, d\mathbf{k}'}{(\omega_{\mathbf{k}'} - \omega_{\mathbf{k}} - \omega_{\varkappa})^2} \right\} \tag{8.2.6}$$

Setting

$$x = \frac{1}{2}(\omega_{\mathbf{k}'} - \omega_{\mathbf{k}} - \omega_{\varkappa})\, t$$

$$\frac{dx}{dk'} = \frac{t}{2} \frac{\partial \omega_{\mathbf{k}'}}{\partial k'} \tag{8.2.7}$$

and

$$\hbar\omega_{\mathbf{k}'} = \frac{\hbar^2 k'^2}{2m_r^*} \tag{8.2.8}$$

we find

$$\frac{dx}{dk'} = \frac{t\hbar k'}{2m_r^*} \tag{8.2.9}$$

where m_r^* is the reduced mass of the electron and its associated hole

$$\frac{1}{m_r^*} = \frac{1}{m_e^*} + \frac{1}{m_h^*} \tag{8.2.10}$$

Using

$$\int_{-\infty}^{+\infty} \frac{\sin^2 x}{x^2} = \pi$$

and

$$(\hbar\omega_\varkappa - E_c)^{1/2} \left(\frac{2m_r^*}{\hbar^2}\right)^{1/2} = k' \tag{8.2.10a}$$

we find

$$P = \frac{-e^2 A^2 \nabla t\,(2m_r)^{3/2}}{4m_e^{*}\hbar^4} \left| \hat{e}_\varkappa \cdot \int_c \psi_{\mathbf{k}}^* \cdot \nabla\psi_{\mathbf{k}}\, d\mathbf{r} \right|^2 (\hbar\omega_\varkappa - E_c)^{1/2} \tag{8.2.11}$$

We may define the absorption coefficient $\alpha(\omega_\varkappa)$ as the probability that a quantum will be absorbed by an electron in the valence band

$$\alpha(\omega_\varkappa) = \frac{|P_{i,j}|^2}{F} \tag{8.2.12}$$

where F is the flux. Thus when direct transitions take place from $\mathbf{k} = 0$, the absorption coefficient may be shown to be[9]

$$\alpha(\omega_\varkappa) = \frac{2e^2(2m_r^*)^{3/2}}{4\hbar^3\omega_{\mathbf{k}} m_e^{*2} n\varkappa c} \left| \hat{e}_\varkappa \cdot \int_e \psi_{\mathbf{k}}^* \cdot \nabla\psi_{\mathbf{k}}\, d\mathbf{r} \right|^2 (\hbar\omega_\varkappa - Eg)^{1/2} \tag{8.2.13}$$

Direct transitions may also take place at points in k space other than at $\mathbf{k} = 0$. In this case we must utilize the second term in (8.2.4) which includes a k^2 term in $P_{\mathbf{k},\mathbf{k}'}$. Thus the absorption coefficient for these so-called "unallowed" direct transitions varies as $(\hbar\omega_\varkappa - Eg)^{3/2}$. The absorption coefficient for single crystal GaAs is shown in Figure 8.4.

It has been found[10] that the energy gap derived from optical absorption measurements on InSb and some highly doped semiconductors seems to be shifted toward higher energies. This was explained by Burstein[11] for InSb

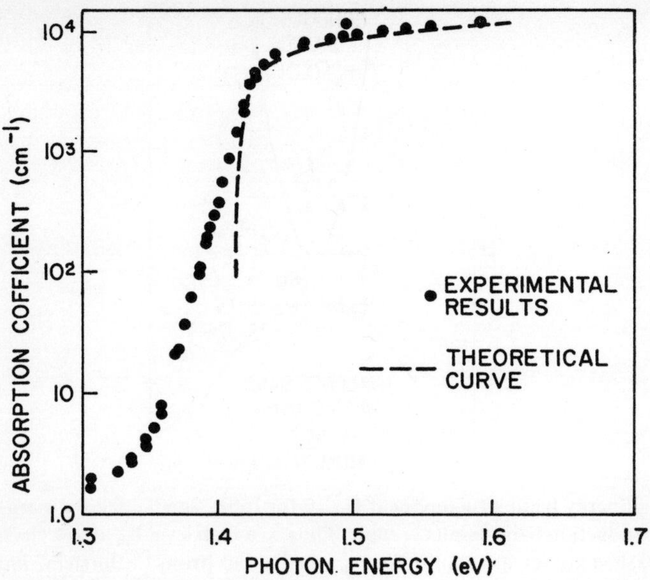

Figure 8.4 Absorption coefficient of single crystal GaAs [from T. S. Moss and T. D. F. Hawkins, *Infrared Phys.* **1**, 111, 1961].

by taking into account the degeneracy produced by the large ratio of hole to electron effective masses [see (6.1.19)] InSb becomes degenerate at electron concentrations of only $1 \times 10^{19}\ \text{cm}^{-3}$ (300°K). Thus the Fermi level is approximately $4k_0T$ above the lowest unfilled energy band level E_m in the conduction band (see Figure 8.5). E_m corresponds to E_c at an electron concentration of $n_m = 8.6 \times 10^{17}\ \text{cm}^{-3}$. Thus, at electron concentrations lower than n_m the optical energy gap corresponds to the true gap E_g. At electron concentrations larger than n_m the optical energy gap is approximately given by

$$E_{g(\text{opt})} = E_g + \left(1 + \frac{m_n}{m_p}\right)(\mu - E_c - 4k_0T) \qquad (8.2.14)$$

For p-type samples the optical energy gap is equal to the true gap at hole concentrations less than $P_m = 1.4 \times 10^{19}\ \text{cm}^{-3}$. For hole concentrations larger than P_m, $E_{g(\text{opt})}$ becomes

$$E_{g(\text{opt})} = E_g + \left(1 + \frac{m_p}{m_n}\right)(E_V - \mu - 4k_0T) \qquad (8.2.15)$$

This apparent shift in the energy gap to higher energies has been called the Burstein shift.

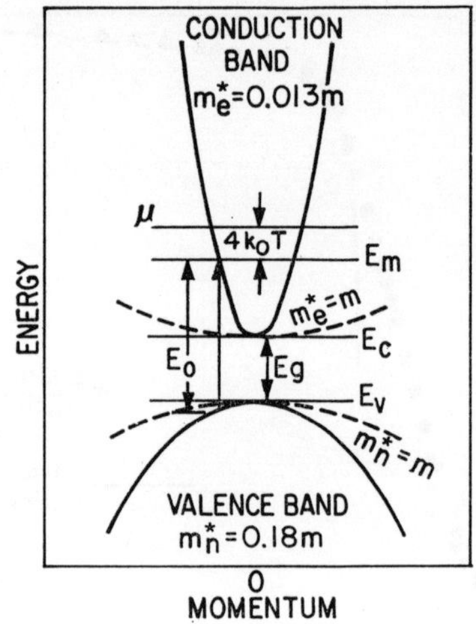

Figure 8.5 Energy band structure near $\mathbf{k} = 0$ for InSb. Small effective mass of electrons in narrow conduction band results in rapid filling to a high level E_m at low electron concentrations. Dashed curves are shown for $m_e^* = m_h^* = m$. [from E. Burstein, *Phys. Rev.* **93**, 632 (1954)].

8.3 ELECTRON-PHONON-PHOTON INTERACTIONS (INDIRECT TRANSITIONS)

As we have seen, the interaction of an electron with a phonon will cause a transition which conserves both energy and momentum. It has been found that there also exists an important interaction between electrons, phonons and photons. This type of interaction explains an important type of optical transition frequently seen in indirect band gap materials. It involves the so-called short-lived "virtual state", transitions to and from which, do not require conservation of energy but only conservation of momentum. This type of interaction gives rise to transitions of the form *ABC* as shown in Figure 8.6. An electron may interact with a photon of energy $\hbar\omega$. Although $\hbar\omega < \Delta E$, this photon may induce a transition to state *B* only if the electron remains in state *B* for a very short time. This time must by short enough that

$$(\Delta E - \hbar\omega)\,\Delta t \lesssim \hbar \tag{8.3.1}$$

i.e., the time that the electron spends in the virtual states must be well within the allowed uncertainty in time prescribed by the uncertainty principle.

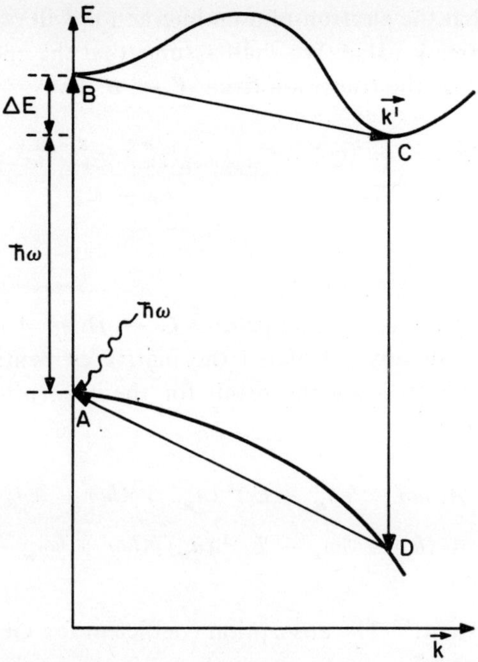

Figure 8.6 Indirect transitions $A \to B \to C$ and $C \to D \to A$.

After spending only a small time in state B, the electron interacts with a photon of wave vector $\mathbf{k}' - \mathbf{k}$ and makes a transition to state C. The transition $B \to C$ requires conservation of momentum but, again, not conservation of energy because the state B is short-lived. The electron ends up in state C and, for the resultant transition $A \to C$, both energy and momentum are conserved. Needless to say, the transition $C \to D \to A$ is also allowed through the emission of a photon of energy $\hbar\omega$ for the transition $C \to D$ and the emission of a phonon of wave vector $\mathbf{k}' - \mathbf{k}$ for the transition $D \to A$ as shown in Figure 8.6.

The application of first order perturbation theory to the electron-phonon-photon interaction results in a zero probability for the transition. Thus we must go to second order (see section 1.4). The total Hamiltonian is given by

$$\mathscr{H} = \mathscr{H}_{\text{êl-phonon}} + \mathscr{H}_{\text{el-photon}} \tag{8.3.2}$$

The first term in (1.4.18) is given by

$$\langle \gamma_{\mathbf{k}}^{C} | \mathscr{H}_{\text{el-phonon}} | \gamma_{\mathbf{k}}^{A} \rangle + \langle \gamma_{\mathbf{k}}^{C} | \mathscr{H}_{\text{el-photon}} | \gamma_{\mathbf{k}}^{A} \rangle = 0 \tag{8.3.3}$$

since we found that the electron-photon interaction will occur only between states with the same **k** value. We shall assume that the energy of a phonon is too small to cause the transition from $A \rightarrow B$. Thus

$$P_{\mathbf{k}-\mathbf{k}'} = \frac{2\pi}{\hbar} \sum_{\substack{|\gamma_B\rangle \\ \neq \\ |\gamma_C\rangle}} \sum_{|\gamma_A\rangle} \left| \frac{\langle \gamma_{\mathbf{k}'}^{C} | \mathscr{H}_{\text{el-photon}} | \gamma_{\mathbf{k}'}^{B} \rangle \langle \gamma_{\mathbf{k}}^{B} | \mathscr{H}_{\text{el-phonon}} | \gamma_{\mathbf{k}}^{A} \rangle}{E_B - E_A} \right|^2 \tag{8.3.4}$$

We must note that the competing process $C \rightarrow D \rightarrow A$ must also be considered. We have already calculated the matrix elements which occur in (8.3.4). We will merely state the result for the absorption coefficient for indirect transitions

$$\begin{aligned} \alpha\omega = A[(\hbar\omega - \hbar\omega_{\varkappa} - E_c)^2 \, (n_{\varkappa+1}) \, \delta(\hbar\omega - \hbar\omega_{x} - E_c) \\ + (\hbar\omega + \hbar\omega_{\varkappa} - E_c)^2 \, (n_{\varkappa}) \, \delta(\hbar\omega + \hbar\omega_{\varkappa} - E_c)] \end{aligned} \tag{8.3.5}$$

where A is a constant.[12] The absorption coefficient for Ge which is indirect is shown in Figure 8.7.

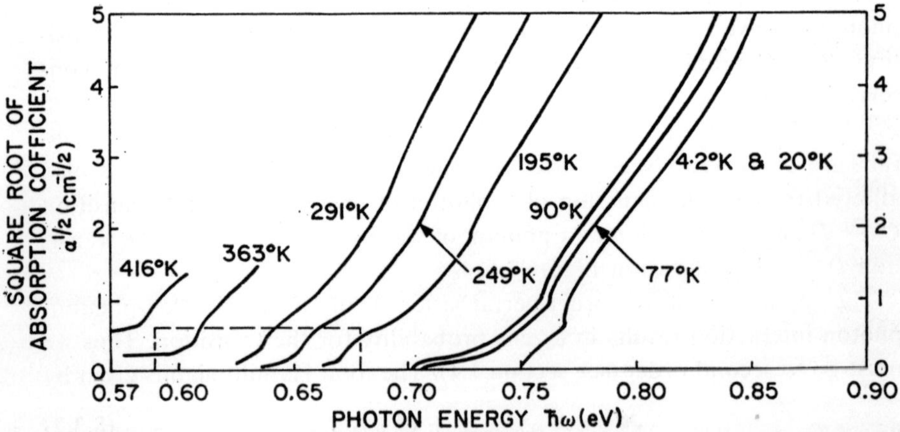

Figure 8.7 Absorption spectrum of Ge. Low and high energy portions are indirect and direct, respectively [from G. G. Macfarlane, T. P. McLean, J. E. Quarrington and V. Roberts, *Jnl. Phys. Chem. Solids*, **8**, 388 (1959)].

Excitons

When an electron makes a band-to band transition, either direct or indirect, it leaves behind a positive hole. Until how we have neglected the coulomb interaction between the electron-hole pair. This stable configuration is called an "exciton" and its inclusion into the theory of solid results in interesting thermal and optical properties of crystals.

When a conduction band electron is created as a result of a direct transition, the momentum vectors of the electron and hole must by equal in magnitude and oppositely directed to conserve crystal momentum. Thus their center of mass must have no net momentum, and the exciton energy level must lie at $\mathbf{k} = 0$. When an exciton is formed through an indirect transition, which involves a phonon, the center of mass may then move with the momentum vector of the phonon involved in the transition. We may treat the electron-hole combination in much the same way that we treated the impurity atom problem in section 5.1. In the present case, however, we replace the electron mass with the exciton reduced mass

$$m_r^* = \frac{m_e^* m_h^*}{m_e^* + m_h^*} \tag{8.3.6}$$

see (8.2.10). We find for the energy of an exciton[13]

$$E = Eg - \frac{m_r e^4}{2\hbar^2 \varkappa^2 n^2} + \frac{\hbar^2 k^2}{2(m_e^* + m_h^*)} \tag{8.3.7}$$

where the last term results from the motion of the center of mass and vanishes for excitons which result from direct transitions. The second term results from the electrostatic attraction between the electron and hole.

For direct transitions the exciton level is quite sharp and lies approximately 0.005 eV below the conduction band edge. The appearance of the principle quantum number n in (8.3.7) indicates that excited exciton states may lie within the conduction band. Because of the large number of allowed lattice wave vectors, the exciton lines due to indirect transitions are not sharply defined, but form bands of states close to the conduction band edge.

The electron-hole separation in a semiconductor are usually quite large, resulting in weakly bound excitons. In molecular crystals, the electron-hole pairs may be quite strongly bound.[14] The effects of exciton absorption are seen superimposed on much optical data as can be seen in Figure 8.8. The left hand side is a detail of the low energy portion of Figure 8.7 and the right is the extension of $\alpha^{\frac{1}{2}}$ to high energies.

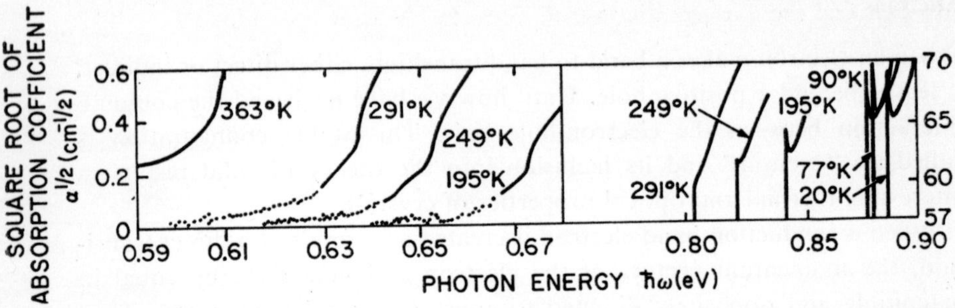

Figure 8.8 Low energy portion and high energy extension of the absorption coefficient of Ge shown in Figure 8.7 [from G. G. Macfarlane, *et al*].

8.4 TRANSPORT THEORY

The Bloch continuity in a crystal is also disturbed by the presence of impurity atoms, dislocations, strains, defects, etc. The interactions of an electron with these regions impede the electron's progress through the crystal and are a major source of electrical and thermal resistance. In this section we will investigate the effect of some specific types of lattices disorder on electrons progressing through a crystalline solid.

We must now find a method by which we will be able to treat the transport of electrons through a lattice which contains perturbations. To a first approximation, we can assume that the collisions an electron makes with the perturbations in Bloch periodicity are random. We will also assume that the interactions are elastic and independent of one another. With these assumptions, the probability P that an electron will suffer a collision is taken to be independent of time.

Thus the total number dn' of collisions that will occur in time dt may be written

$$dn' = Pn'\,dt \tag{8.4.1}$$

Thus

$$n' = n'_0\,e^{tP} \tag{8.4.2}$$

where n' is the number of collisions (or the number of electrons which will suffer collisions) as a function of time. n'_0 is the number of electrons which have suffered collisions up to the time $t = 0$. The number of electrons not having suffered collisions is given by

$$n = n'_0\,e^{t(1-P)} = n_0\,e^{-tP} \tag{8.4.3}$$

We can find the average lifetime $\mathcal{T}$ between electron collisions by

$$\mathcal{T}_e = \frac{\int_0^\infty t\,dn}{\int_0^\infty dn} = -\frac{\int_0^\infty n_0 Pt\,e^{-tP}dt}{\int_0^\infty n_0\,e^{-tP}dt} \tag{8.4.4}$$

$$\mathcal{T}_e = \frac{1}{P} \tag{8.4.5}$$

Thus the average time between collision is equal to the inverse of probability of collision. We can define a so-called average "mean free path" l by

$$l_i = v_i \mathcal{T}_e \tag{8.4.6}$$

where v_i is the average velocity attained between collisions and l_i is the average distance travelled in the ith direction by an electron between collisions. It is a simple matter to argue that we may write the average velocity in the ith direction as

$$\mathbf{v}_i = \frac{e\mathbf{E}_i \mathcal{T}_e}{m_e^*} \tag{8.4.7}$$

where $\mathbf{E}_i$ is the electric field in the ith direction. One usually defines an electron "drift" mobility μ_e as the velocity per unit electric field

$$\mu_e = \frac{\mathbf{v}_i}{|\mathbf{E}|} \tag{8.4.8}$$

Thus from (8.4.7)

$$\mu_e = \frac{e\,\mathcal{T}_e}{m_e^*} \tag{8.4.9}$$

Thus we may write the current density in the i direction as $\mathbf{J}_i = n\,e\mathbf{v}_i$. And

$$\mathbf{J}_i = \frac{n\,e^2 \mathbf{E}_i \mathcal{T}_e}{m_e^*} \tag{8.4.10}$$

Thus, from Gauss' law

$$\mathbf{J}_i = \sigma \mathbf{E}_i \tag{8.4.11}$$

$$\sigma = \frac{n\,e^2 \mathcal{T}_e}{m_e^*} \tag{8.4.12}$$

where we neglect the directional dependence of the conductivity which, in truth, is a tensor quantity.

Similar equations may be written for holes

$$\mu_h = \frac{v_{ih}}{|\mathbf{E}|} = \frac{e\mathcal{T}_h}{m_h^*} \tag{8.4.13}$$

$$\mathbf{J}_{hi} = \sigma_h \mathbf{E}_i \tag{8.4.14}$$

And the total current density may be written

$$\mathbf{J}_i = (n e\mu_{e_i} + P\, e\mu_{h_i})\, \mathbf{E}_i$$

$$= \left(\frac{n e^2 \mathcal{T}_e}{m_e^*} + \frac{p\, e^2 \mathcal{T}_h}{m_h^*}\right) \mathbf{E}_i \tag{8.4.15}$$

And

$$\sigma = n\, e\mu_e + p\, e\mu_h \tag{8.4.16}$$

We shall define a quantity called the differential scattering cross-section $\sigma(\theta, \phi)$ which tells us the probability that an electron will scatter into a unit solid angle in directions θ and ϕ to the incident direction of the electron. The total scattering cross-section σ_T is given by

$$\sigma_\tau = \int_0^\pi \int_0^{2\pi} \sigma(\theta, \phi) \operatorname{Sin}\theta\, d\theta\, d\phi \tag{8.4.17}$$

We shall define

$$\sigma(\theta, \phi) = \frac{k^2}{\text{velocity}} \times (\text{Probability of Transition}) \tag{8.4.18}$$

$$\frac{1}{\mathcal{T}} = 2\pi \int_0^\pi \frac{k^2(P_{\mathbf{k},\mathbf{k}'})}{\text{vel.}} \sin\theta\, d\theta \tag{8.4.19}$$

when $\sigma = \sigma(\theta)$.

Calculations of $\mathcal{T}$ are quite important when one is attempting to determine the relevant charge carrier scattering mechanisms in a solid: this topic will be dealt with in the next section.

The Boltzmann Equation

We have been specifying the distribution of electrons in energy by e.g., the Fermi-Dirac function which is valid only for a system in thermal and energetic equilibrium. In the present study, in which we are investigating the influence of collisions on the transport of electrons through a crystalline solid, our distribution function must be modified to take into account small deviations from equilibrium. We may assume that our new distribution function is a function of **k** vector, position in the crystal and time, and that it may be expanded for small deviations from equilibrium as

$$f(\mathbf{k}, \mathbf{r}, t) = f(\mathbf{k}) + \frac{\partial f}{\partial t}\, dt + \frac{\partial f}{\partial \mathbf{k}}\, d\mathbf{k} + \frac{\partial f}{\partial r}\, dr \tag{8.4.20}$$

We may assume that a force **F** is acting on our crystal to produce small deviations from equilibrium. From Newton's second law

$$\dot{\mathbf{P}} = \mathbf{F} = \hbar\dot{\mathbf{k}} \tag{8.4.21}$$

Thus we may write the ith component of the third term of (8.4.20) as

$$\frac{\partial f}{\partial k_i}\left(\frac{F_i\, dt}{\hbar}\right) \tag{8.4.22}$$

Similarly, the force-produced velocity will give for the fourth term

$$\frac{\Delta f}{\Delta r_i}\,\Delta r_i = \frac{\Delta f}{\Delta r_i}\, v_i\, d\mathscr{T} \tag{8.4.23}$$

Thus the total change Df of the revised distribution function may be written

$$\frac{Df}{dt} = \frac{\partial f}{\partial t} + \frac{1}{\hbar}\,\mathbf{F}\cdot\nabla_k f + \mathbf{v}\cdot\nabla_r f \tag{8.4.25}$$

Since $\dfrac{\partial f}{\partial t}$ can only be brought about by interactions within the crystal, we will define

$$\frac{Df}{dt} \equiv \left[\frac{\partial f}{\partial t}\right]_{\text{interactions}} \tag{8.4.26}$$

Under conditions of dynamic equilibrium, $\dfrac{Df}{\partial t}$ is zero, and we find

$$\left[\frac{df}{dt}\right]_{\text{interactions}} = \frac{1}{\hbar}\,\mathbf{F}\cdot\nabla_k f + \mathbf{v}\cdot\nabla_r f \tag{8.4.27}$$

Note that the left hand side vanishes if no fields exist to produce forces or changes in the distribution function. (8.4.27) is the Boltzmann transport equation.

It is usually assumed that the change in f due to collisions is proportional to both the deviation from equilibrium and the probability of scattering.

$$\left[\frac{df}{dt}\right]_{\text{interactions}} = \frac{f(\mathbf{k}, \mathbf{r}, t) - f(o)}{\mathscr{T}} \tag{8.4.28}$$

Many problems in transport theory are addressed to finding $\frac{df}{dt}$ due to a particular type of interaction. It is unfortunate, however, that only in a very few cases can we define a single $\mathcal{T}$ valid for an entire temperature range. (8.4.13) assumes, of course, that one can define a single $\mathcal{T}$ for the system. This is equivalent to saying that the energy of an electron gained or lost during a collision is negligible with respect to the initial energy of the electron.

Scattering of Charge Carriers

Acoustical Phonons

We have calculated the probability that an electron will interact with an acoustical lattice wave. Using our result (8.1.18) and (8.4.19), we find for the relaxation time for acoustic scattering

$$\frac{1}{\mathcal{T}} = \frac{m_e^* \hbar k E_1^2 k_0 T}{\pi \hbar^4 \varrho v_l^2} \tag{8.4.29}$$

where ϱ is the density and v_l is the longitudinal sound velocity in our crystal. The mean free path l is given by

$$\frac{1}{l} = \frac{1}{\mathcal{T} v_l} \tag{8.4.30a}$$

$$= \frac{m_e^* E_1^2 k_0 T}{\pi \hbar^4 \varrho v^2} \tag{8.4.30b}$$

which yields a mobility of the form

$$\mu = \frac{4e\,l}{3(2\pi m k_0 T)^{1/2}} \tag{8.4.31a}$$

$$= \frac{2(2\pi)^{1/2}\, e\hbar^4 \varrho v^2}{3(m_e^*)^{5/2}\,(k_0 T)^{3/2}\, E_1^2} \tag{8.4.31b}$$

Scattering by acoustic phonons is the dominant scattering mechanism near and above room temperature in the elemental semiconductors. The mobility of electrons in Ge up to room temperature is compared to the $T^{-3/2}$ slope predicted by acoustic scattering in Figure 8.9. As the density of impurities increases, scattering by impurities sets in and destroys the $T^{-3/2}$ dependence of μ.

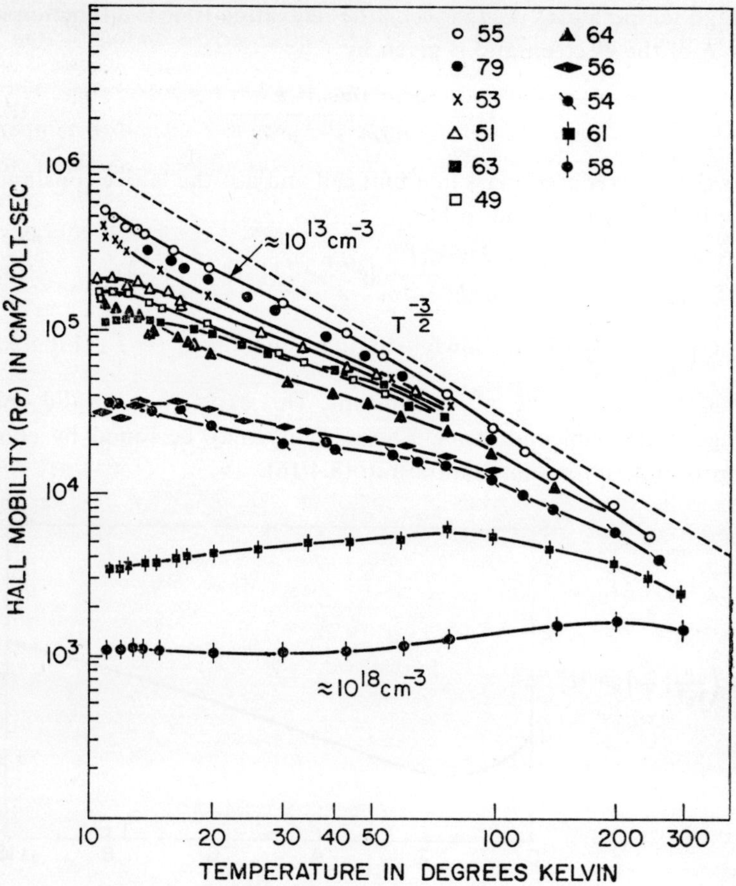

Figure 8.9 Hall mobility in As doped Ge [from P. D. Debye and E. M. Conwell, *Phys. Rev.* **93**, 693 (1954)].

Optical (Polar) Phonons

The scattering of electrons by the optical branch of the vibrational spectrum is an extremely complicated problem. This is because one cannot define a universal relaxation time $\mathcal{T}$ which is valid for all electron densities and at all temperatures. We will merely state the results of Howarth and Sondheimer[15]. We can define a single frequency for optical vibrations ω_0 and call this the fundamental frequency of optical vibrations, its wavelength λ_0 is of the order of magnitude of the interatomic distance, and it is the shortest wavelength that the lattice can support.

At high temperatures ($k_0T > \hbar\omega_0$) the relaxation time is a function of the energy E of the electron and is given by

$$\mathscr{T} = \frac{a^3 m_r^* (\hbar\omega_0)^2 E^{1/2}}{2\pi(2m^*)^{1/2} e^4 k_0 T} \tag{8.4.32}$$

where m_r^* is the reduced mass in a unit cell and a is the lattice constant. The conductivity may be found to be

$$\sigma = \frac{16a^3 M\Theta k_0^3 T^2}{3e^2\hbar^3} [e^{\Theta/T} - 1]\, G\left(\frac{\xi}{k_0T}, \frac{\Theta}{T}\right) \tag{8.4.33}$$

where $G\left(\frac{\xi}{k_0T}, \frac{\Theta}{T}\right)$ is a slowly varying function of (Θ/T) (Figure 8.10), $\Theta = \hbar\omega_0/k_0$ and a is the lattice constant. This expression is valid only for non-degenerate semiconductors. The mobility may be found by choosing the appropriate n for substitution into (8.4.16).

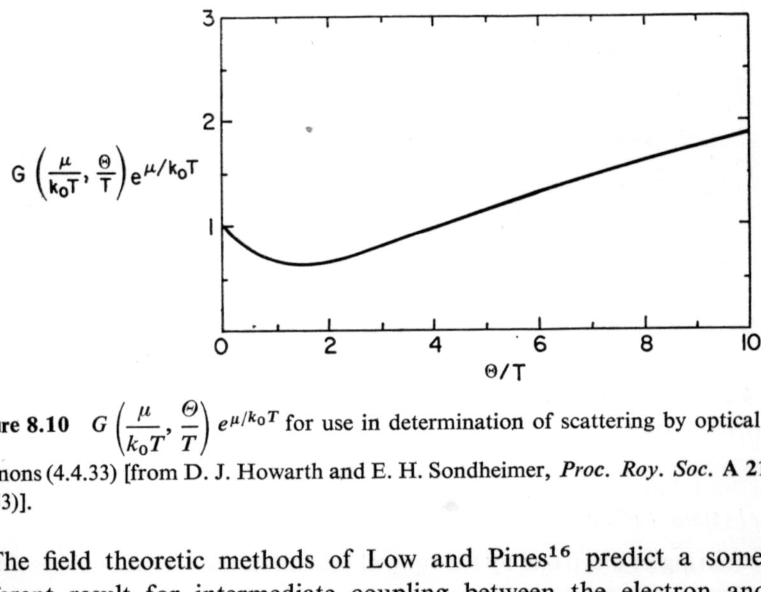

Figure 8.10 $G\left(\frac{\mu}{k_0T}, \frac{\Theta}{T}\right) e^{\mu/k_0T}$ for use in determination of scattering by optical polar phonons (4.4.33) [from D. J. Howarth and E. H. Sondheimer, *Proc. Roy. Soc.* **A 219**, 53 (1953)].

The field theoretic methods of Low and Pines[16] predict a somewhat different result for intermediate coupling between the electron and the lattice

$$\mu = \frac{e}{\alpha\omega_0\hbar^2 m_e^*}\left[1 + \frac{\alpha}{6}\right]^{-3} f(\alpha) \exp\frac{\omega_0\hbar}{k_0T} \tag{8.4.34}$$

where α is given by (8.1.31), and $f(\alpha)$ is a slowly varying function shown in Figure 8.11. A comparison between the mobility predicted by acoustical and polar scattering and experimental data for InSb is shown in Figure 8.12.

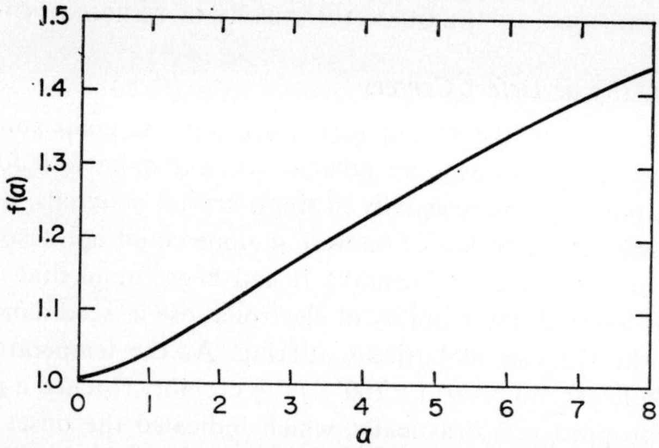

Figure 8.11 $f(\alpha)$ vs. α for Low and Pines derivation of mobility due to scattering by optical polar phonons [from F. E. Low and D. Pines, *Phys. Rev.* **98**, 414 (1955)].

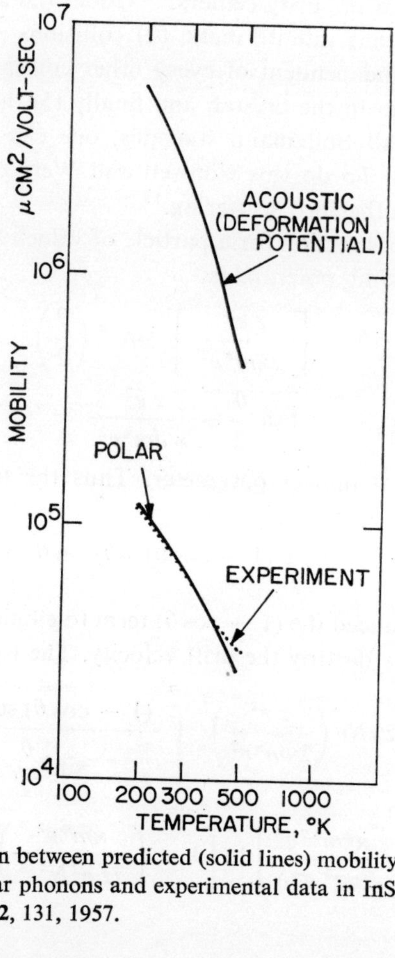

Figure 8.12 Comparison between predicted (solid lines) mobility due to scattering from acoustic and optical polar phonons and experimental data in InSb [from H. Ehrenreich, *Jnl. Phys. Chem. Solids*, **2**, 131, 1957.

Charged Impurity or Defect Centers

Although the previous theories of scattering from the polar and acoustical modes of lattice vibrations were able to account quite well for the high temperature mobility and resistivity in single crystal materials, it was found by many researchers that lattice scattering alone could not account for the low temperature data. Experimentally it had been found that as the temperature was lowered, the mobility of electrons rose in accordance with one or both of the theories of lattice scattering. As the temperature became progressively lower, however, (<100°K) the mobility reached a plateau and there after dropped very drastically, which indicated the onset of another strong scattering mechanism. Conwell and Weiskopf showed that this low temperature behaviour could be predicted if one considered the scattering from localized charged impurity centers.[17] Under the assumptions that (1) the scattering center has infinite mass, (2) collisions are perfectly elastic (3) each collision is independent of every other collision, (4) the impurity, distribution is random in the crystal, and finally (5) that the electrons can be treated by Maxwell Boltzmann statistics, one can calculate the right-hand side of (8.4.17). To do this Conwell and Weiskopf made use of the classical theory of Rutherford scattering.[18]

The differential cross section for a particle of velocity v scattered through an angle θ by a Coulomb potential is

$$\sigma(\theta) = \left[\frac{z\,e^2}{2\varkappa m^* v^2}\right]^2 \sin^{-4}\left(\frac{\theta}{2}\right) \tag{8.4.35a}$$

where

$$\tan\frac{\theta}{2} = \frac{z\,e^2}{\varkappa\, d m^* v^2} \tag{8.4.35b}$$

and d is the classical impact parameter. Thus the total scattering cross section σ_T becomes

$$\sigma_T = 2\pi \int_0^{\pi} (1 - \cos\theta)\,\sigma(\theta)\sin\theta\, d\theta \tag{8.4.35c}$$

where we have introduced the $(1 - \cos\theta)$ term to emphasize those collisions which tend to greatly destroy the drift velocity. The total relaxation time is given by

$$\frac{1}{\mathscr{T}} = 2\pi N v \left(\frac{z\,e^2}{2\varkappa m^* v^2}\right)^2 \int_{\theta_{\min}}^{\pi} \frac{(1 - \cos\theta)\sin\theta}{\sin^4\dfrac{\theta}{2}}\, d\theta$$

$$\mathscr{T} = \frac{\varkappa^2 m^{*2} v^3}{2\pi z^2 e^4 N}\left[\ln\left[1 + \left(\frac{\varkappa m^* v^2}{2z\,e^2 N^{1/3}}\right)^2\right]\right]^{-1} \tag{8.4.36}$$

where N is the number of charged impurities and we have chosen the minimum scattering angle θ_{min} by inserting the largest possible impact diameter $d = \frac{N^{-1/3}}{2}$ within a Wigner-Seitz sphere in the "impurity sublattice" and substituting into (8.4.35b). Taking into account the fact that we must average all allowed electron velocities we find

$$\mathcal{T}_{\text{average}} = \frac{\langle v^2 \mathcal{T} \rangle}{\langle v^2 \rangle} = \frac{\varkappa^2 m^{*2}}{2\pi z^2 e^4 N} \frac{\displaystyle\int_0^\infty \frac{v^7 e^{-\frac{mv^2}{k_0 T}} dv}{\ln\left[1 + \left(\frac{\varkappa m^{*2} v^2}{2ze^2 N^{1/3}}\right)^2\right]}}{\displaystyle\int_0^\infty v^4 e^{-\frac{m^* v^2}{k_0 T}} dv}$$

$$= \frac{8\varkappa^2 (kT)^{3/2} (2m^*)^{1/2}}{\pi^{3/2} z^2 e^4 N} \left[\ln\left(1 + \left\{\frac{7\varkappa k_0 T}{2ze^2 N^{1/3}}\right\}\right)^2\right]^{-1} \tag{8.4.37}$$

and the mobility becomes

$$\mu = \frac{e\mathcal{T}_{\text{average}}}{m^*} = \frac{8\varkappa^2 (k_0 T)^{3/2} \sqrt{2}}{\pi^{3/2} z^2 e^3 m^{*1/2} N} \left[\ln\left\{1 + \left(\frac{7\varkappa k_0 T}{2ze^2 N^{1/3}}\right)^2\right\}\right]^{-1} \tag{8.4.38}$$

The effect of scattering by charged impurities is seen in Figure 8.9 in which the mobility of electrons in Ge deviates more and more strongly from the $T^{3/2}$ law predicted by considering acoustical scattering as the impurity concentration and thus the importance of impurity scattering increases.

Equation (8.4.38) has been modified to account for screening of the field around impurities by electrons (ref. 17, Brooks, Dingle). The screening factor introduces no change in the concentration or temperature dependence of μ but introduces a small change in the slowly varying logarithmic term.

Neutral Impurities

At low temperatures many of the impurities or defects in a lattice are neutralized. They still, however, introduce a deviation from Bloch periodicity due to their size and thus may scatter charge carriers.[19] Erginsoy[20], using the experimental results of scattering from hydrogen atoms[21], has calculated the relaxation time for small angle scattering from these perturbations and found a relaxation time independent of electron energy

$$\frac{1}{\mathscr{T}} = \frac{20\varkappa N n \hbar^3}{m_e^{*2} e^2} \tag{8.4.39}$$

where N_n is the concentration (temperature dependent) of neutral impurities. The mobility arising from scattering by neutral impurities is then given by

$$\mu_{\text{neutral}} = \frac{m_e^* e^3}{20\varkappa N_n \hbar^3} \tag{8.4.40}$$

Dipole Complexes

In materials which contain both donors and acceptors, a density N_{dip} of dipole complexes will be formed. The compound dipole field is much smaller than the fields of the individual impurities and, as a result, scattering due to dipole complexes will result in a mobility which is larger than the mobility due to scattering by the individual ions.[22] Calculation of the relaxation time due to scattering from dipole complexes (neglecting screening effects) yields[23]

$$\frac{1}{\mathscr{T}} = \frac{(\pi)^{3/2} m^{1/2} e^4 l^2 N_{\text{dip}} \sqrt{2}}{\hbar^2 \varkappa^2 (k_0 T)^2} \tag{8.4.41}$$

Combined Scattering Mechanisms

In the above we have neglected carrier-carrier effects in scattering[24], scattering by dislocations[25], high field mobilities[26] and other detailed scattering mechanisms which are necessary refinements to the theory. Since it is clear that different combinations of scattering mechanisms may be dominant in different portions of the temperature spectrum, we may define, in the approximation that one interaction does not affect another, a reduced relaxation time

$$\frac{1}{\mathscr{T}_{\text{red}}} = \sum_{\substack{\text{i=all} \\ \text{interaction}}} \left(\frac{1}{\mathscr{T}_i}\right) \tag{8.4.42}$$

The resultant mobility will, of course, be lower than the lowest predicted by the individual scattering mechanisms.

At moderate temperatures, in both intrinsic and doped materials, conduction of electrons and holes takes place in the conduction and valence bands, respectively. In cases such as GaAs where a subsidiary minimum occurs a few tenths of an eV above the thermal energy gap, conduction in the secondary minimum may be taken into account by using the technique of

Aukerman and Willardson.[27] As the temperature in lowered, the resistivity usually increases since the temperature dependence of the deionization of impurities, or the dropping of electrons from the conduction to the valence bands, is much stronger than the temperature dependence of the mobility, which may rise with decreasing temperature. This typical behaviour is shown in Figure 8.13 for Ge.

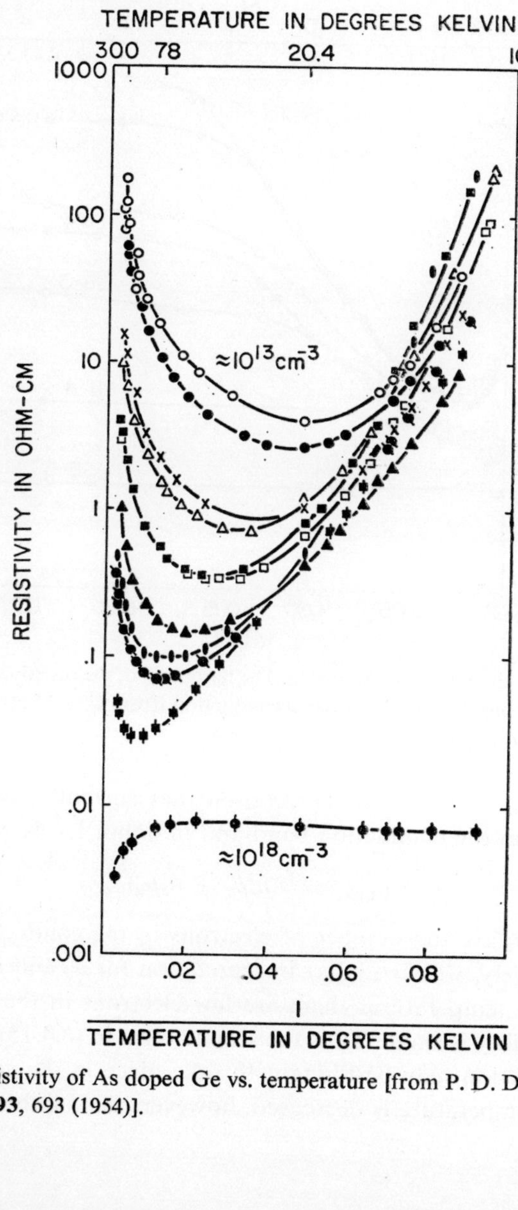

Figure 8.13 Resistivity of As doped Ge vs. temperature [from P. D. Debye and E. Conwell, *Phys. Rev.*, **93**, 693 (1954)].

At very low temperatures, however, the resistivity begins to flatten out instead of continuing to increase, as is shown in Figure 8.14. This results due to the simultaneous conduction of charge carriers (e.g. electrons) in both the conduction band and through the impurity energy levels located within the forbidden energy gap (see section 5.1).[28]

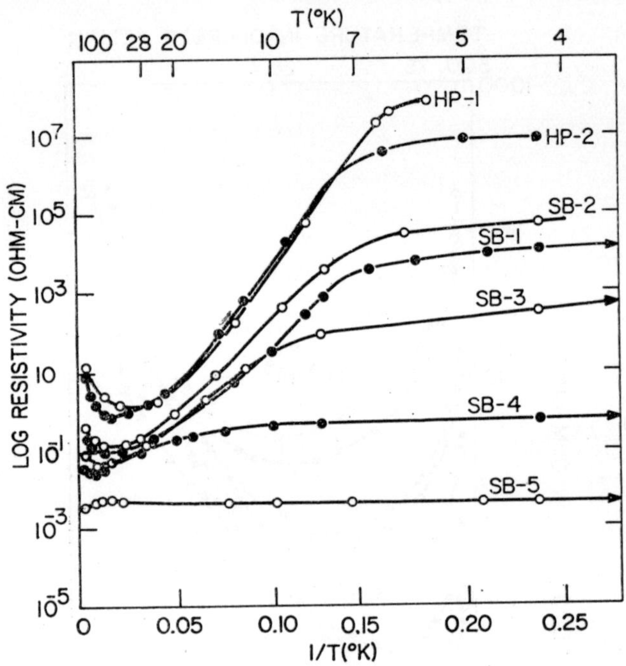

Figure 8.14 Resistivity vs. $1/T°$ K for Ge. The flattening of the resistivity at low temperatures indicates the onset of impurity band conduction [from C. S. Hung and J. R. Gliessman, *Phys. Rev.*, **96**, 1226 (1954)].

Using (8.4.16) with the reinterpretation that current is being carried by electrons both in the conduction band and in donor levels, we may write

$$\varrho_{\text{total}} = e(n_c\mu_c + n_d\mu_d)^{-1} \tag{8.4.43}$$

where n_c and n_d are the number of electrons in the conduction and donor levels, respectively, with a similar interpretation for μ_c and μ_d.

At moderate temperatures there are few electrons in the impurity levels where the mobility is small. Thus the second term in (8.4.15) may be neglected and conduction may be thought to take place in the conduction band only. As the temperature is decreased, however, the number of electrons in the conduction band becomes small as the valence band and impurity levels fill up. At very low temperatures, the second term in (8.4.15) may predominate giving a ϱ similar to that shown in Figure 8.14. This is called impurity band conduction.

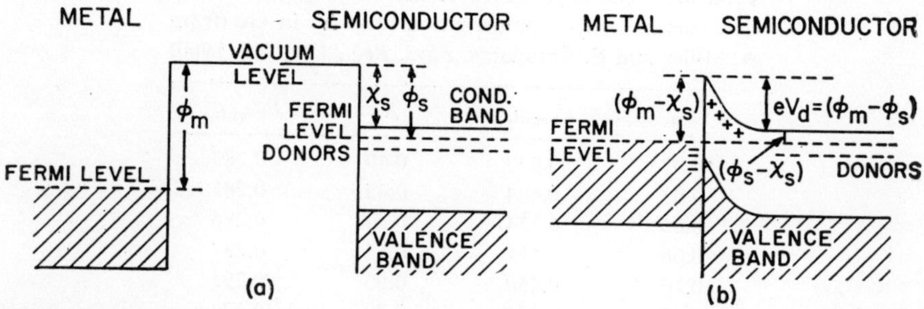

Figure 8.15 Energy level diagram of a metal *n*-type semiconductor contact with $\phi_m > \phi_s$ where ϕ_m and ϕ_s are the work functions of metal and semiconductor, respectively. (a) Energy level diagram before contact, (b) Energy level diagram after contact. The Fermi level is now at equal height; V_d is the diffusion potential. [from A. van der Ziel, *Solid State Physical Electronics*, Prentice-Hall, Inc., N.J., 1961, p. 242].

When the number of impurities is very small with negligible wavefunction overlap impurity band conduction takes place via a hopping-type of mechanism.[29]

When both donors and acceptors are present the combination of an ionized donor and a nearby ionized acceptor may be treated as a singly ionized hydrogen molecule in the presence of local perturbing fields. The electron will hop from one site to another with some preferred frequency. In the absence of any external field gradient, charge transfer will of course be random. The imposition of an electric field across the material produces a slight preference for charge transfer in the direction of lower energy and a small net current will flow.

Miller and Abrahams define an activation energy for the *n*-type resistivity in the region where conduction takes place by hopping as

$$E_{\text{hopping}} = \frac{d[\ln \varrho]}{d[k_0 T]} = \frac{e^2}{\varkappa r_d}[1 - 1.35 N_0^{1/3}] \tag{8.4.44}$$

where N_0 is the compensation ratio ($N_0 < 3\%$), r_d is the average distance between donors, and excited states are neglected. For larger values of N_0 the activation energy must be calculated numerically. Table 8.1 shows the hopping activation energy in units of E_d for various compensation ratios in Ge. Note that the ionization energy reaches a minimum between $0.45 < N_0 < 0.55$. As the concentration is increased, impurity bands are formed and conduction takes place by the usual processes.

Table 8.1 The dependence of the hopping-type activation energy on the compensation ratio N_0 in Ge (from A. Miller and E. Abrahams, *Phys. Rev.*, **120**, 745, 1960)

N_0	E_{hopping}	N_0	E_{hopping}
0.02	0.640	0.40	0.289
0.03	0.604	0.45	0.285
0.05	0.534	0.50	0.286
0.06	0.511	0.55	0.287
0.10	0.450	0.60	0.295
0.20	0.351	0.70	0.323
0.33	0.298	0.80	0.356
		0.90	0.391

8.5 CONTACT PHYSICS

Metal Semiconductor Junctions

In Chapter 2 we found that two systems in contact will reach equilibrium when their Fermi energies are equal if we allow particle flow between them. We will consider the effect this has on the electrical properties of junctions between two materials by first dealing with the metal-semiconductor contact.

The energy band diagrams of a metal and a semiconductor in close proximity, but not in contact, are shown in Figure 8.15a. If the amount of energy necessary to remove an electron from the metal φ_m (the work function) is less than the energy required to remove an electron from the bottom of the conduction band of the semiconductor to the outside world, electrons will flow from the semiconductor (presumed n-type) to the metal until the Fermi energies are aligned as shown in Figure 8.15b. The flow of carriers results in a positive space charge region in the semiconductor due to the ionized impurities left behind, and a negative one in the metal due to the excess electrons. The electron flow results in depressing all of the energy levels of the semiconductor and a potential barrier is set up at the interface between the metal and semiconductor. The height of this potential barrier is given by

$$eV_d = \varphi_m - \varphi_s \tag{8.5.1}$$

where V_d is the diffusion potential.

When a voltage V is placed on the semiconductor, its energy levels will change by an amount $(-eV)$. Thus when V is positive, the barrier height on the semiconductor side will increase according to $e(V_d + V)$. When V is negative, this side of the barrier decreases in height. Thus the current which flows from the semiconductor to the metal is a strong function of the applied voltage. The height of the metal side of the barrier, however, is unaffected by the applied voltage. This is shown in Figure 8.16. The junction between a metal and a semiconductor is thus a rectifying contact. If, however, $\varphi_m > \varphi_s$, no rectification will take place since charge transfer from the metal to the semiconductor raises the energy levels of the semiconductor, producing a very small barrier. The current flowing from the semiconductor to the metal is proportional to N, the number of electrons striking the junction surface and P, their probability of overcoming the barrier. Thus the total current J_τ crossing the junction may be written

$$J_\tau = J_{S\to M} - J_{M\to S} = Ne^{-\frac{eV_d}{k_0T}}\left[e^{-\frac{eV}{k_0T}} - 1\right] \tag{8.5.2}$$

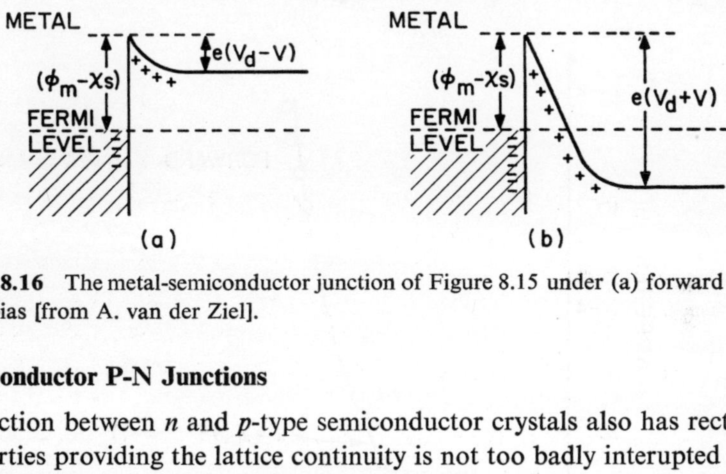

Figure 8.16 The metal-semiconductor junction of Figure 8.15 under (a) forward (b) reverse bias [from A. van der Ziel].

Semiconductor P-N Junctions

A junction between n and p-type semiconductor crystals also has rectifying properties providing the lattice continuity is not too badly interupted at the interface. When contact is made, electrons flow from the n to the p materials. A space charge region is again created, this time by the positively charged donors in the n material and the negatively charged acceptors on the p side as is shown in Figure 8.17.

A negative voltage applied to the n region (forward bias) will decrease the potential barrier and allow current to flow. Backward biasing increases the potential barrier and current soon saturates, as shown in Figure 8.18. The current may again be shown to be[30]

$$J_\tau = \{Ae^{-eV_d/k_0T}\}\,[e^{-eV/k_0T} - 1] \tag{8.5.3}$$

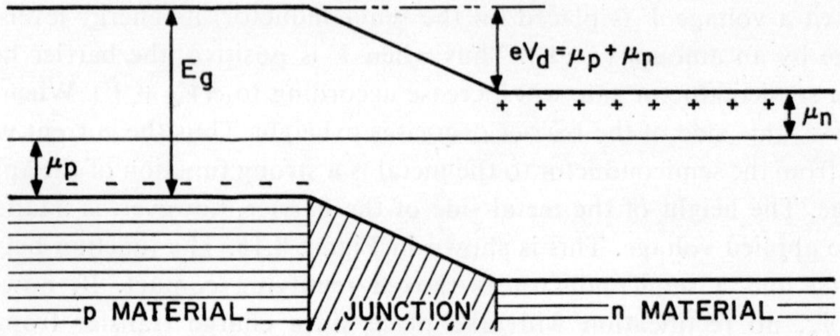

Figure 8.17 Energy band diagram of a *p-n* junction.

where the quantity in curly brackets is called the "saturation current." The width of a *p-n* junction gets smaller as the doping concentration of the *n*- and *p*-type materials increases

$$w = \left[\frac{2\varkappa (V_d + V)(N_a + N_d)}{eN_aN_d} \right]^{\frac{1}{2}} \tag{8.5.4}$$

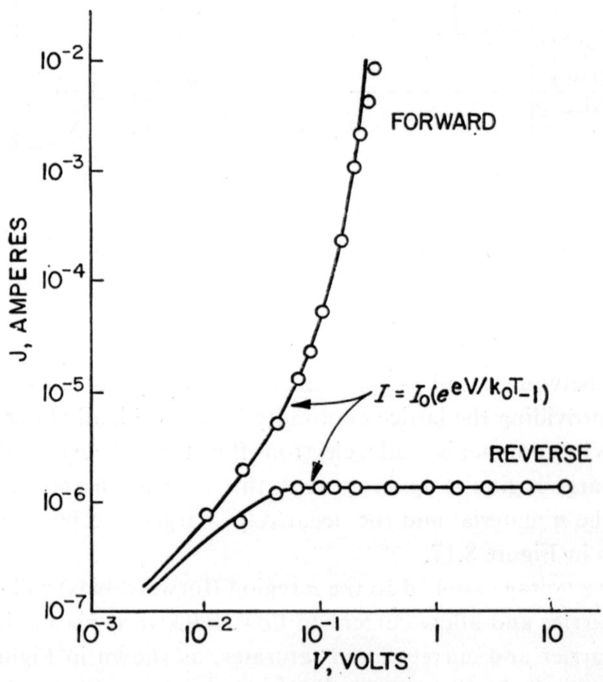

Figure 8.18 Current-voltage curve of a *p-n* junction [from L. V. Azaroff and J. J. Brophy, *Electronic Processes in Materials*, McGraw-Hill Book Co., Inc., New York, 1963, p. 275].

and the maximum electric field is given by

$$E = \frac{2}{w}[V_d + V] \tag{8.5.5}$$

Tunnel Diodes

An interesting effect occurs when the materials used in fabricating a *p-n* junction are highly doped.[31] If the doping density is sufficiently high to force the Fermi level to penetrate into the valence and conduction bands, respectively of the *p-* and *n*-type crystals, the *p-n* junction takes on the character shown in Figure 8.19. Since there are both empty and allowed energy levels located close to the Fermi energy in the conduction band of the *n*-type material and in the valence band of the *p*-type material, there is the possibility of equi-energy tunneling through the thin junction between valence and conduction band states. This tunneling current flows in addition to the usual diffusion *p-n* junction current and results in an *J-V* characteristic such as that shown in Figure 8.20. The low voltage hump comes about in the following way: At zero bias the tunneling current from the conduction band on the *n*-side to the valence band on the *p*-side is just balanced by the oppositely directed current as shown in Figure 8.21 a. As the forward bias voltage is increased (Figure 8.21 b) the barrier on the *n*-side of the junction

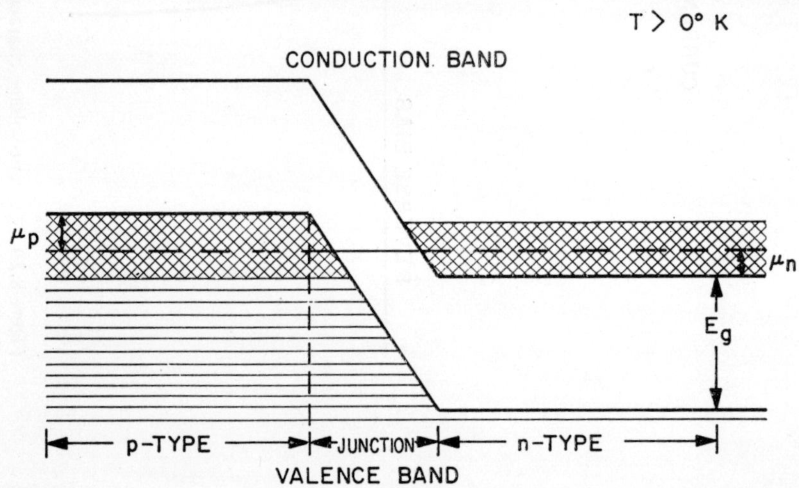

Figure 8.19 Energy band diagram of a *p-n* junction between degenerately doped materials Note the partially occupied levels in the conduction band on the *n* side opposite those in the valence band on the *p*-side.

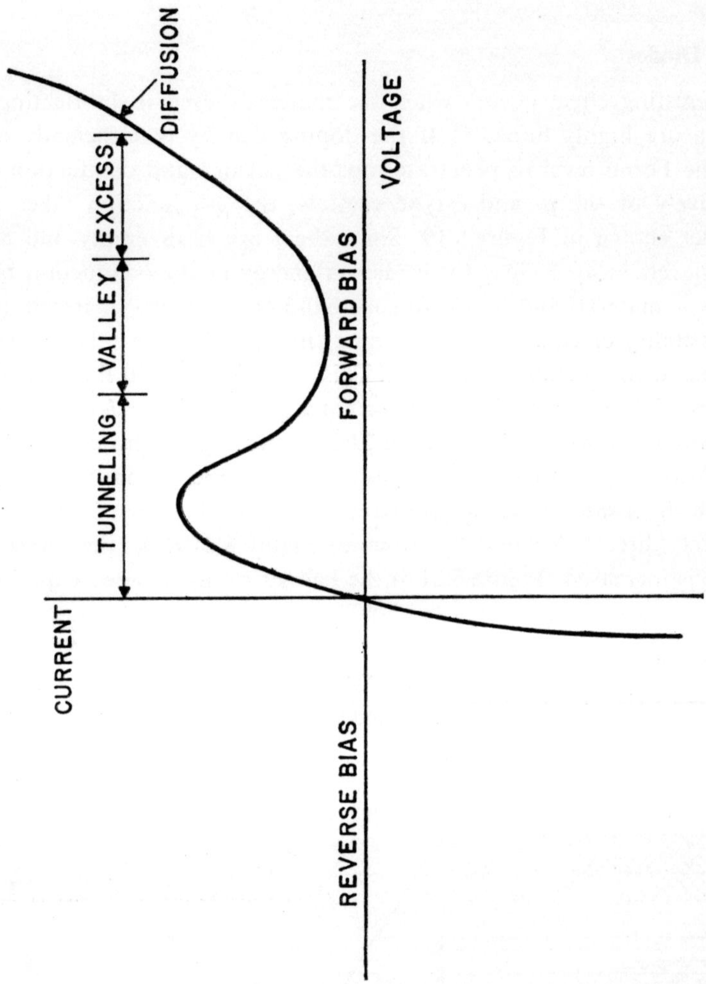

Figure 8.20 Current-voltage characteristic of a tunneling p-n junction diode.

is lowered, however the density of states available for tunneling in the conduction band is decreased. These competing mechanisms combine to produce a peak in the tunneling current which then drops to zero under forward bias sufficiently large to uncross the bands, i.e., $eV = \mu_n + \mu_p$ (Figure 8.21c). Current flows via tunneling in the reverse direction also. The reverse characteristic is very conductive since under reverse bias the energy bands on the *n*-side of the junction are lowered placing more empty conduction band levels opposite the filled valence band.

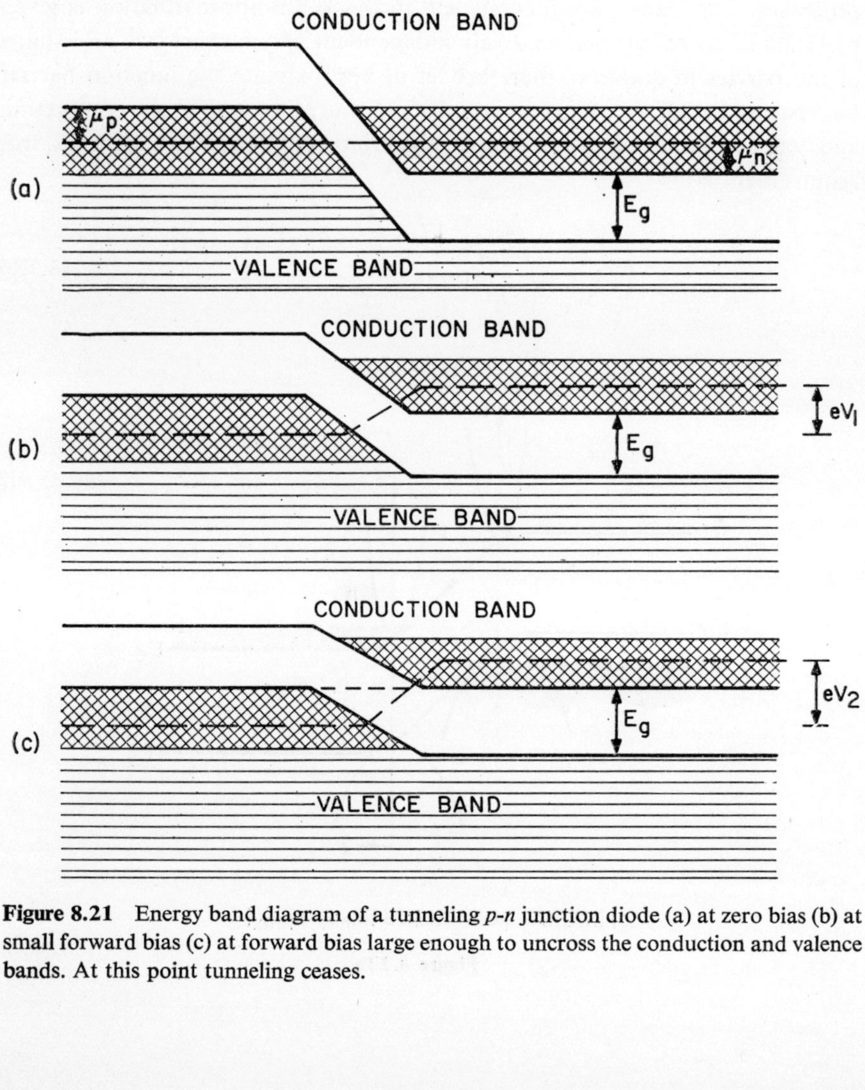

Figure 8.21 Energy band diagram of a tunneling *p-n* junction diode (a) at zero bias (b) at small forward bias (c) at forward bias large enough to uncross the conduction and valence bands. At this point tunneling ceases.

Tunneling may take place in both direct and indirect band gap materials. In direct band gap materials such as GaAs or GaSb, tunneling may take place only between states with the same **k**. This is schematically shown in Figure 8.22a, on which the E vs **k** band diagrams are shown superimposed on the energy band diagram of a tunnel junction. Tunneling may occur in indirect band gap materials also, e.g., from the $\mathbf{k} = 0$ in the valence band of the p-type material to an indirect conduction band minimum if the electron interacts with a phonon and thus exchanges the required momentum. This is shown in Figure 8.22b.

We will now examine the tunneling probability for direct and indirect processes. The transmission coefficient in the WKB approximation is given by (1.3.43). Its results are relatively independent of the exact size and shape of the barrier in question, therefore let us approximate the junction barrier by a parabolic barrier. The barrier must, of course, be zero at the conduction and valence band edges. A simple algebraic function which satisfies this requirement is[32]

$$K_{(x)} = \sqrt{\frac{2m}{\hbar^2}\left\{\frac{\left(\frac{E_g}{2}\right)^2 - (e\varepsilon_x)^2}{E_g}\right\}} \tag{8.5.6}$$

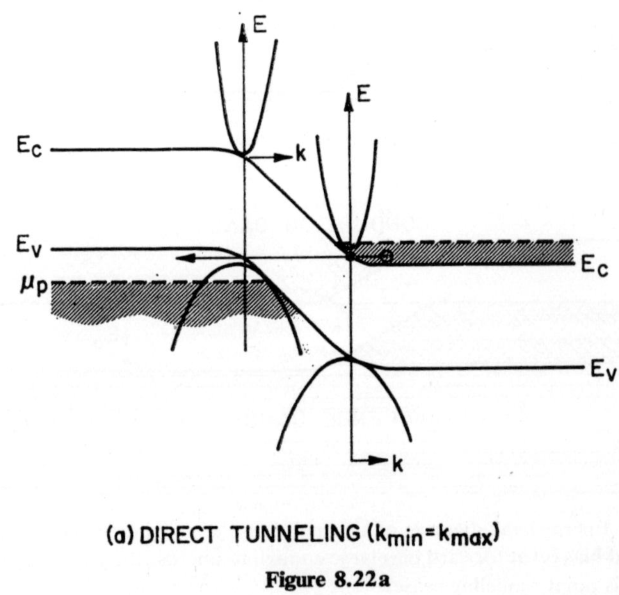

(a) DIRECT TUNNELING ($k_{min} = k_{max}$)

Figure 8.22a

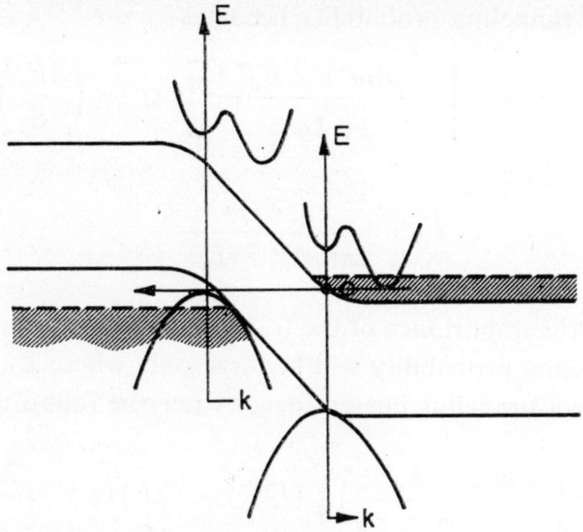

(b) INDIRECT TUNNELING ($k_{min} \neq k_{max}$)

Figure 8.22b

Figure 8.22 E vs k diagram superimposed on the energy band diagram of a tunneling *p-n* junction diode demonstrating (a) direct and (b) indirect tunneling [From S. M. Sze, *Physics of Semiconductor Devices*, Wiley-Interscience, New York (1969), p. 160].

Simple WKB theory for which we require only conservation of energy predicts for the transmission coefficient

$$T = \exp\left[\frac{-\pi(m^*)^{1/2} E_g{}^{3/2}}{2\sqrt{2}e\hbar\,\varepsilon}\right] \tag{8.5.7}$$

Since momentum must also be conserved in the tunneling process, we must also take account of the electron's energy component $E_\perp$ which is associated with its momentum perpendicular to the direction of tunneling. Thus the tunneling probability must be written

$$T = \exp - 2\int\sqrt{\frac{2m^*}{\hbar^2}\,(E_{||} + E_\perp)} \tag{8.5.8}$$

where $E_{||}$ is given by the quantity in brackets in (8.5.6). The classical turning points occur where $K(x) = 0$

$$x = \pm\frac{1}{\varepsilon e}\sqrt{\left(\frac{E_g}{2}\right)^2 + E_g E_\perp} \tag{8.5.9}$$

In this case the tunneling probability becomes

$$T = \exp\left[\frac{-\pi(m^*)^{1/2} E_g^{3/2}}{2\sqrt{2}e\hbar\,\varepsilon}\right] \exp - \left(\frac{2E_\perp}{\bar{E}}\right) \tag{8.5.10}$$

where

$$\bar{E} = \frac{\sqrt{2}\, e\hbar\varepsilon}{2\pi(m^*)^{1/2}(E_g)^{1/2}} \tag{8.5.11}$$

is a measure of the importance of the transverse momentum, i.e., if E is very small, the tunneling probability will be large only where $E_\perp$ is small.

The total direct tunneling current density may be found to be[32]

$$J = \frac{em^*}{2\pi^2\hbar^2} \exp\left(\frac{\pi(m^*)^{1/2} E_g^{1/2}}{2\sqrt{2}e\hbar\varepsilon}\right) \int (f_e(E) - f_v(E) \exp - \frac{2E_\perp}{\bar{E}}\, dE\, dE_\perp \tag{8.5.12}$$

where the limits of integration of $E_\perp$ are determined by

$$0 \leqq E_\perp \leqq E_1$$

$$0 \leqq E_\perp \leqq E_2 \tag{8.5.13}$$

E_1 and E_2 are the electron energy measured from the n and p band edges, respectively. The limits of E are given by the band edges. Integration of (8.5.12) yields

$$J = \frac{em^*}{2\pi^2\hbar^2} \exp\left[\frac{-\pi(m^*)^{1/2} E_g^{3/2}}{2\sqrt{2}\hbar e\varepsilon}\right] \frac{\bar{E}}{2} D \tag{8.5.14}$$

where

$$D \approx \int (f_c(\varepsilon) - f_V(\varepsilon)\left(1 - \exp\frac{-2E_S}{\bar{E}}\right) dE \tag{8.5.15}$$

E_s is the smaller of E_1 and E_2. Thus the shape of the tunneling current curve is determined by the integral D, which is an effective density of states integral.

The transmission coefficient for indirect tunneling is given by

$$T = \exp\left(\frac{-4\sqrt{2}(m_{rx})^{1/2}(E_g - \hbar\omega_\varkappa)^{3/2}}{3e\hbar\varepsilon}\right) \tag{8.5.16}$$

The shape of the tunneling curve predicted by Kane for both direct and indirect tunneling for $\mu_n = \mu_P$ is shown in Figure 8.23.

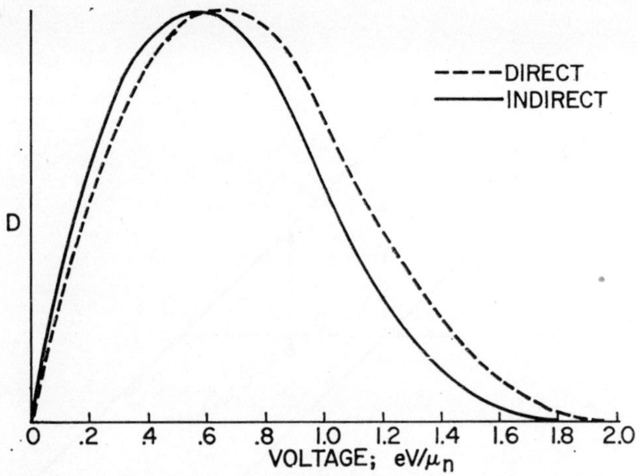

Figure 8.23 Shape of the current tunneling through a *p-n* junction diode. The dashed line shows the predicted curve for direct tunneling and the solid for indirect [from E. O. Kane, *Jnl. App. Phys.* **32**, 83 (1961)].

Photon Assisted Tunneling (*The Franz-Keldysh Effect*)

A type of band-to-band tunneling can occur in which an electron takes advantage of an incident photon to increase its energy and thus decrease the effective width of the barrier it must penetrate. This process is an important one in the theory of optical absorption below the indirect band gap energies.[33, 34]

Under the influence of an electric field, the energy bands are tilted as shown in Figure 8.24. An electron attempting to tunnel from the valence band to the conduction band sees a triangular barrier. If, during the tunneling process the electron interacts with a photon of energy $\hbar\omega_{\varkappa}$ sufficient to raise it to the conduction band, the effective width of the barrier to be traversed has been made smaller. The classical WKB approximation gives the same result as a more advanced treatment, so we will utilize the former. The probability of transmission through the barrier shown in Figure 8.24 is given by

$$T = \exp - 2 \int_A^C \sqrt{\frac{2m_r^*}{\hbar^2} (E_g - e\varepsilon x - \hbar\omega)}\, dx \tag{8.5.17}$$

which yields

$$T = \exp \left\{ \frac{-4\sqrt{2m_r^*}\,(E_g - \hbar\omega)^{3/2}}{3e\hbar\varepsilon} \right\} \tag{8.5.18}$$

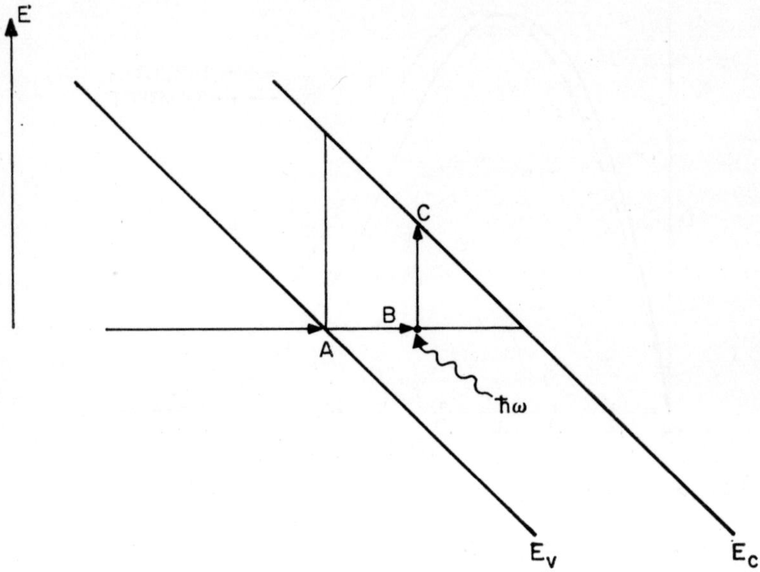

Figure 8.24 Tilted energy bands under the influence of an electric field.

PROBLEM

By assuming that the Fermi functions in the valence and conduction bands may be given by

$$f_c(\varepsilon) \approx \frac{1}{2} - (E - \mu_n)/k_0T$$

$$f_v(\varepsilon) \approx \frac{1}{2} + (\mu_p - E)/k_0T \qquad (8.5.19)$$

show that the tunneling current (neglecting transverse momentum) is given by[35]

$$J = AP\frac{eV}{k_0T}(\mu_n + \mu_p - eV)^2. \qquad (8.5.20)$$

The middle bias region consists of the valley and the excess currents. In these regions the current is larger than that predicted by the sum of the band-to-band tunneling and diffusion currents. The excess current in the valley region is thought to be due to band-to-band tunneling due to exciton or impurity levels and the influence of impurity levels on the band edges and temperature broadening. Kane has shown that the effect of the tailing is to shift the peak tunneling current to larger energies and to increase the energy range over which tunneling takes place as shown in Figure 8.25.[36] The exponential excess current occurs in the voltage region after the bands have uncrossed and no more band-to-band tunneling is possible. This excess current is probably due to a multi-step process in which a conduction band electron satisfies the conservation of energy necessary for tunneling by first dropping to an impurity or defect level located within the energy gap, and then tunnels from this level to an available state in the valence band. This and other similar allowed processes are shown in Figure 8.26. Following Chynoweth, *et al.*[37], under the constraint that the tunneling from the impurity level is the rate-controlling process, the excess current is given by

$$J = Ag_g(\varepsilon)T \tag{8.5.21}$$

where A is a constant, T is the probability of tunneling from the impurity energy level and $g_g(E)$ is the density of impurity energy levels in the energy gap. Franz has shown that the probability of tunneling from an impurity energy level is given by[38]

$$T = \exp\left[\frac{-4\sqrt{2}\,(m^*)^{1/2}\,E^{1/2}}{3e\hbar\varepsilon}\right] \tag{8.5.22}$$

[see (8.5.7)], where we may write

$$E = E_g - eV + \mu_n + \mu_p \tag{8.5.23}$$

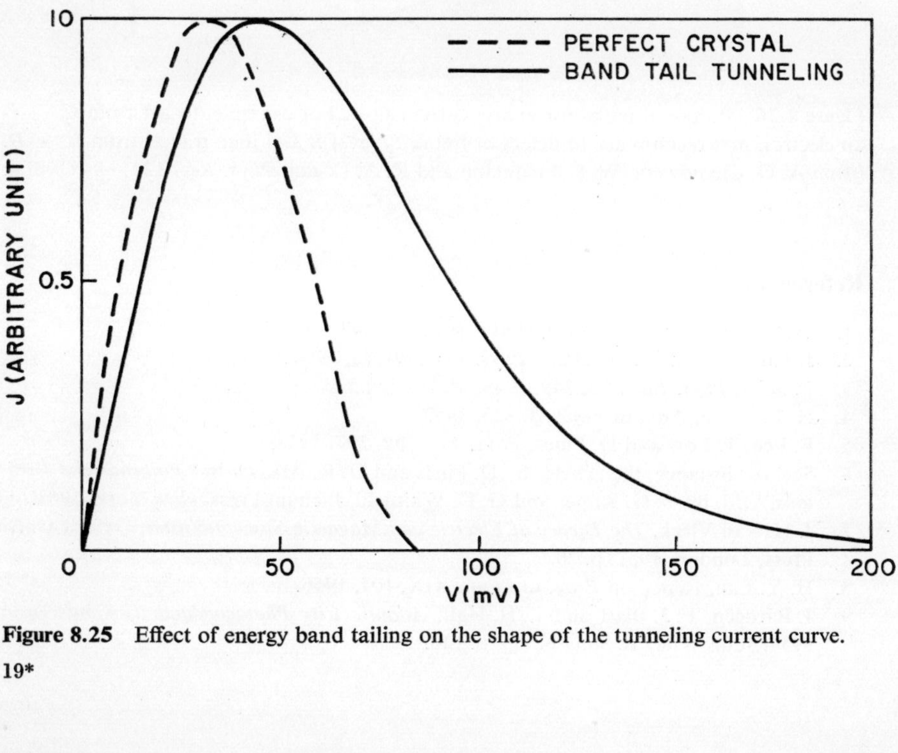

Figure 8.25 Effect of energy band tailing on the shape of the tunneling current curve.

Substitution of (8.5.23) and (8.5.22) into (8.5.21) yields the exponential excess current

$$J = Ag_g(\varepsilon) \exp\left[\frac{-4\sqrt{2}(m^*)^{1/2}}{3e\hbar\varepsilon}(Eg - eV + \mu_n + \mu_p)^{1/2}\right] \quad (8.5.24)$$

Note that the predicted exponential dependence results either from a constant density of levels within the energy gap or one that varies exponentially with gap energy.

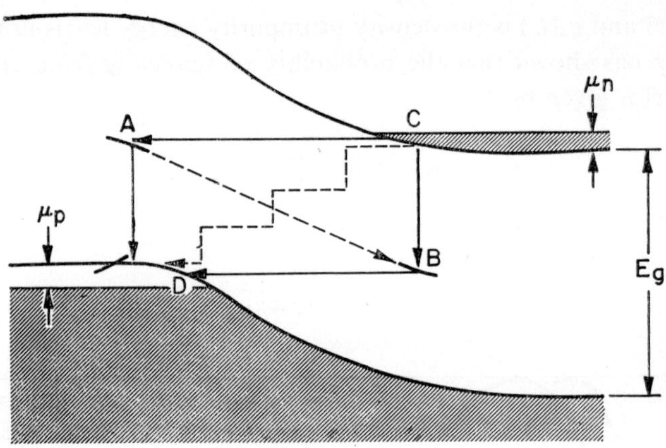

Figure 8.26 Proposed paths for excess current flow. For example, to get from $C \rightarrow D$ an electron first recombines to defect or impurity level B and then travels from $B \rightarrow D$. [from A. G. Chynoweth, W. L. Feldmann and R. A. Logan, *Phys. Rev.* **121**, 684 (1961)].

References

1. W. Shockely and J. Bardeen, *Phys. Rev.*, **77**, 407, 1950.
2. J. Bardeen and W. Shockely, *Phys. Rev.*, **80**, 72, 1950.
3. F. Seitz, *Phys. Rev.*, **73**, 549, 1948.
4. H. Frohlich, Adv. in *Phys.*, **3**, 325, 1953.
5. F. Lee, T. Low and D. Pines, *Phys. Rev.*, **90**, 297, 1953.
6. See, for instance, the articles by D. Pines and G. R. Allcock in "*Polarons and Excitons*," Ed. by C. G. Kuper and G. D. Whitfield, Plenum Press, New York, 1963.
7. J. H. Van Vleck, *The Theory of Electric and Magnetic Susceptibilities*, Oxford Univ. Press, London, 1952, p. 20.
8. H. Y. Fan, Repts, on *Prog. in Phys.*, XIX, 107, 1956.
9. J. Bardeen, F. J. Blatt and L. H. Hall, *Atlantic City Photoconductivity Conference*, 1954, John Wiley & Sons N. Y., p. 146.

10. M. Tannenbaum and H. B. Briggs, *Phys. Rev.*, **91**, 1561, 1953.
11. E. Burstein, *Phys. Rev.*, **93**, 632, 1954.
12. see ref. 9.
13. See the article by R. J. Elliot, *Polarons and Excitons*, Ed. by C. G. Kuper and G. D. Whitfield, Plenum Press, N. Y., 1963, p. 269.
14. D. L. Dexter and R. S. Knox, *Excitons*, Interscience Publishers, N.Y., 1965.
15. D. J. Howarth and E. H. Sondheimer, *Proc. Roy, Soc.* **A 219**, 53, 1953.
16. T. Low and D. Pines, *Phys. Rev.*, **98**, 414, 1955.
17. E. Conwell and V. F. Weiskopf, *Phys. Rev.*, **77**, 388, 1950. H. Brooks, *Phys. Rev.* **83**, 879 (1951). R. B. Dingle, *Phil. mag.* **46**, 831 (1955).
18. R. D. Evans, *The Atomic Nucleus*, McGraw-Hill Book Co., Inc., N.Y., 1955, p. 838.
19. G. L. Pearson and J. Bardeen, *Phys. Rev.*, **75**, 865, 1949.
20. C. Eginsoy, *Phys. Rev.*, **79**, 1013, 1950.
21. H. S. W. Massey and B. L. Moiseiwitsch, *Phys. Rev.*, **78**, 180, 1950.
22. H. Reiss, C. S. Fuller and F. J. Morin, Bell. *Syst. Tech. Jnl.* **35**, 535, 1956.
23. A. G. Samoilovich and M. V. Nitsovich, *Sov. Phys. Solid State*, **5**, 2182, 1964. See also R. Stratton, Proc. Int. Conf. Phys. Semiconductors, Exeter, 1962.
24. T. P. McLean and E. G. S. Paige, *J. Phys. Chem. Solids*, **18**, 139, 1961.
25. D. L. Dexter and F. Seitz., *Phys. Rev.*, **86**, 964, 1952.
26. E. M. Conwell, High Field Transport in Semiconductors, Vol. 9, *Solid State Physics*, Academic Press, N.Y., 1967.
27. L. W. Aukerman and R. K. Willardson, *J. App. Phys.*, **31**, 939, 1960. see also R. S. Algaier, *Jnl. App. Phys.*, **36**, 2429, 1965.
28. C. S. Hung, *Phys. Rev.*, **79**, 727, 1950.
29. A. Miller and E. Abrahams, *Phys. Rev.*, **120**, 745, 1960, N. F. Mott *Can. J. Phys.* **34**, 1356 (1956), E. M. Conwell, *Phys. Rev.* **103**, 51 (1956).
30. A. van der Ziel, *Solid State Physical Electronics*, Prentice-Hall Inc, N.J. 1961, p. 282.
31. L. Esaki, *Phys. Rev.* **2**, 109, 1958. C. Zener, *Proc. Roy. Soc.* **145**, 523 (1934).
32. E. O. Kane, *Jnl. App. Phys.*, **32**, 83, 1961. *J. Phys. Chem. Solids* **12**, 181, 1959.
33. W. Franz, *Z. Naturforsch.*, **13a**, 484, 1958.
34. L. V. Keldysh, *Soviet Phys.* JEPT, **7**, 788, 1958.
35. J. Karlovski, *Phys. Rev.*, **2**, 1070, 1962.
36. E. O. Kane, *Phys. Rev.*, **131**, 79, 1963.
37. A. G. Chynoweth, W. L. Feldmann and R. A. Logan, *Phys.*, *Rev.*, **3**, 684, 1961.
38. W. Franz, *Handbuch der Physik*, Ed. by S. Flugge, Springer-Verlag, Berlin, vol. 17, 1956, p. 155.

Subject Index

Author Index